Developments in Petroleum Science, 10

deconvolution of geophysical time series in the exploration for oil and natural gas

FURTHER TITLES IN THIS SERIES

1 A. GENE COLLINS
GEOCHEMISTRY OF OILFIELD WATERS

2 W.H. FERTL
ABNORMAL FORMATION PRESSURES

3 A.P. SZILAS
PRODUCTION AND TRANSPORT OF OIL AND GAS

4 C.E.B. CONYBEARE
GEOMORPHOLOGY OF OIL AND GAS FIELDS
IN SANDSTONE BODIES

5 T. F. YEN and G.V. CHILINGARIAN (Editors)
OIL SHALE

6 D.W. PEACEMAN
FUNDAMENTALS OF NUMERICAL RESERVOIR SIMULATION

7 G.V. CHILINGARIAN and T.F. YEN (Editors)
BITUMENS, ASPHALTS AND TAR SANDS

8 L.P. DAKE
FUNDAMENTALS OF RESERVOIR ENGINEERING

9 K. MAGARA
COMPACTION AND FLUID MIGRATION

Developments in Petroleum Science, 10

deconvolution of geophysical time series in the exploration for oil and natural gas

MANUEL T. SILVIA
Research Scientist
United States Naval Underwater Systems Center, Newport, Rhode Island

and

ENDERS A. ROBINSON
Distinguished Professor of Mathematics and Geophysics
University of Tulsa, Tulsa, Oklahoma

Adjunct Professor of Electrical Engineering
Northeastern University, Boston, Massachusetts

Consultant
Amoco Production Company, Tulsa, Oklahoma

ELSEVIER SCIENTIFIC PUBLISHING COMPANY
Amsterdam — Oxford — New York 1979

ELSEVIER SCIENTIFIC PUBLISHING COMPANY
335 Jan van Galenstraat
P.O. Box 211, Amsterdam, The Netherlands

Distributors for the United States and Canada:

ELSEVIER NORTH-HOLLAND INC.
52, Vanderbilt Avenue
New York, N.Y. 10017

Library of Congress Cataloging in Publication Data

Silvia, Manuel T
Deconvolution of geophysical time series in the exploration for oil and natural gas.

(Developments in petroleum science ; 10)
Bibliography: p.
Includes index.
1. Seismic reflection method--Deconvolution.
2. Time-series analysis. 3. Petroleum. 4. Gas, Natural. I. Robinson, Enders A., joint author.
II. Title. III. Series.
TN269.S533 622'.18'28 78-4931
ISBN 0-444-41679-X

ISBN 0-444-41679-X (Vol. 10)
ISBN 0-444-41625-0 (Series)

Printed in The Netherlands

CHRISTIAN HUYGENS (1629—1695)

In 1678 Christian Huygens formulated the concept of wave superposition and the principle of regarding each point on a wave front as a source of secondary wavelets. These elegant concepts serve as the foundation for the understanding of all types of wave propagation, and in particular seismic wave propagation.

FOREWORD

The closer you look at something, the more details you will see. This common experience is hardly better exemplified than by seismic exploration of the earth's interior, in particular of its complicated near-surface structure. The seismic waves that leave a wave source, e.g. an explosion, experience lots of influences on their way to the receiver. At the same time as all such effects tend to complicate the records, they also convey the required information on the passed strata. But to disentangle all these various effects, we badly need methods which are capable not only of efficient earth-detective work, but which can also work at high speed — considering the large amount of data generally accumulated in seismic prospecting. The present book by two of the most capable earth detectives deals with these problems, and it is more exciting than any detective story.

But this is not only a game for its own purpose. In order to run and to develop a modern community, in short, for its survival, mankind is nowadays hunting for energy more than ever. Oil and natural gas still constitute some of the most important energy sources. Therefore, we need skilled geophysicists, who are able to extract as much and as accurate information as possible from the records. This is especially important now when we have to explore new areas and can no longer be content with areas where "oil flows like water". We have to search more intensively and to greater depths in the earth. Without efficient methods as developed in this book, there is hardly any chance for success in this hunting for energy.

The book has both theoretical and practical sides, and it can be equally recommended to the university scientist who performs his geophysical work at his desk, in the laboratory or in the lecture room and to the practicing geophysicist who must know how to best perform his seismic prospecting and how to interpret the "text" that Earth writes for him.

September 1977

MARKUS BÅTH
Professor of Seismology
Uppsala, Sweden

PREFACE

In the exploration for oil and natural gas, the geophysicist is confronted with the problem of estimating the structural features of the subsurface to depths of up to 6000 m with an accuracy of a few tens of meters. These estimates must be made over geographic areas covering many hundreds of square miles. The geophysical methods make use of indirect means because all inputs and outputs must be made at or near the surface of the earth. The most successful method is the reflection seismic method. This method was greatly enhanced by the introduction of digital deconvolution in the early 1960's. There has been much practical use of deconvolution in the intervening years as well as a great deal of theoretical work. However, there has never been a unified treatment of the entire subject which actually makes use of the physical properties to simplify and make amenable the mathematical justification.

This book is motivated by the need for a comprehensive treatment of deconvolution which brings out the essential mathematical properties from the physically observed facts. From principles given in this book, the methods of deconvolution can be viewed in a unified way.

A preliminary chapter outlines the field problems in seismic exploration and gives a discussion of the sequence of digital processing techniques applied to the seismic data in order to obtain the final geologic depth section.

Chapter 1 gives a new treatment of seismic wave propagation. This treatment is based on the fact that the direct transmission response of a layered system is minimum-delay. As a result, most practical exploration problems can be reduced to two prototype cases: namely, the case of internal primary reflections and the case of external primary reflections. The characterization of the reflection response of a layered system in terms of these cases is new.

Chapter 2 extends the classic results of the layered earth model. These previous results derived exact expressions for the reflection and transmission response of a stratified system. It is generally observed in the field that the reflection coefficients, which must be necessarily less than one in magnitude, are actually less than 0.1 in magnitude in most physical situations. Use is made of this physical fact to show that the reflection and transmission response can be greatly simplified. This simplification gives the physical insight which is needed by the field geophysicist in order to interpret the results of deconvolution. Furthermore, these simplified responses are precisely the response derived in Chapter 1 on purely physical grounds. Finally, on the basis of these results, a mathematical justification is given for the hypothesis of random reflection coefficients based upon field experience.

The material in Chapter 3 shows the relationship of the kepstrum to the classical logarithmic potential problem. The concept of minimum-advance is introduced. By use of the symmetries between the concepts of minimum-delay and minimum-advance, an arbitrary kepstrum is decomposed into its most basic components. This decomposition is shown to be basic to the theory of deconvolution.

In Chapter 4, unit-step deconvolution is derived from the mathematical and physical models introduced in the previous chapters. A more general type of deconvolution, called α-step deconvolution, is given as a method of removing multiple reflections. The main results show that multiple reflections can be removed by α-step deconvolution, even in the case when the source pulse is not minimum-delay. Kepstral deconvolution is shown to be most effective in the case of detecting a single echo. Finally, state space filtering is related to the foregoing deconvolution problem. State space methods provide a fertile research area for the study of new deconvolution methods.

Chapter 5 gives computer programs in subroutine form which are useful for filtering and spectral analysis. These subroutines provide the link between the mathematical concepts associated with deconvolution and actual numerical implementation.

We wish to express our sincere thanks to Professor Markus Båth of the Seismological Institute of Uppsala University whose encouragement and help made possible the writing of this book. We are grateful to Gerald M. Hill and Frank Spicola of the Naval Underwater Systems Center (Code 352) for their encouragement. We wish to thank Miss Mersina Christopher for her excellent work in typing the manuscript.

MANUEL T. SILVIA
ENDERS A. ROBINSON

CONTENTS

Preliminary Chapter

OVERVIEW OF GEOPHYSICS

1. GEOPHYSICAL EXPLORATION

The three principal geophysical methods used in petroleum exploration are seismic, magnetic, and gravity. The magnetic method is the oldest geophysical method, and is based on the measurement of variations in the magnetic field due to changes of structure or magnetic susceptibility of the rocks. Sedimentary rocks generally have a smaller susceptibility than igneous or metamorphic rocks, so an interpretation of the recorded anomalies can yield the maximum depth values for a sedimentary basin. Today, magnetic surveys for hydrocarbon exploration are usually carried out from the air (aeromagnetics) or from a ship. The gravity method is based on the measurements of the variations in the pull of gravity from rocks in the upper layers of the earth's surface. Denser rocks have greater gravitational attraction than less dense rocks. For example, a structural uplift of a denser rock will appear as an anomaly on the gravity map. Gravity surveys for hydrocarbons are carried out on land, in the air on helicopters, and at sea on ships.

The most widely used geophysical method is the seismic method. For an elementary treatment of seismology the basic text is Båth (1973); for an advanced treatment, Båth (1968) together with his work on spectral analysis (Båth, 1974); for the use of geophysical methods in oil exploration, Dobrin (1976) and Robinson and Treitel (1969), and for an advanced treatment, Claerbout (1976), Kulhánek (1975), and Ricker (1977). These references together with one hundred others are listed at the end of this book.

Exploration seismology is divided into the branches of reflection seismology and refraction seismology. Most petroleum exploration is done by the reflection seismic method. Reflection seismology is a method of mapping the subsurface sedimentary rock layers from measurements of the arrival times of events reflected from the subsurface layers. The technology of collecting and processing reflection seismic data is based on a fundamental concept. By generating seismic energy which penetrates the earth's layered media, reflections of the seismic waves at the interfaces are measured and recorded by receiving devices at or near the earth's surface. Basically, the concept is the same for both land and marine surveys, only the mechanics vary. On land, seismic energy is generated at or near the earth's surface by arrays of small chemical explosions, or vibrating machines, or thumping devices. Preplaced detectors on the surface run in a line as illustrated in Fig. P-1. The seismic waves resulting from the downward propagation of this

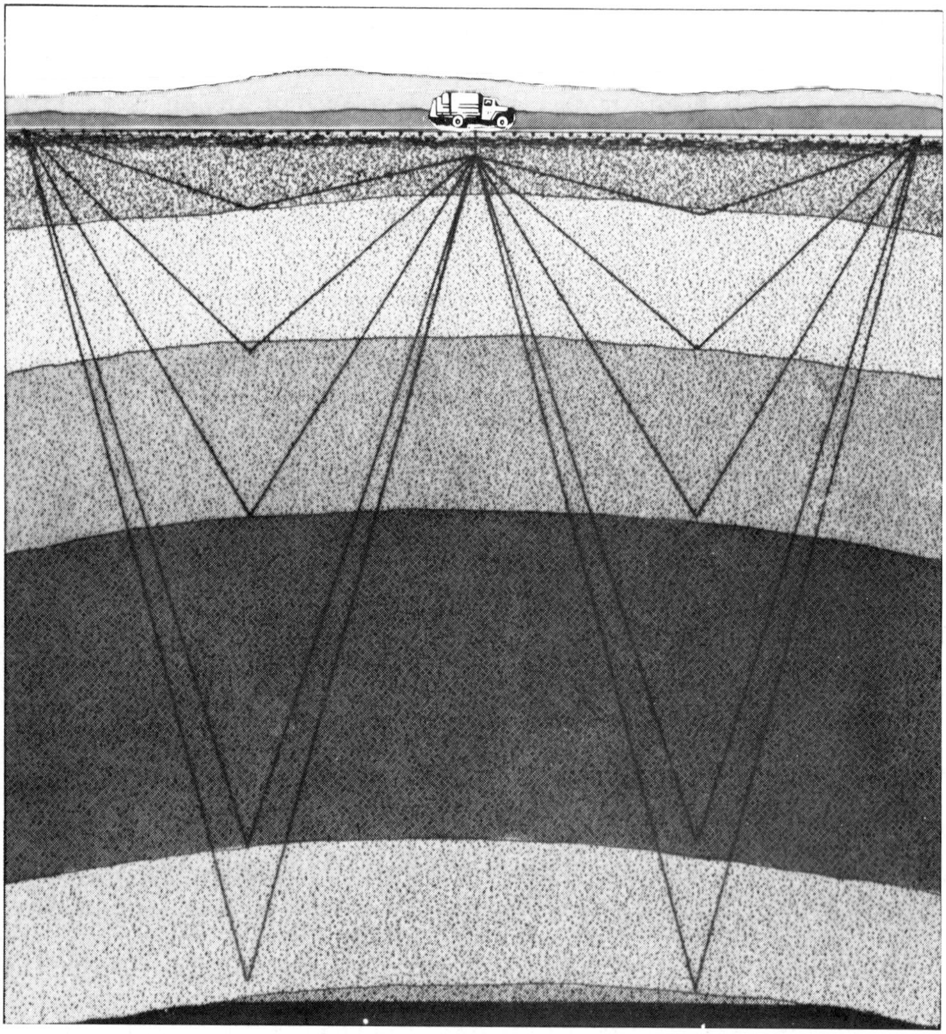

Fig. P-1. Schematic illustration of seismic exploration showing primary reflections. (Oil-bearing layers may be up to 6 km deep.)

source energy are reflected from the various interfaces and received by the detectors. A central recording unit then digitizes the analog signals and records them on magnetic tape for subsequent analysis. At sea, a source such as an array of air guns is actuated every few seconds as the ship moves over a predetermined course. The seismic waves are picked up by detectors embedded in a cable (called a streamer) trailing the ship. As in land surveys, the data are transmitted to a central recording unit and recorded in digital form on magnetic tape.

Fig. P-2. Seismic prospecting ship "Gulf Seal" owned by Digicon, Inc., Houston, Texas.

Let us now describe a typical ship that would be used in a marine seismic survey. The ship would be a steel hull vessel 55 m long with a 12-m beam and a draft of 4 m. There would be twin screw propulsion and a stabilization system. Accommodation would be provided for 24 men (ship's crew and the geophysical crew). The geophysical accommodations would include office space, instrumentation cabins, an enclosed stern deck for the cable reels, and a heliport superstructure. The ship would have a cruising speed of 10 nautical miles per hour and a towing speed of 4—6 nautical miles per hour. Marine navigational facilities would include dual radars with a 24-mile range and a fully integrated satellite navigation system with auto pilot, gyrocompass, sonar doppler, velocimeter, and computer (see Fig. P-2).

In marine seismic work various navigation systems such as Loran C are used. In many surveys a satellite positioning system is used as the primary positioning system. Advantages of the satellite system are 24-hour all-weather operation, real-time statistical filtering to provide "desired track" capabilities, computer-controlled automatic ship steering to minimize on-line course corrections, inclinometers to provide dynamic corrections to improve satellite fix and heading accuracy, and the recording of all sensor data on magnetic tape for rapid post-mission analysis and computer mapping. Normal average deviation from programmed lines in water depths of 200 m or shallower is less than 300 m, and post-plot accuracy of shot points is 100 m; greater accuracy can be attained at a cost premium.

An effective seismic energy source in marine exploration is the Esso sleeve exploder device. Eight air gun units are towed in an array approximately

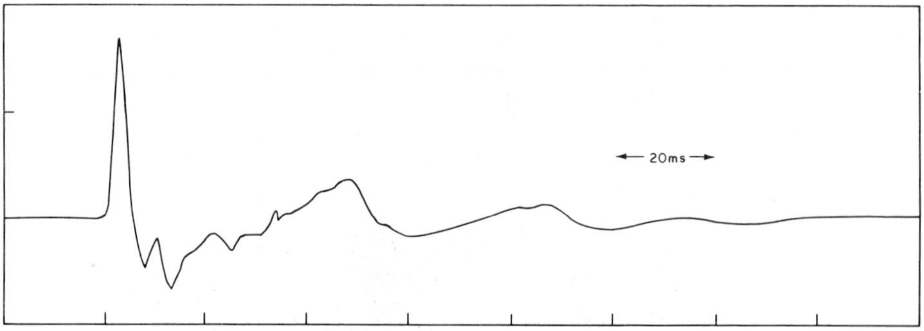

Fig. P-3. Typical air gun signature pulse.

22 m by 12 m at a depth of 6 m. Total air gun performance for this device is continuously monitored by individual gauges at the mechanics' console. The guns are fired by command of the navigation system and a composite gun pulse from each pair of guns is displayed on monitor records and on tape. Fig. P-3 shows a typical source wavelet (i.e. gun signature pulse). Typical quality control standards would be:

(1) Not more than 3 misfires in any 4 consecutive shots, 6 in any 12 shots, 8 in any 20 shots, or 16 in any 100 shots. No more than 6% of any line shall consist of misfires. A misfire is defined as any equipment or operator malfunction that results in an unrecoverable record, or more than 2 dead or excessively noisy traces, or less than 6 guns operating properly.

(2) No new line started with more than 1 dead trace.

(3) Noise level on the main streamer not to exceed 5 microbars rms, except that the use of a radar target towed from the tail buoy increases the allowable noise on channel 1. (Typical streamer noise level is 3 microbars.)

(4) The angle of feathering between the main streamer and the line traverse not to exceed 12°.

Fig. P-4 shows an air gun before lowering into the sea.

The streamer trails below the water's surface behind the ship. A seismic streamer is a high performance and operationally flexible device. Outstanding features are its field repairability, non-rotating stability, quick-coupling connectors, unbreakable bulkheads, high-sensitivity hydrophones, and flexible operational configurations to meet specific recording requirements. Downtime is minimized with quick disconnect sections, a center stress member and a durable, tough cold-resistant vinyl jacket. The center stress cable is a wire rope (0.7 cm outside diameter) of torque-balanced galvanized aircraft steel. It has a minimum breaking strength of 5000 kg. A 0.05-cm layer of high dielectric insulation reduces leakage. A 48-channel streamer has a total length of 2400 m divided into 48 sections (1 section for each channel), each section being 50 m long. There are 40 hydrophones spaced in each of the 50-m-long sections. Each hydrophone is 6 cm by 1.25 cm and weighs 27 g.

Fig. P-4. Air gun before lowering into the sea.

The hydrophones are connected in parallel in each section. The conducting wires are 24 AWG with composite stranding with no duplication of color code. Each section is enclosed in a plastic tube (vinyl jacket) 7 cm in diameter with bulk-heads at each end, so that each section is detachable and interchangeable. The plastic jacket is filled with oil to make the section neutrally buoyant. As a result the streamer is extremely flammable. At intervals along the cable, pressure-sensitive depth controllers keep the cable at the optimum depth. Passive sections 50 m long are available for desired spacing patterns. Fig. P-5 is a picture of the streamer reel on the stern deck of the ship. Fig. P-6 shows the ship dragging the streamer cable.

The 40-unit linear array of hydrophones in each section is called a group, and the signals received from all the hydrophones in a group are composited so as to form one signal, called a seismic trace. As a result the entire streamer of 48 groups yields 48 seismic traces each time the source is activated. These 48 traces are recorded by a 48-channel recorder digitally on magnetic tape inside the ship.

The above description pertains to deep-water exploration. Shallow-water seismic surveys under various conditions are also carried out in all parts of the world. The aim of these special investigations is to link land and marine surveys. Flexibility in the equipment and in the execution of the survey are necessary in order to survey estuaries with heavy traffic or strong currents, areas of mud flats, reefs, and other coastal features, and coastal regions with great tides. Such shallow-water areas require vessels of shallow draught as well as other special vehicles such as pontoons, catamarans, and hovercraft.

Fig. P-5. Streamer cable wound on its reel on afterdeck of ship.

Land exploration ranges over all areas of the world with such extremes as cities such as Los Angeles, the arctic tundra, the tropical jungle. As an example, let us briefly describe oil exploration in the Arabian Desert. Air-conditioned trailers and helicopters have replaced the tents and camel caravans of the early oil explorers in the desert. As a seismic source, arrays of light charges of explosives can be used, but now often the comparatively gentle mechanical jolts of a thumper are used. The thumper, which is on a special vehicle, is a gas-driven chamber that can deliver a sharp thump on the ground. Several such units normally work together and pop simultaneously. Seismic field parties may consist of more than 100 men. Convoys of large air-conditioned trailers for living and office work are towed to camp locations by heavy-duty trucks equipped for desert travel. Much of the desert region is covered with sand. Both heavy and light vehicles can be used if equipped with proper sand tires, but it is often necessary to follow circuitous routes to avoid the worst sand conditions. Helicopters are also used, and personnel and some supplies are often flown to the remote areas.

It may be said that geophysical exploration is pushing more and more into the remote areas of the world which have difficult accesses and operating conditions. Also there are many offshore areas that are coming under active interest. For example, offshore western Florida is a major area of marine data collection. Fig. P-7 depicts the tracts (solid blocks) which the U.S. Department of the Interior offered for lease at sales scheduled in 1973. The seismic results are used by oil companies to evaluate these lease tracts prior to competitive bidding.

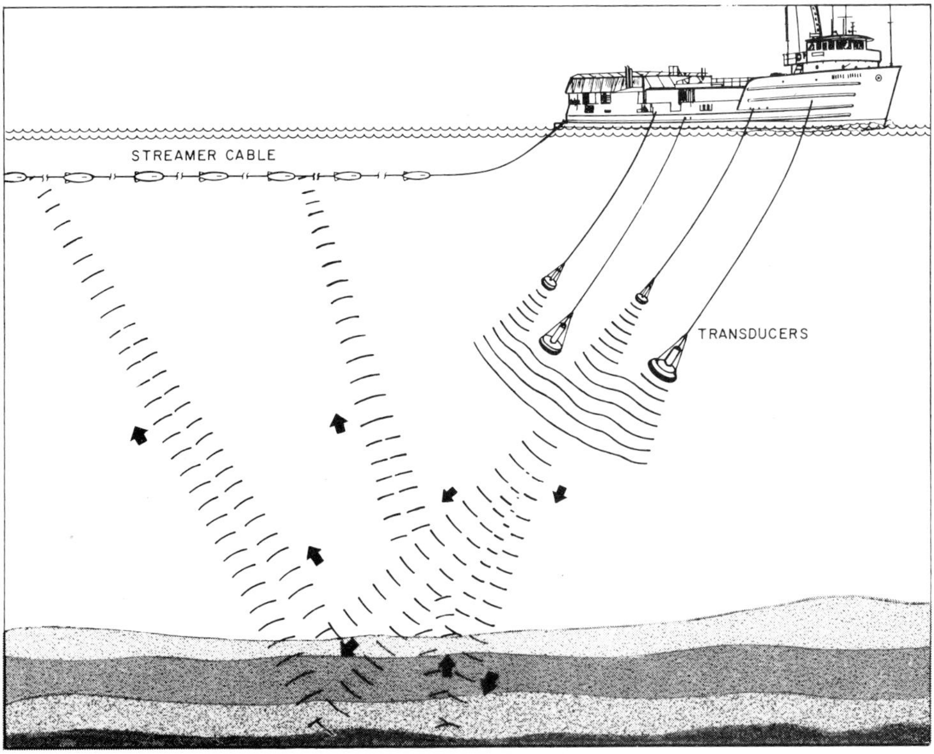

Fig. P-6. Ship dragging streamer cable.

The seismic data are usually processed and interpreted in a computer center in a central location such as Houston, London or Singapore. However, with the advent of small computers, more and more processing is being done in the field. In the acquisition of seismic data, countless confusing signals are recorded simultaneously with the desired primary signal. The unwanted signals are referred to broadly as noise. In a processing center use is made of a software system which enhances the signal while attenuating the noise. A significant and continuing research and development program exists to further improve the signal content of the seismograms. These magnetic tapes are processed according to the specifics dictated by the oil company, and then filed and held as inventory or returned to the oil company.

Fig. P-8 shows the final product in seismic exploration; namely, the final seismic depth section after all the processing methods have been applied. This depth section represents a portion of the geologic structure of offshore Florida indicating the possibility of oil accumulation. On the basis of such depth sections, oil companies submit bids for acreage in government lease sales. Fig. P-9 shows the offshore drilling platform used for drilling the wildcat oil well based on this seismic depth section.

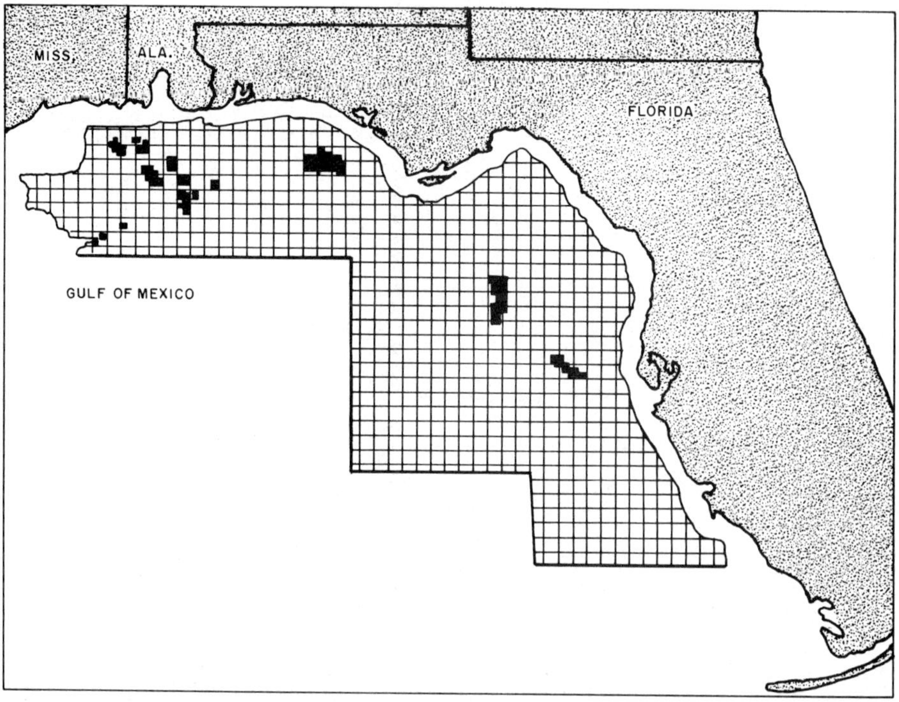

Fig. P-7. Tracts considered for leasing in 1973 in offshore western Florida shown by the solid squares.

2. DIGITAL PROCESSING OF SEISMIC DATA

In a seismic exploration program as conducted by a seismic contractor or an oil company, literally millions of seismic traces are collected. These seismic traces must be analyzed so as to yield contour maps of the subsurface sedimentary structure. These maps are used in order to make decisions as to the locations of where to drill exploratory oil and gas wells. Because of the great costs of drilling and the much greater costs of leasing potential oil producing land or of obtaining oil concessions from foreign governments, the geophysical results are expected to be accurate. These geophysical results must stand up to the test of the drill. The usefulness of the seismic method rests on the fact that it is extremely accurate. For example, a seismic survey can delineate a geologic structure 4000 m under the ground to an accuracy of the order of a few tens of meters. Such accuracy is beyond what reasonably may be expected, in view of facts such as the extreme conditions in which the data are often collected. For example, much arctic exploration has to be done in the winter while the tundra is frozen; temperatures of 50° below zero, storms, and the winter blackness make this work dangerous. The accuracy of a seismic survey depends on solving difficult interpretation

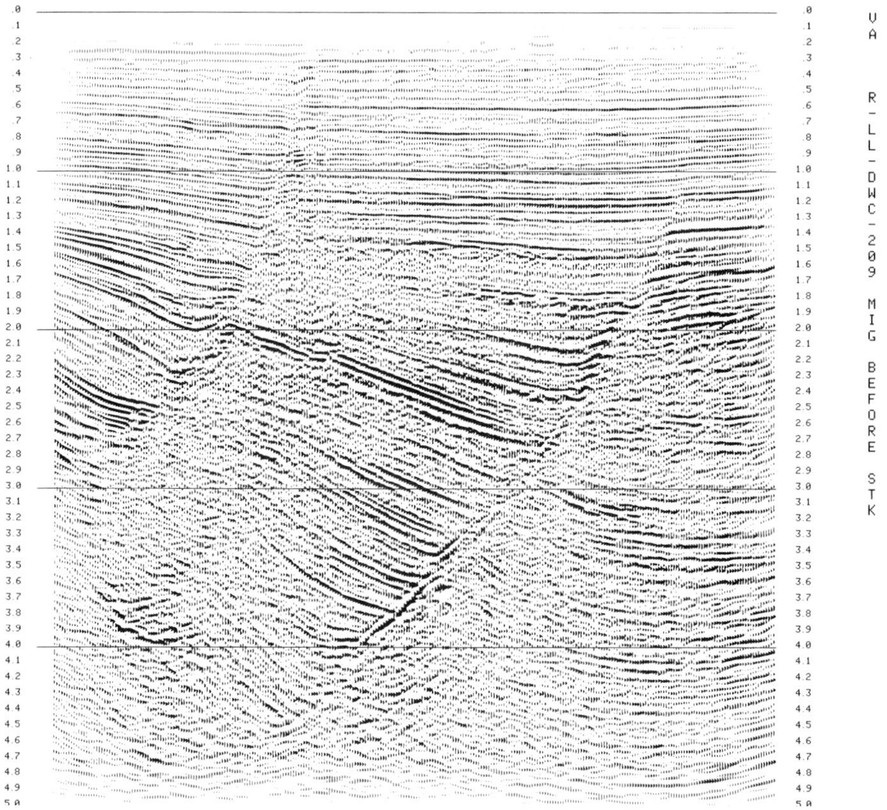

Fig. P-8. Final seismic depth section collected and processed by Digicon, Inc., Houston, Texas.

problems due to the intricate wave patterns which result from the complex configurations of the strata in the upper 6 km of the earth's crust. The purpose of the digital processing of seismic data is to convert these complex wave patterns as recorded on the seismic records into meaningful information that can be used to determine the underground structure.

Ideally it would be nice to be able to put all the data into a computer, fit the necessary wave equations, and come out with the final solution. However, we are far from such an ultimate method. Instead we must process the data through a sequence of operations, each operation involving serious approximations and short-cuts. The seismic analysis is so complex that even an approximate understanding as to what parameters are important, and what are not, in a given situation is worth having. Each of the operations is based on the physical point of view acquired over the years by the field geophysicists who analyze and interpret the reflection seismic records. One of the things learned is that the analysis requires an understanding of the statistical

Fig. P-9. Offshore drilling platform.

make-up of the seismic record section. We do not want to know where every seismic wave is actually moving, but instead we want to know how many move here and there on the average and the mutual buildup and cancellation of effects. This statistical interpretation forms the basis of all the important digital processing techniques used in seismic data analysis; namely, velocity analysis, static and dynamic corrections, stacking, source signature analysis, deconvolution, and migration. We now want to briefly describe these operations, in order to put the technique of deconvolution, with which the present book is concerned, into context with the other techniques in the processing sequence.

First let us discuss *velocity analysis*. The earth's sedimentary layers are approximately horizontal, but they do have features such as anticlines, unconformities, and faults that can serve as traps for petroleum. In order to map the subsurface, the geophysicist must convert the received seismic traces which record the events as a function of time into a function of depth. That is, a time function recorded at the surface must be transformed into a depth function. Unlike radio waves, seismic waves have a velocity which is very much dependent on the medium. Thus the velocity changes as the waves travel into the earth. Generally the velocity increases with depth, although occasionally there may be layers in which a decrease in velocity occurs. For a given surface point, the velocity plotted as a function of depth is called the velocity function. Thus in reflection seismology there are two equally important variables: time of reflected events and velocity. From knowledge of these variables the depth to the reflecting horizons can be determined. Because there are important lateral changes in velocity, that is, because the velocity function varies from one location to another, a given velocity function cannot be assumed to be valid for an entire prospect. As a result the velocity function must be continually corrected from place to place over the area of exploration.

One method of measuring the velocity function is to drill a deep hole, namely an oil well, and determine the velocity by placing seismic detectors in the hole at various depths. However, in most cases, it is necessary to estimate the velocity function by measurements confined to the surface, since oil wells are available only in old prospects. The velocity function can be estimated by considering the time differentials of the same event received by a lateral array of detectors. Any such estimate always depends upon a *ceteris paribus* (other things being equal) assumption. Computers can determine velocities by carrying out calculations based on many intricate time-distance relationships, and the results can present average velocity as a function of travel time (or depth) in a form called a *velocity spectrum*.

Another important digital technique used in the sequence of operations is the determination of *static and dynamic corrections*. Because of lateral variations in the near-surface layers, each trace is corrected by a time shift which has the effect of placing the source and receiver on a fictitious horizontal datum plane. This time shift is additively composed of a source correction and a receiver correction, which together are referred to as static corrections.

The dynamic corrections convert each trace to the equivalent trace that would have been received if the source and receiver were at the same lateral point, namely the point midway between the actual source and receiver locations. In this conversion we are referring to traces made up only of so-called primary reflections. According to ray theory, a primary reflection results from a ray path down from the source to the reflecting horizon and then directly upward to the receiver. Given the velocity function this ray path can be computed by means of Snell's law. If the layers are horizontal, then all the reflection points (or depth points) are always directly beneath the midpoint between source and receiver. If the layers are dipping, then the depth points are offset from the midpoint. Thus the dynamic corrections depend on both the velocity function and the dip of the reflecting beds. The component of the correction due to the separation of source and receiver is called the normal moveout correction and the component due to dip is called the dip correction.

Hence there are four important corrections that must be made to the

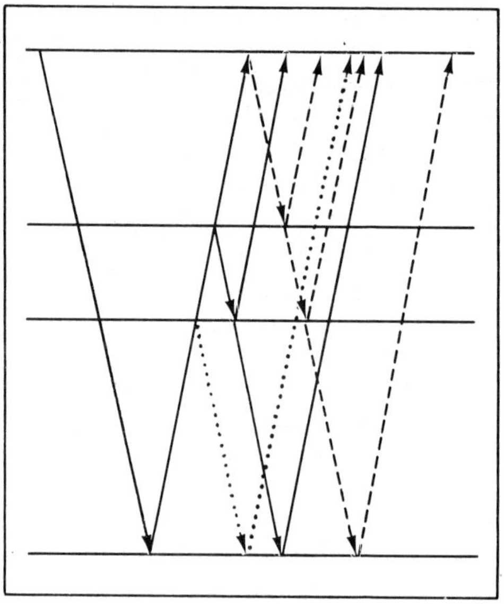

Fig. P-10. Ray paths of some multiple reflections.

recorded traces, namely source corrections, receiver corrections, normal moveout corrections and dip corrections. Each recorded trace is a time series made up of reflected events together with various interfering waves and noise. The desired reflected events are the primary reflections, that is, waves that travel down to a given reflector and then back up to the surface where they are recorded. An important type of undesired interfering wave is the multiple reflection. A multiple reflection has a ray path that goes down to a given reflector, then up to another reflector, then down to still another reflector, then up again, and so on until the multiple reflected event is recorded on the trace (see Fig. P-10). In any layered system there are infinitely many possible multiple reflections. The presence of multiple reflections makes the identification of primary reflections difficult, and thus it is necessary to attenuate the multiples as much as possible on the final display. Hence, in the computer processing, in addition to making the above corrections it is necessary to attenuate the multiple reflections and other noise.

To achieve these goals the seismic data are collected in a special way. For each source a spread (or array) of detectors is laid out. The source is activated (the shot) and the traces are recorded. Then the entire configuration is moved laterally and the process is repeated. By moving the configuration in small enough increments, each depth point is covered several times (multiple coverage). In six-fold coverage six traces will be recorded for each depth point, from the near-trace which has the shortest offset distance between source and receiver to the far-trace which has the longest. Thus the recording of seismic data by a multiple coverage scheme introduces a considerable amount of overlap or redundancy. The redundancies allow us to make the necessary corrections and to attenuate severely the unwanted interference and noise.

For example, all the traces in the prospect can be segregated (or gathered) into sets called gathers such that all traces within each gather have a common mid-point between source and receiver. Under a *ceteris paribus* assumption let us suppress for the time being everything except the normal moveout correction. The appropriate normal moveout correction would convert each trace in this gather into the same equivalent trace, namely that primary-reflection trace which would have been received if source and receiver were directly at the common mid-point in question. In other words, under the appropriate normal moveout correction, the primary reflections of all the traces in the gather will be in phase, thus making the corrected traces coherent. Let $c(n)$ denote a coherency measure of the traces in the common mid-point gather after normal moveout correction n. If we plot $c(n)$ for different corrections n, the appropriate correction would be that n for which $c(n)$ is a maximum, under the conditions of the *ceteris paribus* assumption. In practice a preliminary normal moveout correction is often applied to the data using a preliminary velocity function. In such a case this discussion would then refer to a second-order correction, that is, the residual normal moveout correction which in turn represents a correction of the preliminary velocity function.

The same *ceteris paribus* approach used above for normal moveout can also be used to make appropriate source, receiver, and dip corrections by different gatherings of the traces. On the other hand, instead of making the corrections one at a time under *ceteris paribus* assumptions, a simultaneous approach can be used. Because of the complexity of the problem iterative methods are more readily devised than direct methods.

Let us now discuss the next important technique in the sequence of operations, namely *stacking*. Because of the redundancy inherent in the multiple coverage method of seismic prospecting, unwanted interference and noise can be attenuated by stacking. As we have seen the dynamic corrections put all primary reflections in phase on the traces in each common mid-point gather. Because the ray paths of multiple reflections are different, the dynamic corrections do not put the multiples in phase. Hence, if we add together (i.e. stack) all the corrected traces in a common mid-point gather, we severely attenuate the multiples as well as other incoherent noise. Thus we obtain one output trace for each mid-point which is called the stacked trace for that mid-point. This type of stack is usually called the common depth point (CDP) stack, although the more accurate term common mid-point (CMP) stack is gaining acceptance. The reason for using the term mid-point instead of depth point is that for slanting subsurface horizons the depth points (i.e. points of reflection) do not lie directly under the source/receiver mid-points. The CMP stack is based on the fact that the reflection times of multiple reflections generally increase faster with increasing shot-receiver distances than those of primary reflections, which are superimposed by those multiples. The static corrections necessary before stacking are time-independent time shifts. They are supposed to compensate irregularities in the elevation of the earth's surface and in the character of the so-called weathering layer, at least down to the depth of the groundwater.

Source signature analysis as well as related techniques of digital filtering are essentially the same methods as used in other sciences, so we will not include them in our discussion here which is concerned mainly with methods peculiar to geophysics.

Let us now briefly introduce *deconvolution*, which is the subject matter of this book. A special kind of multiple reflection is the so-called reverberation. Energy trapped in the near-surface layers keeps being reflected back and forth, and this energy becomes attached to the primary reflections as they travel through the near-surface layers. As a result, instead of having sharp clear reflections with good time resolution, one has reflections followed by long reverberation trains. These trains overlap with the trains of succeeding primary reflections, and thus the whole seismic trace is given a ringing or sinusoidal character where it is difficult or impossible to pick out the onset of the primary reflections. The solution of this problem consists in cancelling the energy in each reverberation train but leaving intact the primary reflection, thus increasing the resolution of all the reflected events. This process is called deconvolution and is accomplished in the following way. If we consider the wavelet made up of a primary with its attached reverberation, we know from energy considerations that the wavelet is minimum-delay. Moreover, all such wavelets have approximately the same shape. Because the primaries result from geologic beds laid down with irregular thicknesses, the arrival times of the primaries are effectively random. Hence the autocorrelation function of the trace is the same as the autocorrelation function of the wavelet, and so from this autocorrelation function we can compute the required inverse (or deconvolution) operator. The application of this operator to the trace yields the deconvolved trace, namely, a trace where the reverberation components of the wavelets have been removed thereby increasing the resolution of the primary reflections. The process of deconvolution can also be extended to remove certain long-period multiple reflections as well. This type of deconvolution is called α-step deconvolution. In practice, deconvolution can be applied to the seismic traces either before or after stacking, depending upon cost and other considerations.

3. MIGRATION OF SEISMIC DATA

The final digital technique applied to the data before plotting the final results is *migration*. The word migration as used in seismic data processing has a special meaning and should not be confused with other uses of the word migration, such as the migration of oil from source rocks to an oil trap. Another term for the seismic processing use of the word migration is *depropagation*. That is, migration is the process of propagating waves observed on the surface of the ground backward in time into the earth to the underground structures. In a geographical area where all the subsurface layers are perfectly horizontal, such depropagation would take place on perfectly vertical paths. However, if the subsurface layers are dipping or bent, the depropagation would take place on curved paths turning away from the vertical. The reason for performing migration or depropagation is that it reveals the actual spatial positions of the subsurface reflection points at depth, whereas the unmigrated seismic data observed on the surface of the ground only gives the apparent reflecting positions. Thus migration can be described as the transformation of data observed at the surface to data that would have been observed at depth. That is, migration is the process of mathematically pushing the data back into the ground so as to reveal the true spatial locations of the subsurface structure (Hagedoorn, 1954; Stolt, 1978).

In physics, one does not usually think of migration but of diffraction. *Diffraction* represents the forward process, whereas *migration* represents the inverse process. Suppose that we have a point source at depth. From this point the wave energy spreads in all directions, and in an isotropic homogeneous medium the wave fronts would be spherical. However, in a heterogeneous medium the wave motion is not so simple. For example, if the waves pass the edge of an opaque body, some energy is deflected (or diffracted) into the shadow region. The presence of many refracting layers in the earth means that the wave fronts from the various diffractions will interfere with one another. Out of this complicated mixture a wave pattern will emerge at the surface of the ground and the recording of this wave pattern makes up the seismic data. (Note that in physics the terms interference and diffraction are generally used to describe the same type of wave phenomena. It is a question of usage, as there is no specific physical difference between the two terms. Usually, when there are only a few interfering sources, say two, the result is called interference, but if there is a large number of interfering sources the word diffraction is more often used.)

The common mid-point (CMP) stack record section is made up of all the CMP stacked traces along the survey line. The CMP stack section has approximately the properties of a record section in which each trace has the same source and receiver position. This coincident source/receiver geometry has the following implications: (1) although the energy travel path between the surface source/receiver location and the reflecting interface at depth may be quite complex, the upgoing and downgoing legs must be identical, and (2) the energy travel path strikes the reflecting interface at right angles (see Fig. P-11). The wave equation describes the motion of waves generated by a

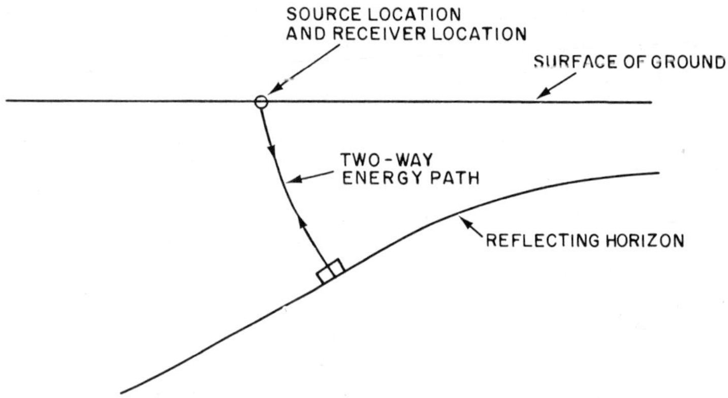

Fig. P-11. Energy path (down and up) for a coincident source and receiver.

physical experiment. However, the stacked record section does not correspond to a wavefield resulting from any single experiment. There were many shots taken sequentially, but the stacked record section gives the appearance that all the shots occurred simultaneously. As a result we hypothesize a theoretical physical experiment to justify the use of the wave equation to operate on the wave motion appearing on the stacked record section. The theoretical experiment may be described in this way. The receivers are located at the surface of the ground, but the sources are no longer on the surface but instead are distributed within the earth. More specifically, along every reflecting interface the sources are positioned with strengths that are proportional to the reflection coefficients. All the sources are fired at the same time instant, namely $t = 0$. We concern ourselves with upward traveling waves only, that is, with waves traveling from the sources at depth to receivers on the surface of the ground. Because the stacking and the deconvolution has effectively removed the multiple reflections, we omit consideration of the multiple reflections in our theoretical experiment. As we know, a seismic trace is recorded in terms of two-way travel time, that is, the time of the round-trip from the surface to depth point. However, in our theoretical experiment, we are only concerned with one-way travel time from the sources at depth to the surface receivers. As a result we must convert our stacked record section from two-way travel time to one-way travel time. We can make this conversion by simply dividing our stacked record time scale by two.

The migration problem can now be stated in the following terms. For simplicity, let us consider two spatial dimensions x, y and time t. The coordinate x is horizontal distance along the surface survey line, and the coordinate y is vertical distance representing depth, measured positively downward in the ground. The full wavefield is the function $u(x, y, t)$. The wavefield $u(x, y, 0)$ at time $t = 0$ represents the wave motion at the time of the initiation of the

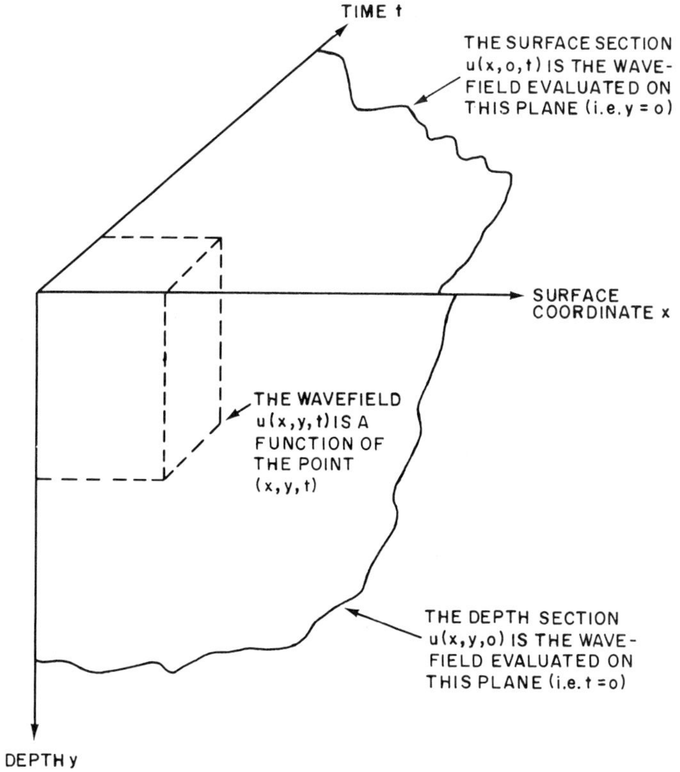

Fig. P-12. The wavefield space showing the plane of the surface section and the plane of the depth section.

sources (i.e. at the time of the simultaneous shooting of all the hypothetical sources distributed within the earth). According to our theoretical experiment, the sources are located at the reflecting interfaces and have strengths proportional to the reflection coefficients. Thus $u(x, y, 0)$ represents the cross-section of the geologic structure of the earth and is called the *depth section*. The wavefield $u(x, 0, t)$ at depth $y = 0$ (i.e. at the surface of the ground) is called the *surface section*. The surface section represents the seismic data collected as a function of time along the surface survey line; the depth section represents the desired geologic structure under the ground. The migration problem can be simply described as follows. Given the surface section $u(x, 0, t)$ find the depth section $u(x, y, 0)$ (see Fig. P-12).

In order to indicate how the migration problem is solved let us assume that the traveling wave velocity c is constant. Then the wave field $u(x, y, t)$ satisfies the wave equation:

$$\frac{\partial^2 u}{\partial t^2} - c^2\left(\frac{\partial^2 u}{\partial x^2} + \frac{\partial^2 u}{\partial y^2}\right) = 0 \qquad [3\text{-}1]$$

Let $U(k, l, \omega)$ be the three-dimensional Fourier transform of the wavefield $u(x, y, t)$; thus:

$$u(x, y, t) = \frac{1}{8\pi^3} \int_{-\infty}^{\infty} dk \int_{-\infty}^{\infty} dl \int_{-\infty}^{\infty} d\omega U(k, l, \omega) e^{i(kx + ly + \omega t)} \qquad [3\text{-}2]$$

Then the three-dimensional Fourier transform of each term in the wave equation gives the equation:

$$\omega^2 - c^2 k^2 - c^2 l^2 = 0 \qquad [3\text{-}3]$$

which is called the *dispersion equation*. If we solve this equation for the angular frequency ω we obtain:

$$\omega = \pm cl\sqrt{1 + k^2/l^2} \qquad [3\text{-}4]$$

At this point we use the fact that we only want to consider upgoing waves (i.e. waves that travel from the underground hypothetical sources up to the surface detectors, i.e. travel in the direction of negative y). Thus we want to retain the plus sign only in the above equation; that is, we want to choose:

$$\omega = + cl\sqrt{1 + k^2/l^2} \qquad [3\text{-}5]$$

Let $V(k, l, t)$ be the two-dimensional Fourier transform of the wavefield $u(x, y, t)$ with respect to the coordinates x and y; thus:

$$u(x, y, t) = \frac{1}{4\pi^2} \int_{-\infty}^{\infty} dk \int_{-\infty}^{\infty} dl \, V(k, l, t) e^{i(kx + ly)} \qquad [3\text{-}6]$$

In particular $V(k, l, 0)$ is the Fourier transform of the depth section $u(x, y, 0)$. If we take the two-dimensional Fourier transform of each term in the wave equation we obtain:

$$\frac{\partial^2 V}{\partial t^2} + c^2(k^2 + l^2)V = 0 \qquad [3\text{-}7]$$

or

$$\frac{\partial^2 V}{\partial t^2} + \omega^2 V = 0 \qquad [3\text{-}8]$$

This equation is an ordinary differential equation. There is a solution corresponding to each of the two ω values (i.e. plus or minus) given in equation [3-4]. However, for upgoing waves we choose only one solution, namely the one corresponding to the value of ω given in equation [3-5]. We write this solution as:

$$V(k, l, t) = A(k, l) e^{i\omega t} \qquad [3\text{-}9]$$

where $A(k, l)$ is a function independent of t. The inverse Fourier transform of $V(k, l, t)$ gives the wavefield:

$$u(x, y, t) = \frac{1}{4\pi^2} \int_{-\infty}^{\infty} \mathrm{d}k \int_{-\infty}^{\infty} \mathrm{d}l\, A(k, l) \mathrm{e}^{i\omega t} \mathrm{e}^{i(kx+ly)} \qquad [3\text{-}10]$$

Note that the function exp $i(\omega t + kx + ly)$ which appears under the integral sign describes a plane wave with angular frequency ω traveling with velocity c and with direction cosines:

$$(-k/[l\sqrt{1+(k/l)^2}], \quad -1/\sqrt{1+(k/l)^2}) \qquad [3\text{-}11]$$

We can obtain an expression for the surface section by letting depth $y = 0$ in equation [3-10]; we obtain:

$$u(x, 0, t) = \frac{1}{4\pi^2} \int_{-\infty}^{\infty} \mathrm{d}k \int_{-\infty}^{\infty} \mathrm{d}l\, A(k, l) \mathrm{e}^{i(kx+\omega t)} \qquad [3\text{-}12]$$

The two-dimensional Fourier transform $B(k, \omega)$ of the surface section $u(x, 0, t)$ is defined as:

$$B(k, \omega) = \int_{-\infty}^{\infty} \mathrm{d}x \int_{-\infty}^{\infty} \mathrm{d}t\, u(x, 0, t) \mathrm{e}^{-i(kx+\omega t)} \qquad [3\text{-}13]$$

and the inverse Fourier transform is:

$$u(x, 0, t) = \frac{1}{4\pi^2} \int_{-\infty}^{\infty} \mathrm{d}k \int_{-\infty}^{\infty} \mathrm{d}\omega\, B(k, \omega) \mathrm{e}^{i(kx+\omega t)} \qquad [3\text{-}14]$$

If we compare equation [3-12] and [3-14] we see that:

$$B(k, \omega)\, \mathrm{d}\omega = A(k, l) \mathrm{d}l \qquad [3\text{-}15]$$

so the function $A(k, l)$ can be obtained from the Fourier transform $B(k, \omega)$ of the surface section by the equation:

$$A(k, l) = B(k, \omega) \frac{\mathrm{d}\omega}{\mathrm{d}l} \qquad [3\text{-}16]$$

which is:

$$A(k, l) = B[k, cl\sqrt{1+(k/l)^2}] \frac{c}{\sqrt{1+(k/l)^2}} \qquad [3\text{-}17]$$

Finally, letting $t = 0$ in equation [3-9], we obtain:

$$A(k, l) = V(k, l, 0) \qquad [3\text{-}18]$$

that is, $A(k, l)$ is the Fourier transform of the depth section. Thus equation [3-17] gives us a means of finding the Fourier transform $A(k, l)$ of the depth section from the Fourier transform $B(k, \omega)$ of the surface section.

In summary, the seismic data can be migrated as follows. First, compute the Fourier transform of the given surface section by equation [3-13]. Next, compute the Fourier transform of the depth section by means of equation [3-17]. Finally, compute the required depth section as the inverse Fourier transform by means of equation [3-10] with $t = 0$, i.e. by means of:

$$u(x, y, 0) = \frac{1}{4\pi^2} \int_{-\infty}^{\infty} dk \int_{-\infty}^{\infty} dl\, A(k, l)\, e^{i(kx+ly)} \qquad [3\text{-}19]$$

In practice these Fourier transforms are computed by the fast Fourier transform algorithm (see Chapter 5).

CHAPTER 1

GEOPHYSICAL MODELING

1.1. MODELS IN SCIENCE AND ENGINEERING

Man creates models of his natural and man-made environments to understand and explain them better and as a prelude to any exploratory action. He models the economy in order to learn about price control and inflation, cost of living, balance of payments, or other factors. He models the solar system for many reasons, some of which are: (1) to understand the interactions between the sun and the celestial bodies that revolve about it; (2) to understand gravitational attraction and tidal phenomena; and (3) to send spacecrafts to the moon and other planets.

He models his bodily processes (e.g. homeostatic mechanisms), organs (e.g. heart and brain), and his entire self for countless reasons, some of which are: (1) to understand the diffusion and administration of drugs in the blood stream; (2) to understand brain waves so that, for example, epileptic patients can be forewarned of oncoming seizures; and (3) to optimize the comfort of passengers in an aircraft.

Modeling in general encompasses four problems: representation, measurement, estimation, and validation. The *representation problem* deals with how something should be modeled. In geophysics exploration, mathematical models play an important role. Within this class of models we need to know whether the model should be static or dynamic, linear or non-linear, deterministic or random, continuous or discrete, fixed or varying, lumped or distributed (continuous-time lumped-parameter systems can be described by ordinary differential equations, whereas continuous-time distributed-parameter systems can be described by partial differential equations), in the time domain or in the frequency domain (a time series is a time-domain representation, whereas a transfer function is a frequency-domain representation).

In order to verify a model, physical quantities must be measured. We distinguish between two types of physical quantities, *signals* and *parameters*. Parameters express a relation between signals. For example, Newton's law states that $F(t) = MA(t)$, where $F(t)$ is force as a function of time t, $A(t)$ is acceleration as a function of time t, and M is mass, which is assumed independent of time. Force and acceleration are signals, both of which are often easily measured, and mass is a parameter. In the differential equation:

$$\ddot{y}(t) + a_1 \dot{y}(t) + a_2 y(t) = bx(t)$$

$y(t)$, its derivative $\dot{y}(t)$ with respect to time t, its second derivative $\ddot{y}(t)$ with respect to time t, and $x(t)$ are signals, and a_1, a_2, and b are parameters. In the algebraic equation:

$$x(\lambda) = \sum_{i=1}^{N} g_i f_i(\lambda)$$

$f_1(\lambda), f_2(\lambda), \ldots f_N(\lambda)$ and $x(\lambda)$ are signals which vary with the continuous independent variable λ, whereas $g_1, g_2, \ldots g_N$ are parameters.

The distinction between what is a signal and what is a parameter is sometimes a matter of personal preference. Often, we decide that a physical quantity is a signal because it is a function of time (space, or some other measureable quantity); however, parameters can also vary with time. For example, consider a single-stage rocket. Its mass decreases as its propellant is burned; hence, for this application the mass parameter is time varying.

Not all signals and parameters are measureable. The *measurement problem* deals with which physical quantities should be measured and how they should be measured. The *estimation problem* deals with the determination of those physical quantities that cannot be measured from those that can be measured. We shall distinguish between the estimation of signals and the estimation of parameters. Because a subjective decision must sometimes be made to classify a physical quantity as a signal or as a parameter, there is some obvious overlap between signal estimation and parameter estimation.

After a model has been completely specified, through choice of an appropriate mathematical representation, measurement of "measurable" signals, estimation of "non-measureable" signals, and estimation of its parameters, the model must be checked out. The *validation problem* deals with demonstrating confidence in the model. In practice, the usual philosophy is that if the results from the model are favorable, then the model is adequate. Hence, based on a given set of operating conditions, i.e. a certain geographical location, signal-to-noise ratio, water-layer depth, etc., if the predictions or results from the model yield satisfying results, then the model is deemed favorable. However, if the operating conditions are modified, i.e. offshore exploration versus deep oil fields on land, degradation in signal-to-noise ratio, etc., does the model still respond favorably? How sensitive is the model to changes or perturbations in the parameters, and over what range of parameter variations does the model provide useful results? These are only a few examples of questions the exploration geophysicist must contend with in order to properly interpret his mathematical model.

After deciding on a mathematical model which best describes the process under investigation (whether it be economic, seismic, chemical, etc.), the next decision is the selection of a mathematical technique which provides the desired parameter estimates or model output quantities. Although the type of model (e.g. linear or non-linear, discrete, random, lumped parameter, etc.) usually dictates the technique, there are many techniques which apply to a given model. For example, the well-known least-squares criterion for parameter estimation was first used by Gauss (1809) in his efforts to determine the orbital parameters of the asteroid Ceres. The original formulation by Gauss assumed a linear, discrete, deterministic, constant-parameter model. Later, these ideas were reintroduced and reformulated in communication and control engineering as the linear least-squares problem for the estimation and prediction of random or stochastic processes. Many different methods evolved. All these mathematical techniques are closely related, and in some cases identical. We hope to bring these mathematical techniques together and present them from a unified point of view, with emphasis on their compatibility with the seismic process.

Least-squares techniques are commonly used in statistics, econometrics, control theory, and geophysics. Classical batch processing estimation algorithms are more familiar to statisticians and econometricians, whereas recursive and sequential estimation algorithms are more familiar to control theory engineers. However, depending upon the quantity of data and available computer storage space, recursive solutions might be more advantageous when processing large amounts of data. Further, the practical consideration of cost-effectiveness might warrant one technique over another. It is not uncommon in geophysics exploration to consider the problem of processing 250,000 seismic records per day. Based on current geophysical data collection procedures, this figure translates into approximately 10^{12} bits of information per day! Thus, the exploration geophysicist is faced with an enormous task. He must find a physically meaningful model of the seismic process, a compatible mathematical technique for computing his model output quantities, and process 10^{12} bits of information per day such that his overall processing operation is cost-effective and accurate enough to insure that *no important information is lost in the process!*

In summary, we will review the petroleum exploration problem and development of the deconvolution process. Our approach will be somewhat general in that we will formulate a general model, examine a wide variety of techniques, and attempt to unify various approaches to this problem. The authors hope that the final contribution of this effort will provide a deeper understanding of the deconvolution method and associated techniques, and provide the exploration geophysicist with a comprehensive summary of related signal-processing methods that will aid him in future work.

1.2 EVOLUTION OF GEOPHYSICAL MODELS

Exploration for oil and natural gas, a vital petroleum industry activity, depends on the collection, signal processing, and geophysical interpretation of data obtained from the seismic process. These data are necessary in order for geophysicists to determine the attitudes, shapes, physical properties, and structural relations of subsurface rock formations. These determinations are of critical importance because they provide otherwise unobtainable evidence that local subsurface conditions may be either favorable or unfavorable for petroleum exploration, i.e. the extraction of possible oil or gas deposits.

The data, consisting of several seismic traces which constitute a seismogram, are obtained as follows:

Typically, a seismic source imparts a pulse of acoustic or vibrational energy into the earth's subsurface. (Sources may be an air-gun blast, gas exploders, or, under appropriate conditions, high-velocity dynamite.) This source pulse travels into the subsurface rock formations where it is split into a large number of waves traveling along various paths determined by the material properties of the medium under observation. Whenever such a wave encounters a change in acoustic impedance, i.e. the product of rock density and propagation velocity, a certain fraction of the incident wave is reflected upwards. Seismic detectors, such as geophones, hydrophones, and seismometers, record the continual impact of seismic waves impinging from below. This reflected energy as a function of time constitutes a *continuous* time series. If the recording is performed digitally at a fixed sampling increment, the resulting set of discrete observations is referred to as a *discrete* time series. Thus, seismic data are in the form of time series, which constitute the set of observations available for geophysical analysis. The procedure just described is referred to as the reflection seismic method. Although exploration geophysicists employ gravity, magnetic, magnetotelluric, and reflection and refraction seismic methods, the most widely used technique in petroleum exploration is the reflection seismic method.

Seismic data in the form of time series may be the only source of information about large-scale earth features such as deep structure of entire basins. Such information is vital to the understanding of the geologic evolution of the basin and its potential for oil. Moreover, the reflection seismic method provides information on hydrocarbon trap geometries and, under favorable conditions, it provides information about gross characteristics of the reservoir rocks and the presence of the hydrocarbons. This is the only direct means, other than costly drilling, that exploration geophysicists have for locating and identifying hydrocarbon accumulations. However, not all recorded reflections are "desirable", i.e. only those reflections which can be identified with any structural features or with hydrocarbon deposits are termed "desired reflections". All other wave motion is therefore regarded as interference. Further, ambient noise will also mask any desired reflections. Thus, the ultimate problem for the exploration geophysicist is to identify and delineate the desired reflections that are concealed in the undesired wave motion and ambient noise of the medium.

If we associate the desired reflections with some desired *signal* and the undesired wave motion as being an *interference*, then the seismic time series can be regarded as the superposition of a signal component, interference component, and a noise component due to the ambient noise of the transmission medium or channel. In the context of signals, interference, and noise, the problem in petroleum exploration appears to be the same problem encountered in electrical engineering, i.e. the detection of a desired signal in noise. Although the communications engineer has developed a wealth of information for solving certain classes of signal detection problems, there are some important assumptions and limitations associated with these solutions which must be carefully examined before any attempt is made to apply electrical engineering theory to the petroleum exploration problem. Thus, it is extremely important that we properly characterize the seismic process. This means that we develop the best possible physical model of the process and then select the available technique which best fits our model. If no technique exists, it then becomes necessary to develop new methods for the problem's solution.

One of the advantages that the communications engineer often has over the geophysicist is that he puts a known signal into his channel and he normally knows what to look for in the received signal, which contains the known (desired) signal plus additive noise. This problem has been solved by electrical engineers and is referred to as the matched-filter solution. The geophysicist, on the other hand, puts into the earth (in the conventional process) a sharp pulse-like time signal, which is characterized by a wide frequency spectrum. Due to the wide frequency band characteristics of this signal and the absorption of high-frequency components by the earth, the received signal cannot be modeled as the original input signal plus noise. In many signal detection situations the desired signal is described by a set of parameters (e.g. amplitude, phase, waveshape) which are not precisely known to the observer. For example, the Bayes likelihood ratio philosophy, developed by electrical engineers, assumes that in most conceivable situations, the detector designer has some knowledge (beyond absolute uncertainty) about the unknown signal parameters. With this a priori information, an appropriate detector may be developed. Thus, since the geophysicist does not know in advance what signal to look for in the received signal, he must find new criteria for detecting (separating) the "signal" components from the non-signal components.

Since the geophysicist works with very little a priori information concerning the desired signal, and the transmission channel of the earth is so difficult to define, the conventional communication theory of the electrical engineer is not directly applicable to the seismic process. Thus, new methods had to be developed which were compatible with the geophysicist's requirements. As late as the early 1950's there was essentially no work being done as to the possible use of communication theory methods in geophysical exploration. Neither geologists nor field geophysicists were using any refined mathematical techniques at the time; perhaps a little statistics in sampling and evaluating ore deposits, some geometry and trigonometry in surveying and crystallography, and a little calculus and some differential equations in representing some simple dynamical situations, but that was about all. Exploration geophysicists had to "eye-ball" their seismic data (records) in order to pick out first arrivals and desired reflections. The problem was to find whether there might be certain kinds of geophysical problems that could be investigated more satisfactorily by using communication theory techniques. The most direct approach was to examine some seismograms to see if they might be amenable to this kind of analysis.

The great increase in demand for petroleum products during the years immediately following World War II presented the petroleum industry with the need for a more extensive exploration program. As this program got underway, it soon became clear that the seismic methods developed during the 1930's and early 1940's were not sophisticated enough to explore successfully many of the potential oil-producing regions of the world. In particular, the existing seismic methods were not very successful in exploring offshore areas because of the water reverberations, and in exploring deep strata because of the reverberations of the near-surface layers. By 1950 highly refined electrical engineering methods had been exploited almost to their fullest extent by the petroleum companies in attempts to filter the seismic data, yet these methods had failed to solve the reverberation problem except for the simplest cases. The situation was well described by one of the pioneers of exploration geophysics when he stated in 1950:

> "We have sharpened our tools about as much as we can; what we need now is a new tool."

Much active research was being carried out in universities and industrial laboratories at that time to find such new tools. However, most of this research was devoted to building more elaborate deterministic physical models of the earth, but these models involved such elaborate mathematics that solutions could only be found for the simplest possible cases. These simple cases were of little use to the geophysicist who analyzed real field data, representative of the practical problems of estimating complicated subsurface structures.

A completely different approach to the reflection seismic method led to the development of the concept of deconvolution in 1954 to solve the reverberation problem and the estimation of the subsurface structure. Deconvolution represented the new tool which the exploration industry required. Whereas the existing seismic methods were analog in nature, the characteristic feature of deconvolution was that it was digital. This digital method required for its successful use the accuracy, storage capacity, and speed of the large digital computer. The philosophy behind deconvolution provided the linkage of the mathematical concepts of communication theory with the physical theory of seismic wave propagation: namely, the transmission characteristic of a waveform propagating through a layered medium is *minimum-delay*,* and as a result, a least-squares *deconvolution* filter can be designed from the observed seismic data to remove the effects of propagation. The application of the deconvolution method to field seismic data showed that this digital method could automatically transform, within the computer, a seismic data record that could not be successfully interpreted by existing methods into a record that would provide the geophysicist with the proper information about the subsurface structure.

Owing to a decline in exploration activity in the late 1950's, due largely to the great Middle East discoveries and also because of the relative expense of vacuum-tube computers, the use of the digital deconvolution method did not become widespread until the 1960's when oil exploration again became very active and transistorized computers were available. The essentially complete conversion of seismic exploration at that time, from analog to digital methods, constitutes the so-called "digital revolution" in geophysical exploration. Since the mid-1960's virtually all seismic exploration for petroleum has made use of digital methods in which every seismic record is deconvolved. The discoveries of offshore oil and natural gas deposits as well as many of the deep oil fields on land made in the last decade represent the fruition of these digital methods.

Today, the multi-billion-dollar industry of seismic exploration for petroleum is one of the leading users of digital computers. As a matter of fact, the petroleum exploration industry makes use of more magnetic computer tape than any other organization in the world: academic, industrial, or governmental!

1.3 STATISTICAL MODELS IN GEOPHYSICS

In all fields of geophysics, and in seismology in particular, much of the basic data collected is in the form of time series. Through the analysis of these time series, the geophysicist attempts to gain a better physical understanding of his environment. In some cases the data are uncomplicated and the physical information they contain may be extracted by straightforward analytical methods. More frequently, however, the data are complicated, and the development of physical models and refined computer analyses are necessary.

Any observed geophysical phenomenon is composed of an abundance of subsidiary physical processes. Depending on which geophysical process is selected as the subject of study, the analyst must somehow separate these subsidiary processes and identify the process of interest. If the process under investigation has some salient features associated with it, he may then

* The concept of minimum-delay will be discussed in Chapter 2.

construct a physical model based on these features which will aid him in his information extraction procedure. For example, in radar and sonar detection problems, a statistical model of the background noise allows the communications engineer to identify which parts of the measured process are "signal" and which parts are "noise". The amplitude of the noise at a particular point in time and space might be characterized by an appropriate probability density function and the autocorrelation description of the noise might have an assumed or known relationship. Thus, we sometimes find that there are central relationships that determine the basic features of geophysical processes. In studying the process, one must identify and take into account the essential features and at the same time disregard the unimportant details caused by subsidiary features. It is not a geophysical phenomenon in all its complexity which is subject to analysis, but a simplified physical model of the desired process whose behavior coincides basically with the behavior of the phenomenon, except for details of a minor or less critical nature. The criterion for the correctness of a model is the agreement between theoretical results and results obtained from field data. Moreover, the model of a geophysical phenomenon should be constructed on the basis of making explicit its connections with related phenomena. The subdivision of factors into essential and non-essential ones depends not only on the specific nature of the geophysical phenomenon itself, but also on the actual problem to be solved. Of course, it is often not possible to find such an ideal model in geophysics, so compromises between the ideal and the obtainable must be made.

In classicial geophysical analysis, a great many models are known in which the behavior of a system or process is fully determined by initial-value and boundary-value conditions. Such is the case in the earth model first treated in a classic paper by Lamb (1904). This model is that of a perfectly elastic, homogeneous, isotropic medium bounded by a free plane surface. Lamb showed that a vertical or horizontal impulse applied along a line on the surface produces a P-pulse, an S-pulse, and a Rayleigh pulse in that order. The physical laws that apply to models of this sort are known as dynamical laws. These laws are characteristic in cases where there is a unique specification of the consequences of a given cause.

In addition to models of geophysical phenomena which lead to the setting up of dynamical laws, other models lead to the formulation of laws of a different nature, namely statistical laws. To clarify this concept, let us consider as an example a model taken from geology. Time plays a peculiar role in geology. Whereas in most sciences time can be taken as an independent variable, which is assumed to be known, the geologist sees time as a dependent variable. Consequently, the geologist is faced with problems unique to his science in his effort to measure time quantitatively in terms of events which have occurred over billions of years. Geologic time over these past eons (a division of geologic time) can be defined only by observations and measurements taken on the earth, moon, and planets, together with astronomical evidence. These measurements, whether they be geological, paleontological, geochemical, radiological, geophysical, or astronomical, are subject to "measurement" errors. These errors lead to statistical fluctuations in the reported measurements of geologic time. As a result, geologic time as we know it is not a perfectly measured or uniform variable, but is subject to "chance" effects, that is, geologic time is a random variable.

Let us consider another example, namely the problem of determining the depths of the stratigraphic layers in the earth by seismic prospecting for oil and natural gas. The deep sedimentary layers were laid down in geologic time in an unsystematic way and thus the seismic events produced by these layers are unsystematic in space and time. If it is possible to have at our disposal unlimited computational means and extremely detailed and accurate data, i.e. no measurement errors, a dynamic model could be constructed which would make use of the laws of wave propagation to describe the seismic wave motion. However, for practical situations, the theoretical difficulties connected with the solution of such a problem are virtually insurmountable. As an alternative, let us investigate the possibility of constructing a statistical geophysical model. In considering such an approach, one might initially pose the following question: Even though the sedimentary layers of the earth's subsurface were laid down in geologic time in an unsystematic manner, are they fixed during the course of a seismic experiment conducted in a specific geographic location? Now, assuming that one could perfectly measure the response of the fixed multi-layered earth to an impulsive seismic source in the presence of no ambient noise, then the resultant time series should represent a deterministic process. Further, if one repeats the experiment over and over again under the same conditions, does one indeed measure the identical deterministic time series? Thus, how does one justify a statistical geophysical model? If we view a random or stochastic process as a process developing in time and controlled by probabilistic laws, then it is not apparent that the seismic experiment discussed above be considered as one realization of a random process. However, let us probe deeper into the geophysical process in order to obtain an answer to this question.

In the context of linear system theory, the time series response of the earth to an impulsive seismic source is effectively the impulse response of the earth. Now if the earth model includes the depths and reflectivities (reflection coefficients) of the geologic beds, then these depths and reflectivities are actually unknown but constant parameters. Assuming that we have knowledge of the seismic source waveshape, then we must identify these parameters from the given observed time series response. This problem, referred to by control engineers as the systems identification problem, assumes some underlying mathematical model of the earth. If we assume that the seismic source can be represented by an ideal impulse, then we are in effect estimating the constant parameters of the earth's impulse response. Although this problem is easily formulated, it is heavily dependent on the assumed mathematical model of the earth, the time waveshape of the seismic source, and for complicated earth models becomes computationally complex. Suppose that all the earth's parameters, i.e. the depths and reflection coefficients of the subsurface layers, could be perfectly identified. For the nth layer there is a corresponding reflection coefficient r_n; and for a total of N layers, there are $r_1, r_2, \ldots r_N$ reflection coefficients. Since the geologic formation of these layers was done in an unsystematic fashion, one might suspect that the sequence of numbers $r_1, r_2, \ldots r_N$ are also unsystematically related. In the field of statistics, one would investigate the correlation of the above sequence in order to determine any predictable, orderly relation between these reflection coefficients. It is also physically intuitive to reason that the reflection sequence is uncorrelated, i.e. there is no systematic relation between r_1 and r_3, r_2 and r_7, etc. Thus, we introduce some statistical considerations into the development of an earth model. Merely to state that the sequence of constant parameters $r_1, r_2, \ldots r_N$ represent an uncorrelated sequence of numbers is not remarkable per se, but this important observation is the key to the development of the statistical geophysical models that we discuss.

1.4. THE CONVOLUTION MODEL

In the reflection seismic method, seismometers record the response of the earth's subsurface to an impulsive-like seismic source, which we shall denote as the source wavelet. This recorded signal is representative of the reflected energy received at the seismometer as a function of time and constitutes a time series. Now let us consider the boundary between two adjacent subsurface layers. When a seismic wave encounters this boundary, it experiences a change in acoustic impedance and a certain fraction of this incident energy is reflected upwards. Also, a certain fraction is transmitted downward through the boundary. As a result of many such encounters, the received time series can be considered as a sum of amplitude-scaled and time-delayed wavelets, with the amplitude scale factors being dependent on the properties of the reflecting layers and the time delays dependent on the depths and wave propagation velocities of the constituent layers. In mathematical terms, if we denote the source wavelet as $s(t)$, then the received time series $y(t)$ is given by:

$$y(t) = \sum_{n=0}^{\infty} f_n s_n(t - \tau_n) + n(t) \qquad [1.4\text{-}1]$$

where f_n = amplitude scale factor, τ_n = time delay, and $n(t)$ = additive noise.

Equation [1.4-1] represents a model for the reflection seismic method, although much research in other areas employs the same model. For example, research in fields such as communications, speech, radar, sonar, and biomedical data processing consider the analysis of time series of multiple overlapping wavelets in a noisy environment. It is important to note that in writing the source wavelet as $f_1 s_1(t-\tau_1), f_2 s_2(t-\tau_2), \ldots f_n s_n(t-\tau_n)$, we are allowing for the fact that the medium offers distortion, i.e. the transmitted wavelet's basic waveshape is altered, for example, by the absorption of high-frequency components by the earth. Thus, somehow the geophysicist must identify those reflections which came from structural features within the earth, say $f_j s_j(t-\tau_j)$, in the presence of numerous interfering reflections and ambient noise. Viewing [1.4-1] as it presently stands, there appears to be too many unknowns. For example, we know that $s(t)$ is a wideband frequency source, but we do not know the true wavelet shape. (In "active" radar and sonar problems, the transmitted waveshape is known.) The quantities f_n and τ_n represent constant but unknown parameters. Moreover, if we do not assume any statistical model for the noise, then the problem in its present form is not amenable to analysis. Let us consider the following assumptions:

(1) Assume that the seismic process satisfies the classical theory of the propagation of elastic waves in homogeneous, isotropic media, i.e. a waveform remains unchanged as it is transmitted through the medium. (Assume a distortionless transmission channel.)
(2) Assume that the continuous representation in [1.4-1] has been properly converted to digital form via the sampling theorem. Hence, $t = k\Delta t$ where Δt is the time between sampled values, and $k = 0, 1, 2, \ldots$
(3) Assume that the time delay τ_n can be represented by an integer multiple of Δt, such that $\tau_n = n\Delta t$, $n = 0, 1, 2, \ldots$

[1.4-2]

With the considerations in [1.4-2], equation [1.4-1] becomes:

$$y(k\Delta t) = \sum_{n=0}^{\infty} f_n s[(k-n)\Delta t] + n(k\Delta t) \qquad [1.4\text{-}3]$$

Defining the quantities $y(k\Delta t) = y_k$, $s[(k-n)\Delta t] = s_{k-n}$, and $n(k\Delta t) = n_k$, then [1.4-3] becomes:

$$y_k = \sum_{n=0}^{\infty} f_n s_{k-n} + n_k \qquad [1.4\text{-}4]$$

With the assumptions of [1.4-2], we recognize the convolution sum expressed by [1.4-4]. Thus, the given seismic model is greatly simplified and states that the seismogram $\{y_k\}$ is the convolution of the source wavelet $\{s_k\}$ with the impulse response of the earth $\{f_k\}$ plus an additive noise sequence $\{n_k\}$. Fig. 1-1 depicts the convolution model.

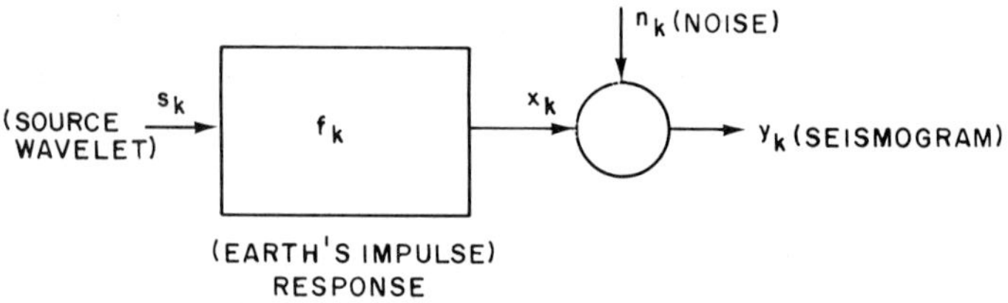

Fig. 1-1. The convolution model.

Thus, [1.4-4] and Fig. 1-1 represent a convolutional model, characteristic of linear time-invariant systems. In this form, the reflection seismic model is more suitable for analysis, since we have at our disposal the well-developed techniques associated with linear system theory. However, the sequence $\{f_k\}$ merely represents a sequence of numbers, which are somehow related to the reflectivities (reflection coefficients) of the subsurface layers. One of our major goals is to find that relationship.

1.5. THE ROBINSON SEISMIC MODEL

In this section, we shall present the work of Robinson (1954), which began a new approach to geophysical modeling. It is important that we review the evolution of this new approach, the limitations and complexity of the classical dynamical models, and the computational simplicity afforded by the formulation of the Robinson model. This model in its final form is simple and, as a result, can be easily evaluated by conventional computational methods. However, without understanding the basic underlying assumptions and physical approximations involved in the model development, one cannot view this simplicity in the proper perspective. Furthermore, we shall give the so-called "statistical" interpretation of the Robinson seismic model so that we can gain deeper physical insight into the theory of deconvolution, which will be discussed in Chapter 4.

Part 1: Transmission of a "deep" source through the earth

Consider an inhomogeneous system (sedimentary layers) bounded by (sandwiched in) two homogeneous infinite half-spaces, the air and basement rock. Fig. 1-2 depicts the situation. Now in earthquake seismology the source is considered to be embedded deep within the earth's subsurface. The classical approach to determining the response of the earth to a deep-source excitation was to solve the governing partial differential equations, which described the wave propagation through the earth into the air, subject to the initial-value and boundary-value conditions. Thus, due to the presence of the governing partial differential equations, the earth was considered as a distributed-parameter system. Research in classical seismology is concerned

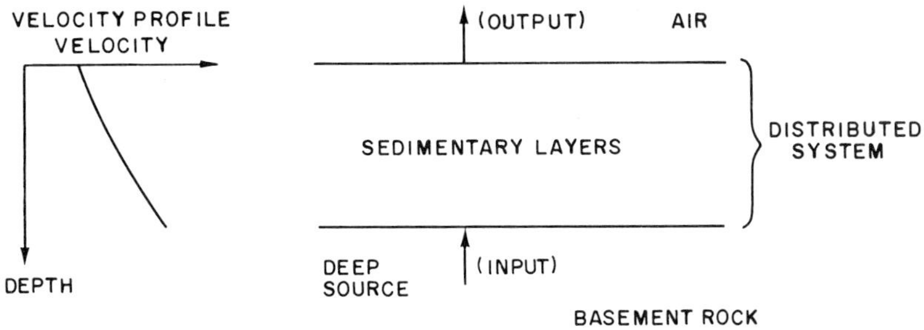

Fig. 1-2. The inhomogeneous earth excited by a deep source and considered as a distributed-parameter system. The associated velocity profile of this system is considered as a continuous function of depth.

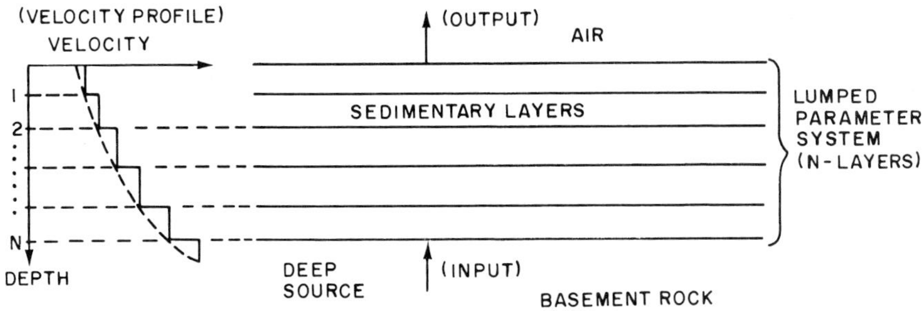

Fig. 1-3. The multi-layered earth excited by a deep source and considered as a lumped-parameter system. The continuous velocity profile is now represented as a discontinuous profile.

with analytic solutions of these partial differential equations, and except in the simplest cases such solutions are mathematically extremely difficult to find.

Consider now the idea of modeling the earth by a lumped-parameter system. For example, we could quantize the continuous velocity profile (distribution) shown in Fig. 1-2. This quantization is equivalent to considering the earth as a multi-layered system, with the velocity of propagation in each layer and the depth of a layer determined by the quantization procedure. A description of the earth as a lumped-parameter system is given in Fig. 1-3.

Now if the time of signal propagation through a layer is short compared with the duration of the signal, then the lumped-parameter assumption is valid. By choosing the thickness of each layer to be very small, i.e. considering many layers, we can satisfy the lumped-parameter conditions. In terms of transmission lines and linear network theory, Fig. 1-2 represents a distributed circuit-element transmission line while Fig. 1-3 represents a lumped circuit-element transmission line. Note that the model for the lumped-parameter system contains a discrete velocity profile distribution, with each discontinuity in the velocity representative of one transmission line section. Thus, the lumped-parameter system is modeled by many sections of transmission lines.

Now if we further assume that the system of Fig. 1-3 is linear and time-invariant, we can expect that the input (ϵ_k) and output (x_k) satisfy the linear difference equation:

$$x_k + a_1 x_{k-1} + \ldots + a_N x_{k-N} = \epsilon_k \qquad [1.5\text{-}1]$$

where the a_n's represent the difference equation operator (i.e. constant parameters associated with our lumped-parameter system which has N degrees of freedom represented by the N layers). (For future reference, the symbols of this section are identical to those of Robinson (1954).) Our reason for considering a lumped-parameter system with the difference equation [1.5-1] is that difference equations are easily solved by digital computers. Thus, the difficulty of analytic solutions of wave motion is eliminated by the numerical solution of equation [1.5-1].

On physical grounds, we know that equation [1.5-1] represents a stable system, i.e. for every bounded input we observe a bounded output. Now, given the sequences $\{x_k\}$ and $\{\epsilon_k\}$, we shall define their respective Laplace z-transforms by:

$$X(z) = \sum_{k=-\infty}^{\infty} x_k z^k, \quad E(z) = \sum_{k=-\infty}^{\infty} \epsilon_k z^k \qquad [1.5\text{-}2]$$

(The definition of the Laplace z-transform and its relationship to the conventional z-transform is discussed in the Appendix.) With the definition in [1.5-2], the Laplace z-transform of each side of equation [1.5-1] gives:

$$X(z)\,(1 + a_1 z + a_2 z^2 + \ldots + a_N z^N) = E(z) \qquad [1.5\text{-}3]$$

or:

$$\frac{X(z)}{E(z)} = \frac{1}{1 + a_1 z + a_2 z^2 + \ldots + a_N z^N} \qquad [1.5\text{-}4]$$

Equation [1.5-4] represents the transfer function of a so-called "all pole" model of a linear time-invariant system because in the (finite) complex z-plane this function has poles but no zeroes. Our physical claim that the lumped-parameter model represents a stable system is equivalent to the mathematical condition that the transfer function $X(z)/E(z)$ contains no poles inside the unit circle. (Using the conventional z-transform, the condition for stability is that $X(z)/E(z)$ have no poles outside the unit circle.) Let us now formulate this property in terms of the difference equation operator $\{1, a_1, a_2, \ldots a_N\}$. The Laplace z-transform of the operator $\{1, a_1, a_2, \ldots a_N\}$ yields the denominator of equation [1.5-4]. Thus, the property that the transfer function $X(z)/E(z)$ have no poles for $|z| < 1$ is equivalent to the property that:

$$1 + a_1 z + a_2 z^2 + \ldots + a_N z^N \neq 0 \quad \text{for } |z| < 1 \qquad [1\text{-}5.5]$$

We can therefore state that the operator $\{1, a_1, a_2, \ldots a_N\}$ yields a stable difference equation if and only if its Laplace z-transform has no poles or zeroes inside the unit circle. Therefore, an important result from Robinson's work is that the Laplace z-transform of the coefficients of the stable difference equation in [1.5-1] has no zeroes within the unit circle.

Now if we replace the input to our lumped-parameter system by a unit impulse, i.e. let $\epsilon_k = \delta_k$ where:

$$\delta_k = \begin{cases} 1, & k = 0 \\ 0, & k \neq 0 \end{cases} \qquad [1.5\text{-}6]$$

then the impulse response, denoted by $\{b_k\}$, is a stable time sequence. This condition follows from physical reasoning. But since:

$$B(z) = \frac{1}{1 + a_1 z + a_2 z^2 + \ldots + a_N z^N} \qquad [1.5\text{-}7]$$

it follows that $B(z)$ has no zeroes or poles inside the unit circle. Now $\{b_k\}$ is the sequence which is transmitted through the earth into the air when a unit impulse (deep source) excites the earth. From actual observation of field seismic data, $\{b_k\}$ has the bulk of its energy concentrated at its beginning, and is rapidly attenuated due to successive reflections and refractions within the earth. Robinson defined sequences that possess this "front-loaded" property as *minimum-delay* sequences.

Now, if $A(z) = 1 + a_1 z + a_2 z^2 + \ldots + a_N z^N$ is the Laplace z-transform of the operator $\{1, a_1, a_2, \ldots a_N\}$, then from [1.5-7] we observe that:

$$B(z)\, A(z) = 1 \qquad [1.5\text{-}8]$$

The inverse Laplace z-transform of [1.5-8] yields:

$$\sum_{n=0}^{\infty} b_n a_{k-n} = \delta_k, \quad k \geqslant 0 \qquad [1.5\text{-}9]$$

which shows that the operator $\{a_k\}$ and transmission impulse response $\{b_k\}$ are mutually inverse sequences. Thus, the linear operator $\{1, a_1, a_2, \ldots a_N\}$ is the sequence which reduces the impulse response $\{b_k\}$ to a unit impulse δ_k, defined by [1.5-6]. The unit impulse δ_k is sometimes referred to as a unit spike in geophysical signal-processing terms. Let us summarize our conclusions thus far:

(1) From purely physical reasoning, we know that the layered system is stable. This physical conclusion led us to the mathematical relation in [1.5-5], i.e. that the operator $\{1, a_1, a_2, \ldots a_N\}$ has no zeroes or poles inside the unit circle.

(2) From actual physical observations, we can observe the fact that $\{b_k\}$ is

minimum-delay, that is, the bulk of the energy of $\{b_k\}$ is concentrated at its beginning.

(3) Now since $\{a_k\}$ and $\{b_k\}$ are mutually inverse sequences, then we can conclude that $\{a_k\}$ is a stable difference equation operator (i.e. the Laplace z-transform of $\{a_k\}$ has no poles or zeroes within the unit circle) if and only if $\{b_k\}$ is minimum-delay (i.e. the energy of $\{b_k\}$ is front-end loaded). That is, from purely physical grounds, we arrive at the important result:

> $\{a_k\}$ *is a stable difference equation operator if and only if* $\{b_k\}$ *is a minimum-delay response*

Now we defined $\{b_k\}$ as the transmission impulse response of the layered (lumped-parameter) model. The sequence $\{a_k\}$ is the difference equation operator defined in [1.5-1]. From physical arguments, the operator $\{a_k\}$ is stable and the transmission impulse response $\{b_k\}$ is minimum-delay. But the convolution sum defined in [1.5-9] and also expressed by $a_k * b_k = \delta_k$, indicates that $\{a_k\}$ is the operator which converts $\{b_k\}$ to δ_k. Thus, $\{a_k\}$ is the deconvolution operator which removes the effect of transmission. Fig. 1-4 describes the deconvolution operation.

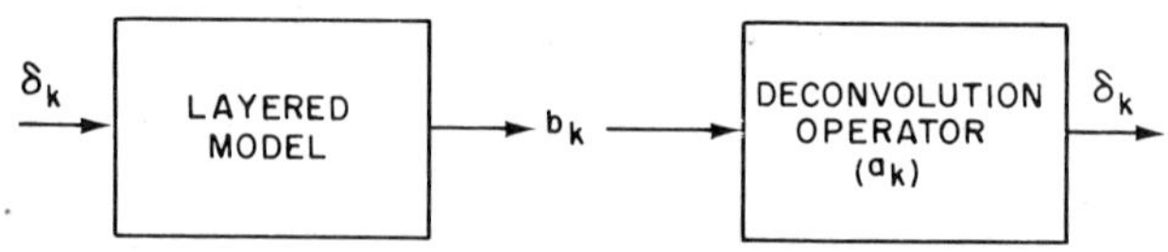

Fig. 1-4. The sequence $\{a_k\}$ viewed as the basic decomposition (deconvolution) operator which reduced $\{b_k\}$ to a unit spike.

Because $\{b_k\}$ represents all the multiple reflections and refractions resulting from an impulsive input, $\{b_k\}$ also represents the system reverberation; $\{b_k\}$ = transmission (impulse) response = layered-system reverberation wavelet. Thus, $\{a_k\}$ can also be viewed as the deconvolution operator which removes the layered system reverberation. We therefore claim the following:

Part 2: Reflections caused by a surface source

Next, let us consider the N-layered (lumped-parameter) system of Fig. 1-3 with the "deep" source replaced by a downgoing unit impulse (source) initiated at the surface. In the following discussion we shall consider two types of reflections: (1) internal primary reflections, and (2) external primary reflections.

Internal primary reflections

We shall consider the case of a downgoing unit impulse initiated at the surface. This situation, along with a pictorial definition of an internal primary reflection, is described in Fig. 1-5.

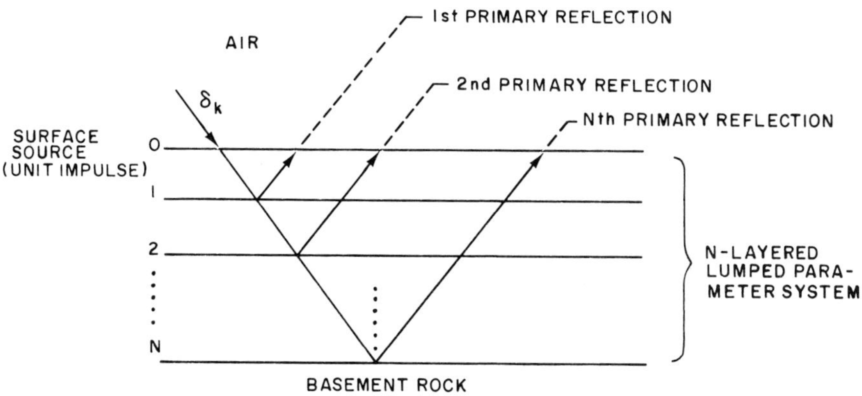

Fig. 1-5. Internal primary reflections caused by a downgoing unit impulse applied at the surface. For clarity, ray paths are drawn at oblique incidence but wave motion analysis is for normal incidence.

Now in Part 1, we claimed that an upgoing unit impulse (source) incident on the lowest interface gave rise to the transmitted (i.e. the system reverberation) waveform $\{b_k\}$. If we consider this situation in the context of linear time-invariant system theory, then the unit impulse δ_k is the *cause* and the system reverberation $\{b_k\}$ is the *effect*. Now if the cause is delayed by m units of time, i.e. δ_{k-m}, then the effect is also delayed by m units of time, i.e. $\{b_{k-m}\}$. Also, if the cause is scaled by a constant C, i.e. $C\delta_k$, then the effect is also scaled by the constant C, i.e. $C\{b_k\}$. These properties, together with the property of superposition, characterize linear time-invariant systems. Thus, instead of considering the downgoing unit impulse (source), let us replace it by a set of hypothetical (i.e. mathematical) upgoing sources at the reflecting interfaces of our N-layered system. Since we know the "effect" $\{b_k\}$ due to the upgoing "cause" δ_k, we can then apply the linear time-invariant system properties to deduce the effect due to a downgoing unit impulse at the surface. Fig. 1-6 shows the downgoing source impulse replaced

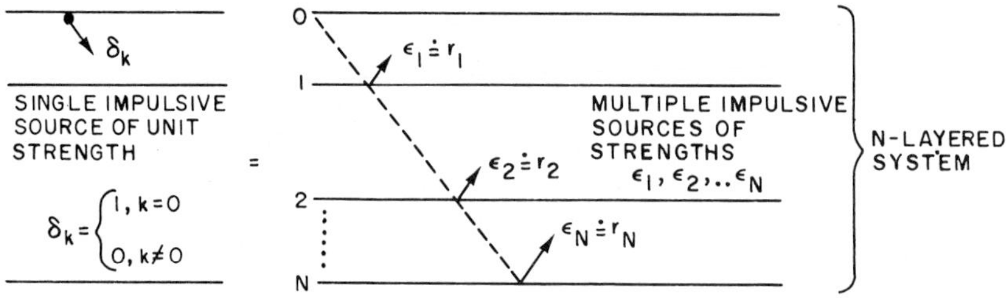

Fig. 1-6. The downgoing unit impulse surface source δ_k replaced by hypothetical (i.e. mathematical) upgoing impulsive sources of strength $\epsilon_n \doteq r_n$.

by upgoing sources of strength $\epsilon_n \doteq r_n$, where r_n is the reflection coefficient of interface n. From physical experience gained in seismic prospecting, it is known that generally these reflection coefficients are small in magnitude, that is, $|\epsilon_n| = |r_n| < 0.1$.

Each of these hypothetical sources gives rise to a reverberation wavelet due to the layering. Robinson made the hypothesis that each reverberation wavelet is the same shape, namely the system reverberation wavelet $\{b_k\}$. In Chapter 2 we shall examine this hypothesis in some detail and show the precise mathematical conditions under which it is a good approximation to the exact model we will develop. Taking into account the time delays associated with the source locations (Fig. 1-6), we reason as follows:

If an upgoing unit spike $\delta_k \xrightarrow[\text{(gives rise to)}]{}$ transmission impulse-response b_k then:

$$\begin{aligned}
\epsilon_1 \delta_{k-1} &\longrightarrow \epsilon_1 b_{k-1} \\
\epsilon_2 \delta_{k-2} &\longrightarrow \epsilon_2 b_{k-2} \\
&\cdot \\
&\cdot \\
\epsilon_N \delta_{k-N} &\longrightarrow \epsilon_N b_{k-N}
\end{aligned}$$

Thus, using the superposition property of linear systems, we can add the responses due to all the hypothetical sources to obtain the total response $\{x_k\}$. Doing so, we get:

$$x_k = \epsilon_1 b_{k-1} + \epsilon_2 b_{k-2} + \ldots + \epsilon_N b_{k-N} \qquad [1.5\text{-}10]$$

which is in the form of the convolution (summation) model discussed earlier. Equation [1.5-10] represents the surface seismogram resulting from a downgoing unit impulse initiated at the surface. Further, $\{x_k\}$ is the resultant of all the system reverberations (reverberation wavelets b_k) due to the hypothetical upgoing sources with strengths given by the reflection coefficient sequence $\{r_k\}$. Rewriting [1.5-10] as a convolution sum, we get:

$$x_k = \sum_{n=1}^{N} \epsilon_n b_{k-n} = \epsilon_k * b_k \qquad [1.5\text{-}11]$$

The Laplace z-transform of [1.5-11] gives:

$$X(z) = E(z)\,B(z) \qquad [1.5\text{-}12]$$

where:

$$E(z) = \epsilon_1 z + \epsilon_2 z^2 + \ldots + \epsilon_N z^N \qquad [1.5\text{-}13]$$

Substitution of [1.5-7] and [1.5-13] into [1.5-12] yields:

$$X(z) = \frac{\epsilon_1 z + \epsilon_2 z^2 + \ldots + \epsilon_N z^N}{1 + a_1 z + a_2 z^2 + \ldots + a_N z^N}$$

or: [1.5-14]

$$\frac{X(z)}{E(z)} = \frac{1}{1 + a_1 z + a_2 z^2 + \ldots + a_N z^N}$$

where the reflection coefficient sequence $\{\epsilon_1, \epsilon_2, \ldots \epsilon_N\}$ represents the strengths of the hypothetical sources and $\{1, a_1, a_2, \ldots a_N\}$ the stable difference equation operator discussed in Part 1.

Let us now consider the case in which our source excitation is not an ideal unit impulse δ_k. Thus, instead of δ_k, we shall consider an arbitrary surface source $\{s_k\}$ which can be represented in terms of δ_k by:

$$s_k = \sum_{n=0}^{\infty} s_n \delta_{k-n}, \quad k \geqslant 0 \qquad [1.5\text{-}15]$$

Recalling the properties of linear time-invariant systems, we can arrive at the response $\{x_k'\}$ due to an arbitrary source $\{s_k\}$ as follows:

If a downgoing unit impulse $\delta_k \xrightarrow[\text{(gives rise to)}]{} x_k = \epsilon_k * b_k$ then:

$$s_1 \delta_{k-1} \longrightarrow s_1 x_{k-1}$$

$$s_2 \delta_{k-2} \longrightarrow s_2 x_{k-2}$$

$$\cdot \qquad \cdot$$

$$\cdot \qquad \cdot$$

Using the superposition property of linear systems, we can add the responses $s_1 x_{k-1}, s_2 x_{k-2}, \ldots s_n x_{k-n}$, which gives:

$$x_k' = \sum_{n=0}^{\infty} s_n x_{k-n} = s_k * x_k, \quad k \geqslant 0 \qquad [1.5\text{-}16]$$

Substitution of [1.5-11] into [1.5-16] gives:

$$x'_k = s_k * \epsilon_k * b_k = \epsilon_k * (b_k * s_k) \qquad [1.5\text{-}17]$$

Defining $w_k \equiv b_k * s_k$ as the composite wavelet consisting of the source wavelet $\{s_k\}$ and the system reverberation wavelet $\{b_k\}$, we get:

$$x'_k = \epsilon_k * w_k \qquad [1.5\text{-}18]$$

Now, from physical reasoning, we concluded that $\{b_k\}$ is minimum-delay. Further, by proper design of the field data collection procedures, the source wavelet $\{s_k\}$ can be made minimum-delay, or approximately so. Thus, we can assume that $\{w_k\}$ is a minimum-delay wavelet.

Let us now introduce the statistical aspect of Robinson's model. From geological considerations, Robinson hypothesized that sections of the sequence of reflection coefficients $\{\epsilon_1, \epsilon_2, \ldots \epsilon_n\}$ can be considered to be uncorrelated. Specifically, if we examine a section of a seismic record from time k to $k + L$, then the sequence of hypothetical sources $\epsilon_k, \epsilon_{k+1}, \ldots \epsilon_{k+L}$ is assumed uncorrelated over this time interval. Let us summarize the Robinson seismic model for the case of internal primary reflections (total system reverberation):

(1) The model is a convolutional model represented by:

$$x'_k = \epsilon_k * w_k$$

with $\{\epsilon_k\}$ = reflection coefficients at the N reflecting interfaces; $\{w_k\}$ = composite wavelet = $s_k * b_k$; $\{s_k\}$ = source wavelet; $\{b_k\}$ = system reverberation wavelet; and $\{x'_k\}$ = observed seismic record (time series).

(2) The composite wavelet $\{w_k\}$ is minimum-delay. This follows from the physical fact that $\{b_k\}$ is minimum-delay and $\{s_k\}$ can be approximately minimum-delay by properly designed field procedures.

(3) The reflection coefficients $\epsilon_n = r_n$ are small in magnitude, a fact generally observed in seismic prospecting.

(4) The reflection coefficients $\{\epsilon_1, \epsilon_2, \ldots \epsilon_N\}$ are uncorrelated over various intervals $[k, k + L]$ of the seismic record. This hypothesis added statistical considerations into geophysical modeling. In this sense, the model is sometimes referred to as the Robinson "statistical" seismic model.

Now, we previously showed that the operator $\{a_k\}$ was a deconvolution operator in the sense that it reduced the system reverberation $\{b_k\}$ to a unit spike δ_k (see Fig. 1-4). Thus, if we desire to eliminate the reverberation component $\{b_k\}$ from the observed seismic record $\{x'_k\}$, then we would convolve the operator $\{a_k\}$ with the observation $\{x'_k\}$. Using the Robinson seismic model, the observation $\{x'_k\}$ is modeled as:

$$x'_k = b_k * s_k * \epsilon_k = w_k * \epsilon_k \qquad [1.5\text{-}19]$$

Performing the convolution of $\{x'_k\}$ with $\{a_k\}$, we get:

$$a_k * x'_k = a_k * b_k * s_k * \epsilon_k \qquad [1.5\text{-}20]$$

However, since $\{a_k\}$ removes the effect of the system reverberation $\{b_k\}$, i.e. $a_k * b_k = \delta_k$, then we write [1.5-20] as:

$$a_k * x'_k = \delta_k * s_k * \epsilon_k = s_k * \epsilon_k \qquad [1.5\text{-}21]$$

Similarly, we can remove the effect of the source wavelet $\{s_k\}$ from our observation $\{x'_k\}$ by convolving $\{x'_k\}$ with the operator $\{s_k^{-1}\}$, which is the inverse sequence associated with $\{s_k\}$ and defined by the relation:

$$s_k * s_k^{-1} = \delta_k \qquad [1.5\text{-}22]$$

By convolving $a_k * x'_k$ with the inverse source operator $\{s_k^{-1}\}$, we get:

$$s_k^{-1} * a_k * x'_k = s_k^{-1} * s_k * \epsilon_k = \epsilon_k \qquad [1.5\text{-}23]$$

In Chapter 4 we discuss how the deconvolution operator $s_k^{-1} * a_k$ can be found. Inspection of [1.5-23] indicates that the operator $s_k^{-1} * a_k$ removes the effects of the system reverberation $\{b_k\}$ and source wavelet $\{s_k\}$ from the observed time series $\{x'_k\}$, and reduces $\{x'_k\}$ to the hypothetical source sequence $\{\epsilon_1, \epsilon_2, \ldots \epsilon_N\}$ associated with each interface of our layered system (refer to Fig. 1-6). Although the reflection coefficient sequence $\{\epsilon_1, \epsilon_2, \ldots \epsilon_N\}$ was constructed from intuitive reasoning, i.e. the concept of hypothetical sources, we shall derive the physical justification of this hypothesis in Chapter 2. Now, if we define the inverse sequence $\{w_k^{-1})$ by the relation:

$$w_k^{-1} * w_k = \delta_k \qquad [1.5\text{-}24]$$

then the convolution of $\{w_k^{-1}\}$ with $\{x'_k\}$ in [1.5-19] gives:

$$w_k^{-1} * x'_k = \epsilon_k \qquad [1.5\text{-}25]$$

Comparison of equations [1.5-23] and [1.5-25] reveals that:

$$w_k^{-1} = s_k^{-1} * a_k \qquad [1.5\text{-}26]$$

can be viewed as the deconvolution operator which removes the effects of system reverberation $\{b_k\}$ and the source wavelet $\{s_k\}$ from the observed time series $\{x'_k\}$, leaving the reflection coefficient sequence $\{\epsilon_1, \epsilon_2, \ldots \epsilon_N\}$. The determination of the reflection coefficient sequence $\{\epsilon_k\}$ from the observed seismic record $\{x'_k\}$ is the fundamental purpose of deconvolution. Fig. 1-7 provides a description of the decomposition (deconvolution) operation involving $\{w_k^{-1}\}$.

External primary reflections

In exploration geophysics, external primary reflections are often referred to as "deep" reflections arising from "deep" subsurface reflecting layers. External primary reflections are modeled as follows: After excitation of the earth by a surface source, the resulting wave motion penetrates a surface-layered system, propagates with no reflections through an intermediate "free space", then is reflected upwards by a deep-layered system. On its upward

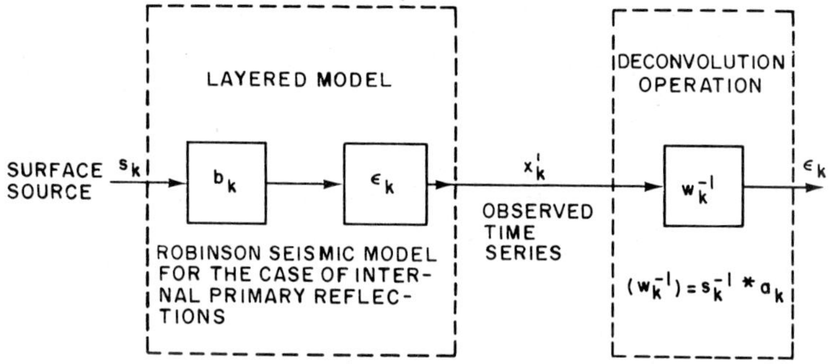

Fig. 1-7. The extraction of the hypothetical source sequence $\{\epsilon_k\}$ from the Robinson seismic model by the deconvolution (decomposition) operator $\{w_k^{-1}\}$.

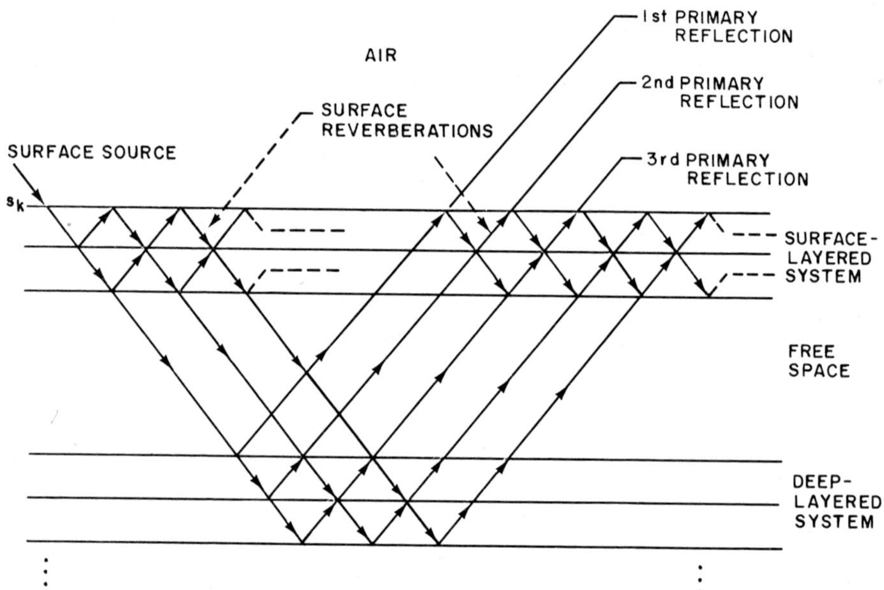

Fig. 1-8. External primary reflections resulting from a source excitation at the surface. As in Fig. 1-7 ray paths are drawn at oblique incidence for clarity.

motion, the waveform propagates the intermediate free space and re-enters the surface-layered system. Waveforms following the above propagation path are referred to as external primary reflections. The complete system, along with a pictorial definition of an external primary reflection, is described in Fig. 1-8. The assumption that the intermediate layer is a "free" space with no impedance discontinuities is equivalent to the assumption of neglecting all multiple reflections that would normally occur in the intermediate layer.

Unlike the internal primary reflections, which are generated "internal" to the layered system, the external primary reflections are generated "external" to the surface-layered system, i.e. from an independent "deep"-layered system. However, these "deep" reflections, which are of interest to the exploration geophysicist, are often masked by unwanted surface reverberations when they re-enter the surface-layered system from below. These surface reverberations, sometimes referred to as "ringing", have a pronounced effect on the seismic record and conceal the "deep" reflection information. The term "ringing" evolved from cases where observed periodicities on a suite of seismograms had a "ringing" or sinusoidal nature.

From Part 1, we saw that $\{b_k\}$ is the response of an N-layered system, measured at the surface, to an upgoing unit impulse applied from below. Now, if we assume that the layered system is also passive, i.e. contains no internal sources, then we can treat the layered system as a reciprocal linear network. Due to the principle of reciprocity, a downgoing unit impulse applied at the surface will give rise to the response $\{b_k\}$ measured from below (see Fig. 1-9a). Thus, if $\{s_k\}$ is a downgoing source wavelet applied at the top of our surface-layered system, then the output of this layered system, measured from below, is given by $s_k * b_k$ (see Fig. 1-9b). Let us consider the signal $s_k * b_k$ as the input to our deep-layered system. We can now use the results of Part 1, namely the case of internal primary reflections as applied to our deep-layered system. Thus, the upgoing response of our deep-layered system to the downgoing excitation $s_k * b_k$ is given by $s_k * b_k * \epsilon_k$, where $\{\epsilon_k\}$ is the reflection coefficient sequence of the deep layers (see Fig. 1-9c). We can now consider the upgoing signal $s_k * b_k * \epsilon_k$ as the input to the surface-layered system, i.e. we are considering $s_k * b_k * \epsilon_k$ as a "deep" source. Thus, the output of the complete system to a downgoing surface source s_k is given by $\{x'_k\}$ (see Fig. 1-9d) where:

$$x'_k = s_k * b_k * \epsilon_k * b_k \qquad [1.5\text{-}27]$$

If we define $w_k = b_k * b_k * s_k$ as a composite wavelet consisting of the surface-layer reverberations $\{b_k\}$ and source wavelet s_k, then the total response $\{x'_k\}$ can be expressed by:

$$x'_k = w_k * \epsilon_k \qquad [1.5\text{-}28]$$

By our previous reasoning, we assume that $\{w_k\}$ is minimum-delay and $\{\epsilon_k\}$ is assumed to be an uncorrelated sequence over the time interval k to $k + L$. Let us summarize the Robinson seismic model for the case of external primary reflections (i.e. deep reflections with surface-layer reverberation):

(1) The model is a convolutional model represented by:

$$x'_k = \epsilon_k * w_k$$

with $\{\epsilon_k\}$ = reflection coefficient sequence of the deep-layer reflecting interfaces; $\{w_k\}$ = composite wavelet = $b_k * b_k * s_k$; $\{b_k\}$ = surface-layer impulse response, i.e. surface-layer reverberation; $\{s_k\}$ = source wavelet; and $\{x'_k\}$ = observed seismic record (time series).

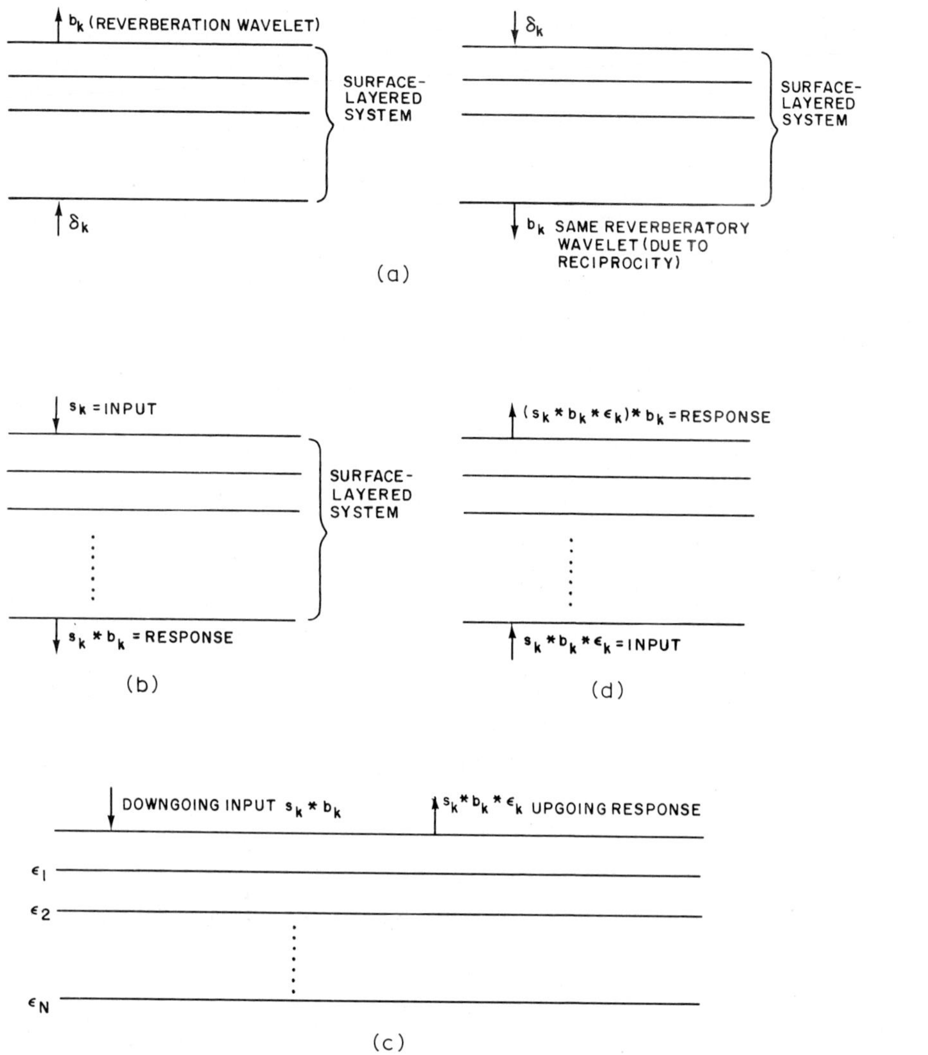

Fig. 1-9. (a) The principle of reciprocity applied to the surface-layered system. (b) Response of the surface-layered system to a downgoing source s_k. (c) Response of the deep-layered system to a downgoing source $s_k * b_k$, based on the results of Part 1 for the case of internal primary reflections. (d) Total response due to surface- and deep-layered systems.

(2) The composite wavelet $\{w_k\}$ is minimum-delay, under the condition that $\{s_k\}$ is minimum-delay.

(3) The reflection coefficients are uncorrelated over a time interval $[k, k + L]$ of the seismic record.

Now we know that $\{a_k\}$ is the linear operator that reduces $\{b_k\}$ to a unit

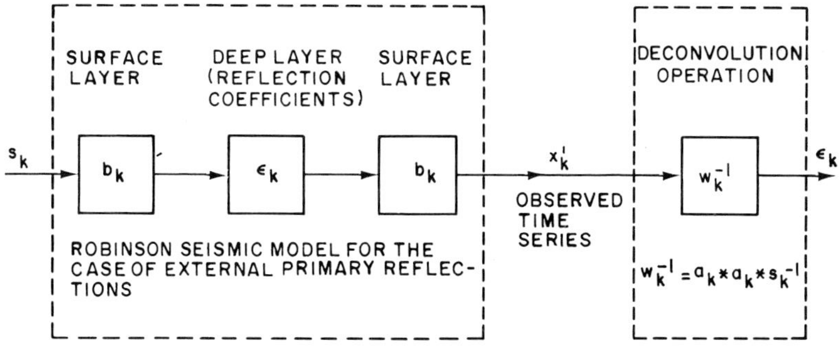

Fig. 1-10. The deconvolution operator $\{w_k^{-1}\}$ for the case of deep reflections masked by surface reverberations.

spike. Similarly, $\{s_k^{-1}\}$ is the linear operator that reduces $\{s_k\}$ to a unit spike. In Chapter 4, we will discuss the existence of $\{s_k^{-1}\}$ in terms of the minimum-delay condition on $\{s_k\}$. Thus:

$$(a_k * a_k * s_k^{-1}) * w_k = \delta_k \qquad [1.5\text{-}29]$$

where:

$$w_k^{-1} = a_k * a_k * s_k^{-1} \qquad [1.5\text{-}30]$$

The convolution of $\{w_k^{-1}\}$ with [1.5-28] gives:

$$w_k^{-1} * x'_k = \epsilon_k \qquad [1.5\text{-}31]$$

where the existence of $\{w_k^{-1}\}$ depends on the condition that $\{s_k\}$ be minimum-delay. Hence, $w_k^{-1} = a_k * a_k * s_k^{-1}$ is viewed as the deconvolution operator which removes the surface reverberations $(b_k * b_k)$ from the observed time series $\{x'_k\}$ and the effect of the source wavelet $\{s_k\}$. A complete description of this procedure is given in Fig. 1-10. In the next chapter we will develop the layered earth model in its mathematical form so as to justify the approach given here.

Chapter 2

THE LAYERED EARTH MODEL

2.1. MINIMUM-PHASE AND MINIMUM-DELAY

An important physical concept often encountered in the study of linear systems is the concept of minimum-phase. The concept of minimum-phase and its relationship to feedback systems was introduced by Bode (1945), who worked in the domain of continuous-time linear systems. Working with the Laplace complex frequency variable $s = \sigma + i\omega$ associated with continuous time, Bode originally stated that a transfer function, derived from a linear differential equation with constant coefficients, is minimum-phase if it contains no zeroes or poles in the right-half s-plane. Systems having poles and/or zeroes in the right-half s-plane are called non-minimum-phase systems. To apply the theory of minimum-phase to discrete-time systems, we first define the Laplace z-transform (see Appendix) of the sequence $\{x_k\}$ by $X(z)$ where:

$$X(z) \equiv \sum_{k=-\infty}^{\infty} x_k z^k \qquad [2.1\text{-}1]$$

and the transformation from the s-plane to the z-plane is accomplished by defining the complex variable

$$z \equiv e^{-s\Delta t} \qquad [2.1\text{-}2]$$

with Δt defined in assumption (2) of equation [1.4-2]. Under these conditions, Bode's definition of minimum-phase in the context of discrete time is as follows:

Definition of minimum-phase. A stable causal sequence is called minimum-phase if its Laplace z-transform has no zeroes within the unit circle in the z-plane.

For completeness, we add that a sequence is stable if its Laplace z-transform contains no poles inside the unit circle. A sequence is causal if it satisfies the condition $x_k = 0$ for $k < 0$. Bode's main theorem on minimum-phase becomes in the context of discrete time:

Theorem. For a given frequency content, a stable causal time sequence is minimum-phase if and only if the sequence has a minimum amount of net phase change in the negative direction (phase lag) over the frequency range from zero to π. A non-minimum-phase sequence is one that does not have this property.

By examining the response of a digital filter to the digital phasor $e^{-i\omega k}$, we can, by definition, determine the frequency response of the filter over the range $0 \leqslant \omega \leqslant \pi$. (Without loss of generality, we can let the time between samples $\Delta t = 1$ discrete time unit. Thus, ω has the dimensions of radians per sample interval.) This procedure is equivalent to evaluating the transfer function of the filter at values of z on the unit circle, i.e. $z = e^{-i\omega}$. Similarly, the frequency content of the sequence $\{x_k\}$ is found by evaluating $X(z)$ on the unit circle, where $X(z = e^{-i\omega}) = X(\omega)$ is a complex-valued quantity with a magnitude $|X(\omega)|$ and argument (phase) $\theta(\omega)$ related by:

$$X(\omega) = |X(\omega)|\, e^{i\theta(\omega)} \qquad [2.1\text{-}3]$$

One important consequence of a minimum-phase sequence is that if the magnitude of the frequency response $|X(\omega)|$ is specified for $0 \leqslant \omega \leqslant \pi$, then the corresponding phase $\theta(\omega)$ is also specified. Conversely, if the phase is specified for $0 \leqslant \omega \leqslant \pi$, then the magnitude is also specified. Thus, the magnitude and phase of the frequency response of a minimum-phase sequence are uniquely related. Moreover, they form a Hilbert transform pair.

The concept of minimum-delay was introduced by Robinson (1954, 1962). Any causal linear system can be described by its gain and its delay. Gain is a measure of the increase or decrease in the magnitude of the output as compared to the magnitude of the input. Delay is a measure of the time delay from input to output. It is possible to have many different causal systems each with the same gain, but each with a different delay. In fact, it is always possible to have causal systems with very large delays, as there is no theoretical limit to the largeness of delay that can be incorporated into a causal system. On the other hand, there is a limit to the smallness of the delay that a causal system can possess. The reason is that it always takes some time for a causal system to respond significantly to an input. Robinson defined the minimum-delay system as the one with the smallest possible delay for its gain. The concept of minimum-delay applies to both discrete- and continuous-time systems: single-channel, multi-channel, or multidimensional systems.

Robinson's approach was to consider the class of all stable causal systems with the same gain, i.e. magnitude or amplitude spectrum. He then showed that each of the following ten properties is necessary and sufficient that a given member of the class be minimum-delay:

(1) The *phase delay* (i.e. the negative of the phase spectrum divided by frequency) is a minimum over the entire frequency range.

(2) The *group delay* (i.e. the negative of the frequency derivative of the phase spectrum) is a minimum over the entire frequency range.

(3) The *energy delay* (i.e. the sum or integral of the magnitude-squared values of the system's impulse from time t) is a minimum.

(4) The *spike delay* (i.e. the time of occurrence of the output spike produced by a causal inverse system) is a minimum (and this minimum is zero).

(5) The *information delay* (i.e. the delay of information from input to output) is a minimum.

(6) The *propagation delay* (i.e. the waveform energy still within the system at time t as compared with the waveform energy that has already been transmitted through the system) is a minimum.

(7) The *phase-lag* (i.e. the negative of the phase spectrum) is a minimum.

(8) The *partial energy* (i.e. the partial sum or integral of the magnitude-squared values of the system's impulse response up to time t (or k)) is a maximum.

(9) The *inverse system* is causal.

(10) For discrete-time systems, the zeroes of the Laplace z-transform of the impulse response lie outside the unit circle. For *continuous-time systems*, the zeroes of the Laplace transform of the impulse response lie in the left-half s-plane.

Let us now express some of these properties in mathematical form. A causal system is characterized by its one-sided impulse response, denoted by the sequence $\{a_k\}$ where $a_k = 0$ for $k < 0$. The *system transfer function* is defined as the Laplace z-transform of the impulse response $\{a_k\}$. In symbols:

$$A(z) = \sum_{k=-\infty}^{\infty} a_k z^k \qquad [2.1\text{-}4]$$

and for causal systems, $A(z)$ becomes:

$$A(z) = \sum_{k=0}^{\infty} a_k z^k \qquad [2.1\text{-}5]$$

Now, given the system transfer function defined by the Laplace z-transform in [2.1-5], we obtain the *frequency response* of the causal system by letting $z = e^{-i\omega}$. Thus, substitution of $z = e^{-i\omega}$ in equation [2.1-5] gives the frequency response $A(e^{-i\omega})$ where:

$$A(e^{-i\omega}) = \sum_{k=0}^{\infty} a_k e^{-i\omega k} \qquad [2.1\text{-}6]$$

Now, [2.1-6] can be written as:

$$A(e^{-i\omega}) = \sum_{k=0}^{\infty} a_k \cos \omega k - i \sum_{k=0}^{\infty} a_k \sin \omega k \qquad [2.1\text{-}7]$$

or in polar form as:

$$A(e^{-i\omega}) = |A(e^{-i\omega})|\, e^{i\theta(\omega)} \qquad [2.1\text{-}8]$$

where the *amplitude spectrum* is:

$$|A(e^{-i\omega})| = \left[\left(\sum_{k=0}^{\infty} a_k \cos \omega k\right)^2 + \left(\sum_{k=0}^{\infty} a_k \sin \omega k\right)^2\right]^{\frac{1}{2}} \qquad [2.1\text{-}9]$$

and the *phase spectrum* is:

$$\theta(\omega) = \tan^{-1} \frac{-\sum_{k=0}^{\infty} a_k \sin \omega k}{\sum_{k=0}^{\infty} a_k \cos \omega k} \qquad [2.1\text{-}10]$$

In the amplitude spectrum representation, we shall frequently refer to $|A(e^{-i\omega})|$ as simply $|A(\omega)|$.

If we consider the class of causal sequences, with members of the class denoted by $\{a_k^{(0)}\}$, $\{a_k^{(1)}\}$, ... $\{a_k^{(j)}\}$, $\{a_k^{(j+1)}\}$, ..., in which each member of the class has the same gain $|A(\omega)|$, then each member of the class has the same total energy. This fact follows from Parseval's relation which states that the energy is:

$$\text{energy} = \sum_{k=0}^{\infty} |a_k^{(j)}|^2 = \frac{1}{2\pi} \int_{-\pi}^{\pi} |A(\omega)|^2 d\omega = \text{constant, since } |A(\omega)| \text{ is the same for each member of the class} \qquad [2.1\text{-}11]$$

Suppose that $\{a_k^{(0)}\}$ represents a minimum-delay member of the class and $\{a_k^{(j)}\}$ represents a non-minimum-delay member. Although these two members have the same total energy, property (8) states that the partial energy of the minimum-delay member $\{a_k^{(0)}\}$ exceeds the partial energy of the non-minimum-delay member $\{a_k^{(j)}\}$. That is (see Robinson, 1962):

$$\sum_{n=0}^{k} |a_n^{(0)}|^2 \geqslant \sum_{n=0}^{k} |a_n^{(j)}|^2 \qquad [2.1\text{-}12]$$

As a result, the energy of the minimum-delay member $\{a_k^{(0)}\}$ is more concentrated at the "front-end" than any other member $\{a_k^{(j)}\}$ of the class. That is, the major concentration of energy in a minimum-delay sequence occurs as early as possible in the sequence and is not delayed any more than necessary to fit the given amplitude spectrum. For that reason, a minimum-delay sequence is often called a "front-loaded" sequence. The fact that a minimum-delay sequence has this early distribution of energy with time is an important physical property. For example, observation that reverberations in seismic prospecting were front-loaded provided the insight that led to the construction of a mathematical model for the generating system.

Comparing the definition of minimum-phase with minimum-delay property (10), we conclude that a linear system is minimum-phase if and only if the system is minimum-delay. Thus, Bode's concept of minimum-phase and

Robinson's concept of minimum-delay are in fact identical. Moreover, minimum-delay property (7) shows that the phase-lag of a minimum-phase system is a minimum over the entire frequency range. As a result of property (7), we see that the concept of minimum-phase should more precisely be called *minimum-negative-phase* or "minimum-phase-lag". That is, the term "minimum-phase" is actually a misnomer, for a so-called "minimum-phase" system actually has the "maximum-phase" of all systems with the same gain. For this reason we prefer to use the term "minimum-delay" instead of "minimum-phase".

Let us now look again at Bode's theorem on minimum-phase. We recall that the theorem states that a minimum-phase system has a minimum amount of net negative phase change over the frequency range zero to π. That is:

$$-[\theta(\pi)-\theta(0)] = \text{minimum for the minimum-phase system} \qquad [2.1\text{-}13]$$

Property (7) states that a minimum-delay system has minimum-negative-phase at every frequency. That is,

$$-\theta(\omega) = \text{minimum for the minimum-delay system} \qquad [2.1\text{-}14]$$

In the application of this property, we usually tie down the phase curves of the systems we are comparing by requiring each phase curve to pass through the origin, i.e. by requiring that $\theta(0) = 0$. Then, letting $\omega = 0$ and π respectively, we have:

$$\begin{aligned} \theta(0) &= \text{same for each member of the class of causal sequences} \\ &\quad\ \text{with the same gain} \\ -\theta(\pi) &= \text{minimum for the minimum-delay system} \end{aligned} \qquad [2.1\text{-}15]$$

from which it follows that:

$$-[\theta(\pi)-\theta(0)] = \text{minimum for the minimum-delay system} \qquad [2.1\text{-}16]$$

In summary, Bode's theorem and the minimum-delay property (7) describe the same concept, as we would expect.

2.2. TRANSMISSION AND REFLECTION RESPONSE FOR A SINGLE LAYER

To gain some physical insight into the concept of minimum-delay, let us consider the mathematical model of an ideal horizontally layered system. To avoid getting lost in mathematical details, we shall consider a single-layered system, which illustrates the physical properties of minimum-delay. Such a system is shown in Fig. 2-1.

Let us assume that the layered system shown in Fig. 2-1 is excited by a downgoing source pulse s_k, the seismic detector (located just beneath the surface) responds only to upward wave motion in layer 1, and the source is initiated at the detector. Further, we restrict ourselves to plane wave motion

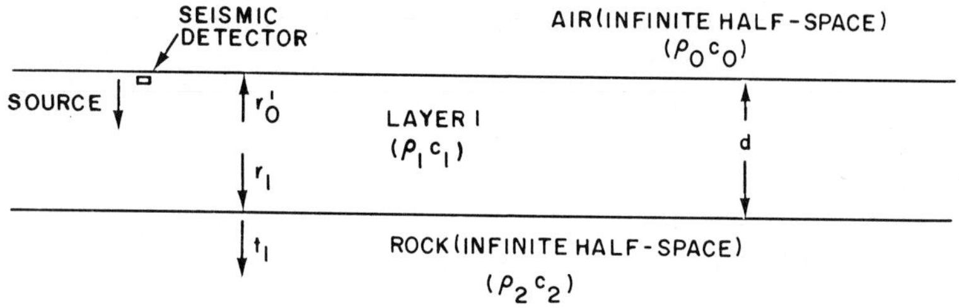

Fig. 2-1. Schematic diagram of a horizontal single-layered system. Symbols as follows: ρ_i = density of layer i ($i = 0, 1, 2$); c_i = propagation velocity of layer i ($i = 0, 1, 2$); $\rho_i c_i$ = characteristic impedance of layer i ($i = 0, 1, 2$); r_1 = reflection coefficient = $(\rho_2 c_2 - \rho_1 c_1)/(\rho_2 c_2 + \rho_1 c_1)$ at layer 1—rock interface; r'_0 = reflection coefficient = $(\rho_0 c_0 - \rho_1 c_1)/(\rho_0 c_0 + \rho_1 c_1)$ at layer 1—air interface ($r'_0 = -r_0$); t_1 = transmission coefficient = $1 + r_1$; d = thickness of layer 1. r_1, r'_0, and t_1 are real numbers such that $|r_1| \leqslant 1$, $|r'_0| \leqslant 1$, and $0 \leqslant t_1 \leqslant 2$.

at normal incidence to the horizontal interfaces and we adhere to the assumptions in [1.4-2] for our convolutional model. We choose the time origin ($k = 0$) as the time of the source excitation, or in geophysical terms, the time of the shot. Assuming that the system of Fig. 2-1 is a lumped-parameter system, there is no additive noise in the measurement process, and the recorded samples are spaced at $\Delta t = 2d/c_1$ seconds apart (two-way travel time in layer 1), let us consider the following situations.

Transmitted response $x_k^{(1)}$ due to a unit impulse excitation ($s_k = \delta_k$)

We shall first examine the waveform which is transmitted to the top of the rock infinite half-space when a unit impulse is applied at the surface. This waveform will be measured at the bottom of layer 1 as depicted in Fig. 2-2.

The sequence $x_k^{(1)}$, measured at the bottom of layer 1, as a function of time is given by:

$$x_k^{(1)} = \delta_{k-\frac{1}{2}} + r_1 r'_0 \delta_{k-\frac{3}{2}} + (r_1 r'_0)^2 \delta_{k-\frac{5}{2}} + (r_1 r'_0)^3 \delta_{k-\frac{7}{2}} + \ldots \qquad [2.2\text{-}1]$$

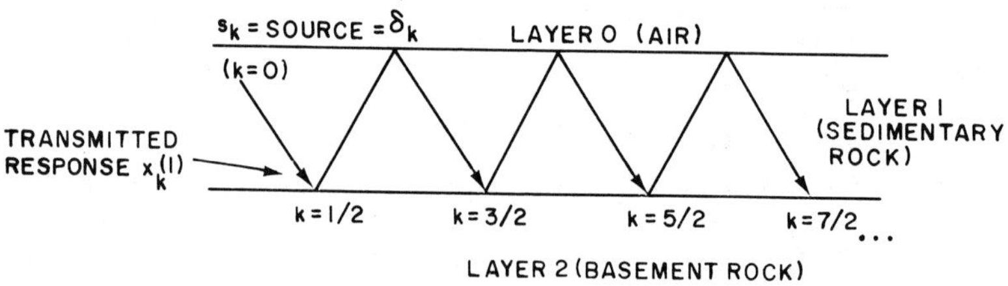

Fig. 2-2. Transmitted response due to surface impulse excitation, measured at the bottom of layer 1. For clarity, ray paths are drawn at oblique incidence but wave motion analysis is for normal incidence.

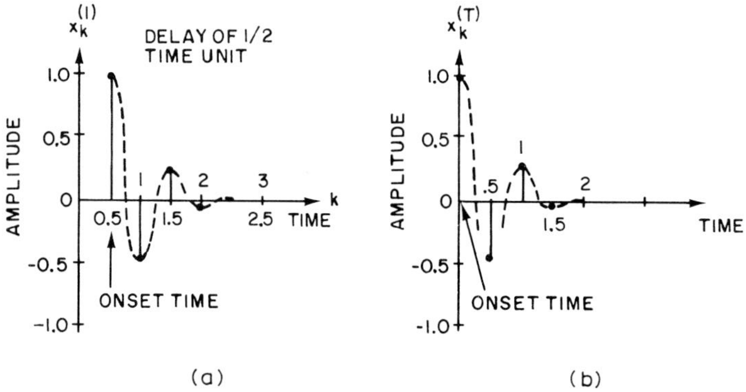

Fig. 2-3. (a) Transmission response ($r_0' = -1$, $r_1 = 0.5$). (b) Transmission response with corrected origin (delay removed).

Taking the Laplace z-transform of [2.2-1], we get:

$$X^{(I)}(z) = z^{1/2} + r_1 r_0' z^{3/2} + (r_1 r_0')^2 z^{5/2} + (r_1 r_0')^3 z^{7/2} + \ldots \qquad [2.2\text{-}2]$$

or:

$$X^{(I)}(z) = z^{1/2}[1 + r_1 r_0' z + (r_1 r_0' z)^2 + (r_1 r_0' z)^3 + \ldots] \qquad [2.2\text{-}3]$$

Now the term in brackets represents a geometric series which converges for $|r_1 r_0' z| < 1$ or $|z| < 1/|r_1 r_0'|$. Thus, [2.2-3] may be rewritten as:

$$X^{(I)}(z) = \frac{z^{1/2}}{1 - r_0' r_1 z}, \quad |z| < \frac{1}{|r_0' r_1|} \qquad [2.2\text{-}4]$$

Because reflection coefficients cannot exceed one in magnitude, it follows that $|r_0 r_1| < 1$. (Note: $|r_0 r_1| = |r_0' r_1|$. Thus, the Laplace z-transform $X^{(I)}(z)$ has no poles inside the unit circle, i.e. $x_k^{(I)}$ is a stable sequence. However, the presence of the factor $z^{\frac{1}{2}}$ in the numerator of [2.2-4] is due to a delay of $\frac{1}{2}$ time units in our measurement, i.e. the one-way transmission delay between shot onset time and measurement time. Thus, $X^{(I)}(z)$ contains a zero at the origin, and according to our definition, is not minimum-delay.

The transmitted response $x_k^{(I)}$ is illustrated in Fig. 2-3a. However, it is useful to separate the pure delay $z^{\frac{1}{2}}$ from the rest of $X^{(I)}(z)$. Doing so, we obtain:

$$X^{(T)}(z) \equiv z^{-\frac{1}{2}} X^{(I)}(z) = \frac{1}{1 - r_0' r_1 z} \qquad [2.2\text{-}5]$$

which is the Laplace z-transform of the transmitted response $x_k^{(T)}$ as measured from its first break (i.e. its onset time; see Fig. 2-3b). Now, the transmitted response $x_k^{(T)}$ as *measured* from its first break is *minimum-delay*. Thus, whenever we deal with transmitted responses, we want to let its time origin be its first break, so that they will have the minimum-delay property. Our different

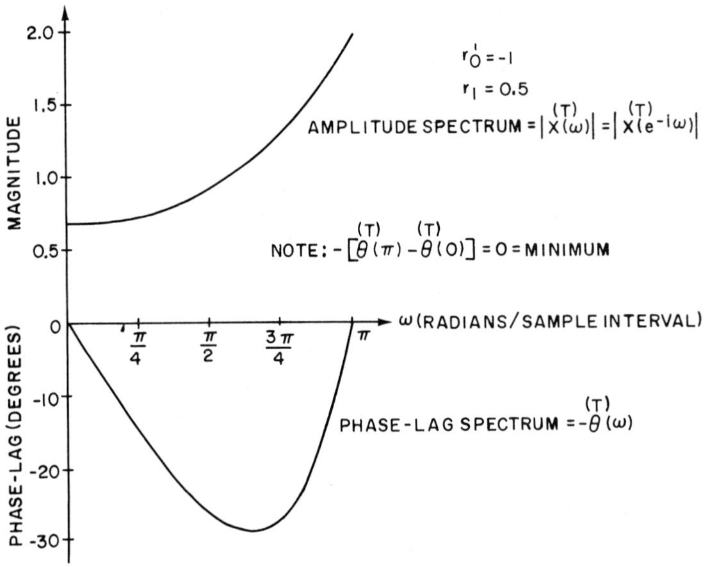

Fig. 2-4. Amplitude and phase-lag spectrum of the minimum-delay transmission response $x_k^{(T)}$; $-\theta^{(T)}(\omega)$ is the minimum-phase-lag spectrum for all the causal stable sequences possessing the gain $|X^{(T)}(\omega)|$.

measurement-time origin has removed the one-way transmission delay associated with $x_k^{(1)}$ by shifting (advancing) it by $\frac{1}{2}$ time units. Thus, $X^{(T)}(z)$ has no zeroes or poles inside the unit circle and is by definition minimum-delay. The frequency response $X^{(T)}(e^{-i\omega})$, denoted by $X^{(T)}(\omega)$, is found by evaluating $X^{(T)}(z)$ on the unit circle. Substituting $z = e^{-i\omega}$ into [2.2-5], we obtain:

$$X^{(T)}(\omega) = |X^{(T)}(\omega)|\, e^{i\theta^{(T)}(\omega)} \qquad [2.2\text{-}6]$$

where:

$$|X^{(T)}(\omega)| = \frac{1}{[1 + (r_0' r_1)^2 - 2 r_1 r_0' \cos\omega]^{\frac{1}{2}}}$$

$$-\theta^{(T)}(\omega) = \tan^{-1}\frac{r_1 r_0' \sin\omega}{1 - r_1 r_0' \cos\omega}$$

The amplitude spectrum $|X^{(T)}(\omega)|$ and phase-lag spectrum $-\theta^{(T)}(\omega)$, associated with the minimum-delay transmission response $x_k^{(T)}$, are plotted in Fig. 2-4 for the values $r_0' = -1$ and $r_1 = 0.5$ over the range $0 \leqslant \omega \leqslant \pi$.

The "reflection response" due to a downgoing unit impulse excitation applied at the surface

Let us now consider the case of the reflection sequence $x_k^{(R)}$ which is

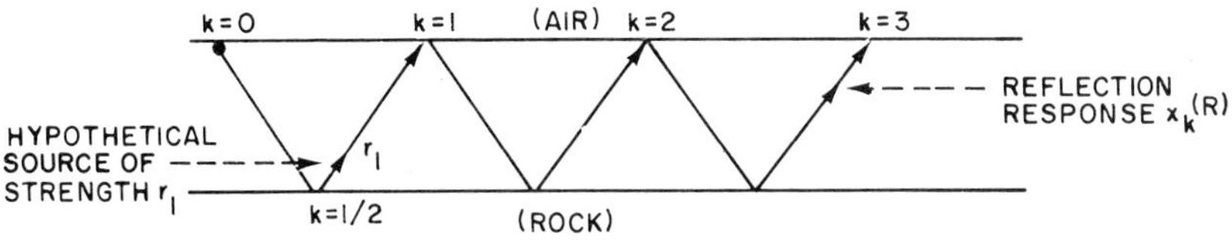

Fig. 2-5. Downgoing unit impulse surface source replaced by an upgoing hypothetical source of strength r_1. The reflection response $x_k^{(R)}$ is measured at the top of layer 1.

recorded at the top of layer 1 and results when a downgoing unit impulse is applied at the surface. Let us first replace the downgoing unit impulse source by an upgoing hypothetical source of strength r_1. Fig. 2-5 depicts the situation.

We shall assume that the configuration of Fig. 2-5 represents a lumped-parameter, linear, passive, time-invariant system. Comparison of Figs. 2-2 and 2-5 indicates that the system of Fig. 2-5 can be considered as the reciprocal of the system in Fig. 2-2. Since the system is linear, time-invariant and passive, then we can use the principle of reciprocity in developing the sequence $x_k^{(R)}$. Now, if the hypothetical source of Fig. 2-5 was a unit impulse, then from reciprocity considerations we could conclude that $x_k^{(R)} = x_k^{(T)}$ = the transmitted response. Further, the linear and shift-invariant assumptions imply that if $\delta_k \xrightarrow[\text{(gives rise to)}]{} x_k^{(T)}$ then:

$$\delta_{k-\frac{1}{2}} \longrightarrow x_{k-\frac{1}{2}}^{(T)} \text{ and } r_1\delta_{k-\frac{1}{2}} \text{ (hypothetical source)} \longrightarrow r_1 x_{k-\frac{1}{2}}^{(T)}$$

But if we introduce a delay of $\frac{1}{2}$ time units, then we can state:

$$r_1\delta_{k-1} \longrightarrow r_1 x_{k-1}^{(T)} \text{ (reflection response)}$$

Thus, we conclude that the reflection response $x_k^{(R)}$ can be expressed as:

$$x_k^{(R)} = r_1 x_{k-1}^{(T)} \qquad [2.2\text{-}7]$$

The Laplace z-transform of [2.2-7] yields:

$$X^{(R)}(z) = (r_1 z)X^{(T)}(z) \qquad [2.2\text{-}8]$$

which implicitly states that the reflection response is the convolution of the delayed hypothetical source of strength r_1 with the transmitted response $x_k^{(T)}$. Substitution of [2.2-5] in [2.2-8] gives:

$$X^{(R)}(z) = (r_1 z) \cdot \frac{1}{1 - r_0' r_1 z}, \quad |z| < \frac{1}{|r_0' r_1|} \qquad [2.2\text{-}9]$$

or:

$$X^{(R)}(z) = r_1 \cdot z \cdot \frac{1}{1 - r_0' r_1 z}, \quad |z| < \frac{1}{|r_0' r_1|} \qquad [2.2\text{-}10]$$

Inspection of [2.2-10] shows that $X^{(R)}(z)$ is composed of a scale factor r_1, a pure delay of 1 time unit due to the factor z, and a minimum-delay factor given by $1/(1-r_0'r_1z)$, which is the transmitted response Laplace z-transform. We see that the reflection response $X^{(R)}(z)$ has the form:

$$X^{(R)}(z) = E(z)/A(z) \qquad [2.2\text{-}11]$$

where $E(z)$ is the Laplace z-transform of the reflection coefficient sequence, i.e. $E(z) = r_1z$, and $A(z)$ is the Laplace z-transform of the difference equation operator $A(z) = 1 + a_1z$, where $a_1 = -r_0'r_1$. The reverberation is the inverse of $A(z)$, i.e. $1/A(z)$, and is given by:

$$1/A(z) = 1 + r_0'r_1z + (r_0'r_1)^2z^2 + \ldots \qquad [2.2\text{-}12]$$

That is, the reverberation is the time sequence $\{1, r_0'r_1, (r_0'r_1)^2, (r_0'r_1)^3, \ldots\}$. Now the amplitude spectrum of the reflection response $x_k^{(R)}$ is equal to the amplitude spectrum of the reflection coefficient sequence $\{0, r_1\}$ *times* the amplitude spectrum of the transmission response $x_k^{(T)}$. In symbols:*

$$|X^{(R)}(\omega)| = |r_1|\,|X^{(T)}(\omega)| \qquad [2.2\text{-}13]$$

However, the phase-lag spectrum of $x_k^{(R)}$ is equal to the phase-lag spectrum of the reflection coefficient sequence $\{0, r_1\}$ *plus* the phase-lag spectrum of the transmission response $x_k^{(T)}$. That is:

$$-\theta^{(R)}(\omega) = \omega + [-\theta^{(T)}(\omega)] \qquad [2.2\text{-}14]$$

For the gain $|X^{(T)}(\omega)|$, we found that $-\theta^{(T)}(\omega)$ was the minimum-phase-lag spectrum. Although in this case the amplitude spectrum of $x_k^{(R)}$ represents the same gain except for a scale factor, i.e. $|r_1|\,|X^{(T)}(\omega)|$, we note that the phase-lag spectrum $-\theta^{(R)}(\omega)$ differs from the minimum-phase-lag spectrum $-\theta^{(T)}(\omega)$ by the linear phase-lag ω. The phase-lag spectrum $-\theta^{(R)}(\omega)$ is shown in Fig. 2-6 for the values $r_0' = -1$ and $r_1 = 0.5$. We conclude that excess phase-lag in the reflection response can be introduced only by the reflection coefficient sequence and never by the reverberations.

We arrived at the reflection response $x_k^{(R)}$ by considering the layered system of Fig. 2-1 as a lumped-parameter, linear, time-invariant system and using the notion of hypothetical sources. Let us now proceed to calculate $x_k^{(R)}$ by a different approach, i.e., we shall not consider the idea of a hypothetical source and will analyze all wave motion due to the downgoing unit impulse δ_k. The resulting reflection response will be denoted by $x_k^{(2)}$ with a description of our procedure shown in Fig. 2-7.

The sequence $x_k^{(2)}$ due to the excitation δ_k in Fig. 2-7 is given by:

$$x_k^{(2)} = r_1\delta_{k-1} + r_1^2r_0'\delta_{k-2} + r_1^3r_0'^2\,\delta_{k-3} + \ldots \qquad [2.2\text{-}15]$$

Taking the Laplace z-transform of 2.2-15, we get:

---

* Recall that $X(\omega) = X(z = e^{-i\omega}) = X(e^{-i\omega})$ when using the Laplace z-transform.

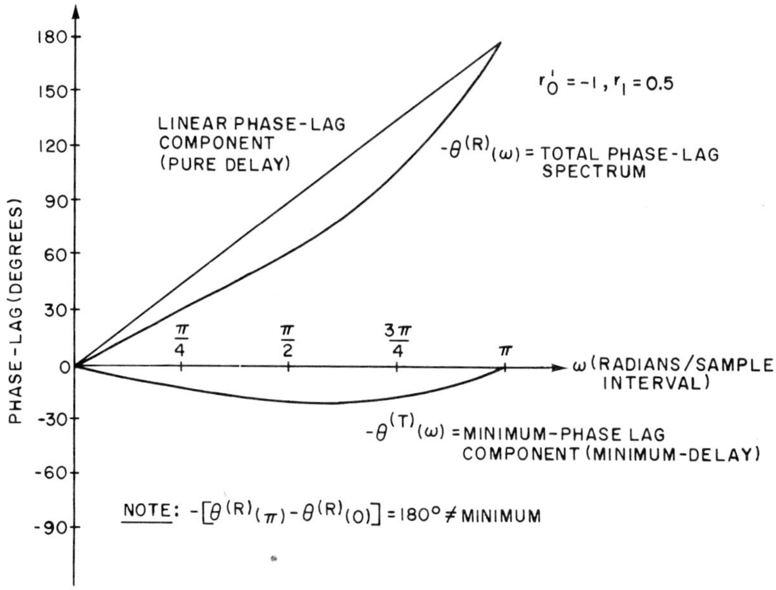

Fig. 2-6. Phase-lag spectrum $-\theta^{(R)}(\omega)$ associated with the reflection response $x_k^{(R)}$. For the gain $|X(\omega)|$, note the difference between the minimum phase-lag $-\theta^{(T)}(\omega)$ and the non-minimum phase-lag $-\theta^{(R)}(\omega)$.

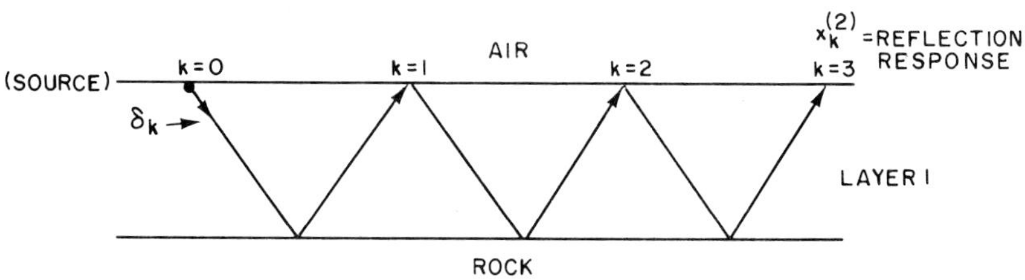

Fig. 2-7. The reflection response produced by a downgoing unit impulse applied at the surface. The reflection response is the waveform consisting of the upgoing pulses at the top of layer 1.

$$X^{(2)}(z) \; = \; r_1 z + r_1^2 r_0' z^2 + r_1^3 r_0'^2 z^3 + \ldots \qquad [2.2\text{-}16]$$

or:

$$X^{(2)}(z) \; = \; r_1 z[1 + r_1 r_0' z + (r_1 r_0' z)^2 + \ldots] \qquad [2.2\text{-}17]$$

which is equivalent to:

$$X^{(2)}(z) \; = \; \frac{r_1 z}{1 - r_1 r_0' z}, \quad |z| < \frac{1}{|r_0' r_1|} \qquad [2.2\text{-}18]$$

Thus, [2.2-18] is identical to [2.2-9], which was derived by replacing the downgoing unit impulse source δ_k by an upgoing hypothetical source of strength r_1. For this particular example, the two approaches yield the same result, i.e. $x_k^{(2)} = x_k^{(R)}$ as we would expect.

The reflection response due to an arbitrary downgoing source pulse s_k applied at the surface

Let us now replace the unit impulse source δ_k in Fig. 2-7 by an arbitrary downgoing source s_k. If we denote the resulting reflection response by $x_k^{(3)}$, then this sequence is given by:

$$x_k^{(3)} = r_1 s_{k-1} + r_0' r_1^2 s_{k-2} + r_0'^2 r_1^3 s_{k-3} + \ldots \qquad [2.2\text{-}19]$$

The Laplace z-transform of [2.2-19] gives:

$$X^{(3)}(z) = r_1 z S(z) + r_0' r_1^2 z^2 S(z) + r_0'^2 r_1^3 z^3 S(z) + \ldots \qquad [2.2\text{-}20]$$

or:

$$X^{(3)}(z) = r_1 z S(z)[1 + (r_0' r_1 z) + (r_0' r_1 z)^2 + \ldots] \qquad [2.2\text{-}21]$$

Simplification of [2.2-21] yields:

$$X^{(3)}(z) = \frac{(r_1 z)S(z)}{1 - r_0' r_1 z}, \quad |z| < \frac{1}{|r_0' r_1|} \qquad [2.2\text{-}22]$$

Thus far, we have shown that the factor $1/(1 - r_0' r_1 z)$ (associated with the reverberation transmission response $x_k^{(T)}$) is minimum-delay. The reflection response $x_k^{(3)}$ is not minimum-delay because of $(r_1 z)S(z)$ in the numerator of [2.2-22], which represents the reflection coefficient sequence and the source. In addition, the source pulse s_k could contain zeroes inside the unit circle. Under this condition, $S(z)$ is not minimum-delay. Equation [2.2-22] indicates that the reflection response $x_k^{(3)}$ can be considered as the convolution of the source pulse s_k, the reflection coefficient sequence $\{0, r_1\}$, and the reverberation transmission response $x_k^{(T)}$. Now, $x_k^{(T)}$ is minimum-delay (refer to Fig. 2-4). If we assume that s_k is also minimum-delay, then we observe that the reflection response is the convolution of the reflection coefficient sequence $\{0, r_1\}$ with a minimum-delay wavelet w_k. The wavelet w_k is the composite wavelet of the minimum-delay source pulse s_k with the minimum-delay reverberation $x_k^{(T)}$.

In the context of linear systems, equation [2.2-9] or [2.2-18] represents the transfer function of the single-layered system shown in Fig. 2-1. In particular, [2.2-9] is the Laplace z-transform of the impulse response $x_k^{(R)}$ of this system, which related the source pulse s_k to the reflection response $x_k^{(3)}$. However, due to the reflection coefficient sequence, the transfer function $X^{(R)}(z)$ of our single-layered system is not minimum-delay. Inspection of Fig. 2-6 indicates the excessive phase-lag $(-\theta^{(R)}(\omega))$ is introduced exclusively by the reflection coefficient sequence. This excessive phase-lag is characteristic of reflection responses or reflection seismograms. The nice feature here is that we have isolated the factor giving the excess phase-lag. The remaining factors (source wavelet and reverberation) are minimum-delay. The fact that we can associate all excess delay with the reflection coefficient sequence provides the physical reason why we can deconvolve a reflection seismogram to yield the reflection coefficients.

Transmission lines, piping-flow delays, transportation lags, and semiconductor diffusion all have transfer functions with z^m components, which characterizes the delay associated with these systems. Successful analysis of these systems would depend on isolating the delay-producing components.

From a physical point of view, our single-layered model (refer to Fig. 2-1) could represent an ideal layered earth model, and in general, such multi-layered models characterize the layered earth. In the layered earth situation, the reflection coefficients are equal to or less than unity in magnitude. In the transmission of energy from one side to the other side of a layered system, a seismic pulse suffers multiple reflections and refractions within the layers, which result in delay and attenuation. Thus, the concentration of energy in a transmitted waveform, say x_k, appears at its beginning rather than at its end. Since this is the condition that a sequence be minimum-delay, then we can always expect that the transmitted output (measured from its first break) of a multi-layered system is minimum-delay, provided the source pulse s_k is minimum-delay.

In terms of information transmission, the concept of minimum-delay has an interesting interpretation. For example, consider the flow of information through a system. A system can destroy information in the sense that the flow of information out of the system is less than the flow of information into the system. In such a case there is a net loss of information; this lost information cannot be recovered. Another situation, and one which is most favorable, is the case where a system does not destroy information but only delays it. In this case, the total information that comes out of the system over all time is the same as the total information that went into the system, but at any given instant, the system withholds information in the sense that information has gone in but has not yet come out. If we wait long enough, we can ultimately obtain this retained information at the system output terminal. The cost to us is delay; we must wait for information until the system is ready to give it to us. The most favorable situation is the one found in the layered earth, namely the case where the system neither destroys nor delays information in transmission. In this case, the total information that has come out of the system up to any given time instant is the same as the total information that has gone into the system up to the same time instant. Thus, such a system does not withhold any information from us; it merely converts the information into a different form with neither loss of time nor content. Certainly, this is the most efficient system; it is the minimum-delay system. The minimum-delay system is the one that produces optimum information transfer. The minimum-delay system is a system which is isomorphic to the layered earth in the transmission of information from one face to the other, provided that we remove the one-way pure delay of the transmission of the direct wave through the layered earth. The recognition of the minimum-delay property of the earth makes possible the deconvolution of seismic records.

2.3. TRANSMISSION AND REFLECTION RESPONSE FOR MULTIPLE LAYERS

In Chapter 1, we introduced the *convolution* model consistent with the assumptions of [1.4-2]. We showed that the seismic record represents the output of a linear time-invariant system with impulse response $\{f_k\}$ (see Fig. 1-1). However, the sequence $\{f_k\}$ is not physically meaningful as it stands, since we have not identified the reflection coefficients $r_0, r_1, r_2, \ldots r_N$ of the layered earth with the earth's transfer function $F(z)$, the Laplace z-transform of $\{f_k\}$. For example, f_1 could correspond to r_1, f_2 to $-r_0r_1^2 + (1-r_1^2)r_2$, f_3 to $r_0^2r_1^3 - r_2(1-r_1^2)(2r_0r_1 + r_1r_2)$, etc., but at the moment we do not have such a dependence. Thus, we must relate $\{f_k\}$ to the reflection coefficients $r_0, r_1, r_2, \ldots r_N$ of the N layers of the desired subsurface structure. Our next task is to find a layered earth model which will provide the necessary relationships and complete the convolution model.

The layered earth model considered here is the familiar horizontally layered elastic medium, as earlier developed by Goupillaud (1961), Kunetz (1964), Robinson (1967a, 1975), and many others. Each layer is homogeneous and isotropic, and subject to the analysis of plane wave motion at normal incidence. The layers in this model are numbered from top to bottom, with layer 0 denoting the infinite air half-space and layer $N+1$ denoting the lowermost layer which is also an infinite half-space. For example, in the case of marine geophysical exploration, we could associate layer 1 with the water layer. Thus, the earth model consists of N layers, $N+1$ interfaces, and $N+1$ reflection and transmission coefficients (see Fig. 2-8).

We restrict ourselves to plane wave motion at normal incidence to the horizontal interfaces. The wave motion itself can be measured in terms of any number of related quantities consistent with the plane wave equation, i.e. acoustic pressure, particle velocity, or particle acceleration. For example, seismometers convert particle velocity to electrical voltage for land surveys, whereas in marine work they convert acoustic pressure variations to voltage. Whatever quantity is used, reflection and transmission coefficients can be defined for each interface.

If a downgoing unit impulse is incident on the top of interface n, then the reflection coefficient r_n is equal to the resulting upgoing impulse reflected from the top of interface n, and the transmission coefficient t_n is equal to

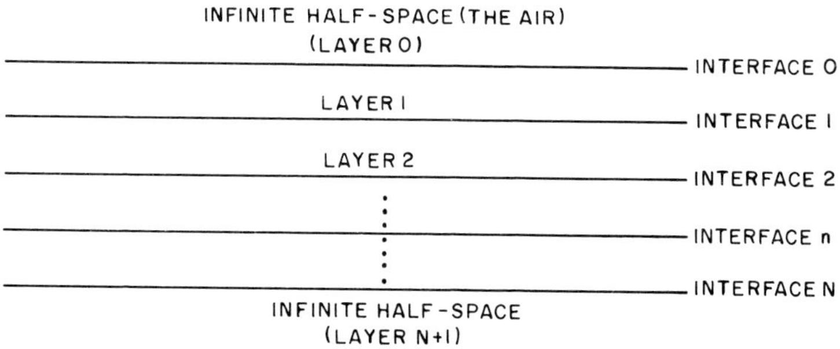

Fig. 2-8. The layered earth model.

the resulting downgoing impulse transmitted through interface n. Similarly, if an upgoing unit impulse is incident on the bottom of interface n, then the reflection coefficient r'_n is equal to the resulting downgoing impulse reflected from the bottom of interface n, and the transmission coefficient t'_n is equal to the resulting upgoing impulse transmitted through interface n (see Fig. 2-9).

The reflection and transmission coefficients at an interface are dependent on the characteristic impedance of the adjacent layers. These coefficients, and some useful relationships resulting from the boundary conditions at an interface, are (where ρ_n = density of layer n ($n = 0, 1, 2, \ldots N$) and c_n = propagation velocity of layer n ($n = 0, 1, 2, \ldots N$)):

$$r_n = \frac{\rho_{n+1}c_{n+1} - \rho_n c_n}{\rho_{n+1}c_{n+1} + \rho_n c_n}, \; t_n = \frac{2\rho_{n+1}c_{n+1}}{\rho_{n+1}c_{n+1} + \rho_n c_n} \quad (n = 0, 1, 2, \ldots N)$$

$$t_n = 1 + r_n; \quad r'_n = -r_n; \quad t'_n = 1 - r_n; \quad t_n t'_n = 1 - r_n^2 \qquad [2.3\text{-}1]$$

In our treatment of wave propagation in a layered medium, the reflection and transmission coefficients are real numbers with $|r_n| \leqslant 1$ and $0 \leqslant t_n \leqslant 2$.

For mathematical simplicity, it is convenient to add hypothetical (i.e. mathematical, not geological) interfaces where necessary, so as to make the

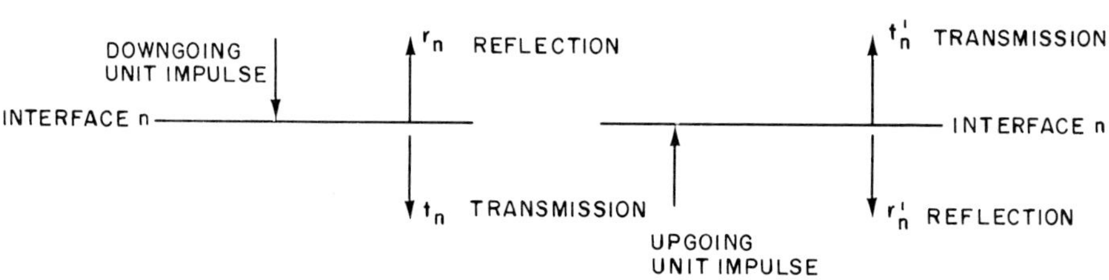

Fig. 2-9. Schematic diagram illustrating convention of reflection and transmission coefficients at an interface.

two-way travel time in each layer equal to the same quantity.* The reflection coefficients are zero and the transmission coefficients are one for any such hypothetical interfaces that are added. As a practical consideration, we suppose that a source is located just below interface zero and gives rise to a downgoing unit impulse. We further suppose that the seismometer (seismic receiver) responds only to upward wave motion and is also located in layer one just below interface zero. We choose our time origin ($k = 0$) as the time of the shot, i.e. the time of source excitation. With this convention, the reflected sequence $\{x_k\}$ recorded by the seismometer begins at $k = 1$ and $x_0 = 0$. Further, we choose the time between samples as one unit, corresponding to the two-way travel time in a layer. Thus, for the case of "no layers", i.e. an infinite half-space below interface zero, the received sequence $\{x_k\}$ is zero for all time. For the case of a single layer ($n = 1$), which we discussed in an example contained in the previous section, we found that the transfer function was:

$$F(z) = \frac{r_1 z}{1 + r_0 r_1 z} = \frac{X(z)}{S(z)} = \frac{E(z)}{A(z)\,S(z)} \qquad [2.3\text{-}2]$$

where $S(z)$ is the Laplace z-transform of the source wavelet and $X(z)$ is the Laplace z-transform of the reflection seismogram. From [2.3-2], we deduce that the reflection sequence $\{x_k\}$ can be modeled as the output of a linear shift-invariant recursive digital filter described by the difference equations:

$$x_k + a_1 x_{k-1} = \epsilon_k \quad \text{or } x_k + r_0 r_1 x_{k-1} = r_1 s_{k-1} \qquad [2.3\text{-}3]$$

Thus, the parameters of the transfer function $F(z)$ and difference equations [2.3-3] take on physical meaning; they are functions of the reflection coefficients of the two interfaces bounding layer 1. In the following discussion, we will generalize this result by finding the impulse response of the earth for the case of N layers. Since we are exciting the N-layered system with a unit impulse, then by definition the received sequence $\{x_k\}$ is the desired impulse response. Our end result will be a rational function $F(z)$ which will contain physically meaningful parameters derived from our layered earth model.

Let us now derive the transfer function of the N-layered earth, denoted by $F^{(N)}(z)$, by first considering the situation at interface n at some time k after the source excitation. Fig. 2-10 depicts an expanded version of Fig. 2-8 at the nth interface. It is convenient to draw the ray paths (see Fig. 2-10) in a diamond-grid configuration, so that the rays appear at non-normal incidence, in order to clearly illustrate the possible ray paths. However, the reader is reminded that the analysis is for the normal-incidence wave propagation. Let

---

* We note that the two-way travel time $= 2d/c$, where $d =$ the layer thickness and $c =$ the propagation velocity in a layer, is constant for all layers. As c differs from layer to layer, this would then mean that layer thickness also has to vary from layer to layer so as to make d/c constant.

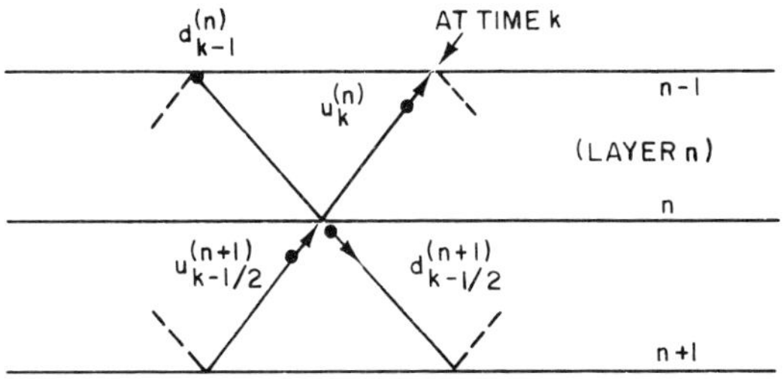

Fig. 2-10. Schematic diagram of reflected and transmitted waves at interface n.

$u_k^{(n)}$ denote the upgoing waveform measured at the top of layer n at time k, and let $d_{k-\frac{1}{2}}^{(n+1)}$ denote the downgoing waveform measured at the top of layer $n+1$ at time $k-\frac{1}{2}$. Then the upgoing and downgoing waves in the adjoining layers of Fig. 2-10 are related by:

$$\begin{aligned} u_k^{(n)} &= r_n d_{k-1}^{(n)} + t_n' u_{k-\frac{1}{2}}^{(n+1)} \\ d_{k-\frac{1}{2}}^{(n+1)} &= r_n' u_{k-\frac{1}{2}}^{(n+1)} + t_n d_{k-1}^{(n)} \end{aligned} \qquad [2.3\text{-}4]$$

where $u_k^{(n)}$ = upgoing waveform measured at the top of layer n at time k; $d_{k-\frac{1}{2}}^{(n+1)}$ = downgoing waveform measured at the top of layer $n+1$ at time $k-\frac{1}{2}$; $d_{k-1}^{(n)}$ = downgoing waveform measured at the top of layer n at time $k-1$; and $u_{k-\frac{1}{2}}^{(n+1)}$ = upgoing waveform measured at the top of layer $n+1$ at time $k-\frac{1}{2}$.

Rearrangement of [2.3-4] by solving for $u_{k-\frac{1}{2}}^{(n+1)}$ and $d_{k-\frac{1}{2}}^{(n+1)}$ in terms of $u_k^{(n)}$ and $d_{k-1}^{(n)}$ yields:

$$\begin{aligned} u_{k-\frac{1}{2}}^{(n+1)} &= \frac{1}{t_n'} u_k^{(n)} - \frac{r_n}{t_n'} d_{k-1}^{(n)} \\ d_{k-\frac{1}{2}}^{(n+1)} &= -\frac{r_n}{t_n'} u_k^{(n)} + \frac{1}{t_n'} d_{k-1}^{(n)} \end{aligned} \qquad [2.3\text{-}5]$$

where we have used the relation $r_n' = -r_n$ given in [2.3-1]. Taking the Laplace z-transform of [2.3-5] we get:

$$\begin{aligned} z^{\frac{1}{2}} U^{(n+1)}(z) &= \frac{1}{t_n'} U^{(n)}(z) - \frac{r_n}{t_n'} z D^{(n)}(z) \\ z^{\frac{1}{2}} D^{(n+1)}(z) &= -\frac{r_n}{t_n'} U^{(n)}(z) + \frac{1}{t_n'} z D^{(n)}(z) \end{aligned} \qquad [2.3\text{-}6]$$

Multiplication of [2.3-6] by $z^{-\frac{1}{2}}$ gives:

$$\begin{aligned} U^{(n+1)}(z) &= \frac{z^{-\frac{1}{2}}}{t_n'} U^{(n)}(z) - \frac{r_n}{t_n'} z^{\frac{1}{2}} D^{(n)}(z) \\ D^{(n+1)}(z) &= -\frac{r_n}{t_n'} z^{-\frac{1}{2}} U^{(n)}(z) + \frac{1}{t_n'} z^{\frac{1}{2}} D^{(n)}(z) \end{aligned} \qquad [2.3\text{-}7]$$

Writing [2.3-7] in matrix form we get:

$$\begin{bmatrix} U^{(n+1)}(z) \\ D^{(n+1)}(z) \end{bmatrix} = \frac{z^{-\frac{1}{2}}}{t'_n} \begin{bmatrix} 1 & -r_n z \\ -r_n & z \end{bmatrix} \begin{bmatrix} U^{(n)}(z) \\ D^{(n)}(z) \end{bmatrix} \qquad [2.3\text{-}8]$$

Now, $U^{(1)}(z)$ is the Laplace z-transform of the sequence $u_0^{(1)}, u_1^{(1)}, u_2^{(1)}, \ldots$ and represents the waveform measured at the seismometer. Hence, $U^{(1)}(z)$ is the transfer function of our N-layered model and we will refer to the sequence $\{u_k^{(1)}\}$ as the *reflection response* of the N-layered earth. Let us proceed to relate the waveform at the top of layer $N+1$ to the reflection response $u_k^{(1)}$. For $n = 1$, [2.3-8] becomes:

$$\begin{bmatrix} U^{(2)}(z) \\ D^{(2)}(z) \end{bmatrix} = \frac{z^{-\frac{1}{2}}}{t'_1} \begin{bmatrix} 1 & -r_1 z \\ -r_1 & z \end{bmatrix} \begin{bmatrix} U^{(1)}(z) \\ D^{(1)}(z) \end{bmatrix} \qquad [2.3\text{-}9]$$

Similarly, for the case $n = 2$ we obtain the result:

$$\begin{bmatrix} U^{(3)}(z) \\ D^{(3)}(z) \end{bmatrix} = \frac{z^{-1}}{t'_1 t'_2} \begin{bmatrix} 1 & -r_2 z \\ -r_2 & z \end{bmatrix} \begin{bmatrix} 1 & -r_1 z \\ -r_1 & z \end{bmatrix} \begin{bmatrix} U^{(1)}(z) \\ D^{(1)}(z) \end{bmatrix} \qquad [2.3\text{-}10]$$

Simplification of [2.3-10] gives:

$$\begin{bmatrix} U^{(3)}(z) \\ D^{(3)}(z) \end{bmatrix} = \frac{z^{-1}}{t'_1 t'_2} \begin{bmatrix} (1 + r_1 r_2 z) & -z(r_1 + r_2 z) \\ -(r_2 + r_1 z) & z(r_1 r_2 + z) \end{bmatrix} \begin{bmatrix} U^{(1)}(z) \\ D^{(1)}(z) \end{bmatrix} \qquad [2.3\text{-}11]$$

If we continue this procedure for $n = 3, 4, \ldots N$, we obtain the final result:

$$\begin{bmatrix} U^{(N+1)}(z) \\ D^{(N+1)}(z) \end{bmatrix} = \frac{z^{-N/2}}{t'_1 t'_2 \ldots t'_N} \mathbf{M}^{(N)} \cdot \mathbf{M}^{(N-1)} \ldots \mathbf{M}^{(1)} \begin{bmatrix} U^{(1)}(z) \\ D^{(1)}(z) \end{bmatrix} \qquad [2.3\text{-}12]$$

where we define the square matrix $\mathbf{M}^{(n)}$ by:

$$\mathbf{M}^{(n)} = \begin{bmatrix} 1 & -r_n z \\ -r_n & z \end{bmatrix}, \quad n = 1, 2, \ldots N \qquad [2.3\text{-}13]$$

From [2.3-11] we observe that the matrix product $\mathbf{M}^{(2)}\mathbf{M}^{(1)}$ has a set of polynomials as its elements. The product $\mathbf{M}^{(2)}\mathbf{M}^{(1)}$ may be expressed as:

$$\mathbf{M}^{(2)}\mathbf{M}^{(1)} = \begin{bmatrix} 1 + r_1 r_2 z & -z(r_1 + r_2 z) \\ -(r_2 + r_1 z) & z(r_1 r_2 + z) \end{bmatrix} = \begin{bmatrix} m_{11}^{(2)}(z) & m_{12}^{(2)}(z) \\ m_{21}^{(2)}(z) & m_{22}^{(2)}(z) \end{bmatrix} \qquad [2.3\text{-}14]$$

Now, the polynomial $m_{11}^{(2)}(z) = 1 + r_1 r_2 z$ also represents the Laplace z-transform of a minimum-delay sequence $\{m_k\} = \{1, r_1 r_2\}$ because the zeroes of $m_{11}^{(2)}(z)$ lie outside the unit circle. (Note that the zero of $m_{11}^{(2)}(z)$ is at $z = -1/(r_1 r_2)$. Now, in general, $|r_1 r_2| < 1$. Thus, $z = -1/(r_1 r_2)$ lies outside

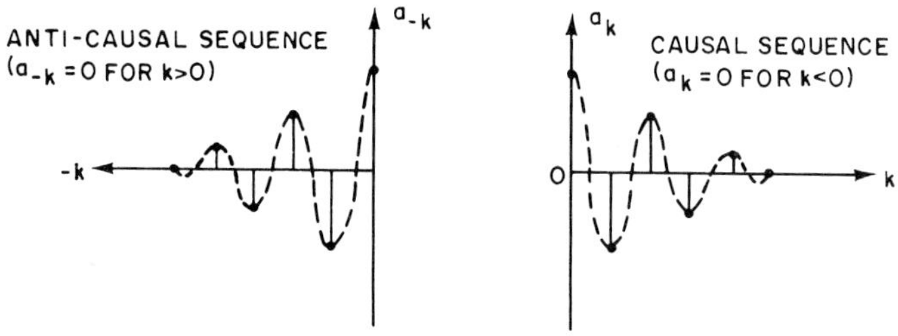

Fig. 2-11. Causal and anti-causal sequences.

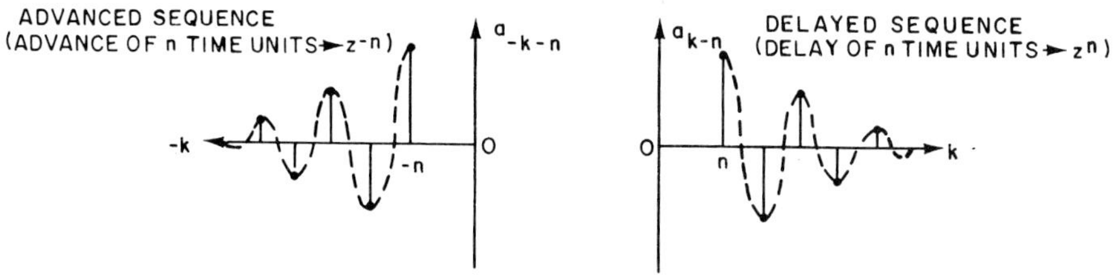

Fig. 2-12. Delayed and advanced sequences.

the unit circle.) When we reverse the direction of time (i.e. when we change k to $-k$), a causal sequence $\{a_k\}$ becomes an anti-causal sequence $\{a_{-k}\}$ (see Fig. 2-11). When we reverse the direction of time, a delay of n units becomes an advance of n units (i.e. $z^n \to z^{-n}$; see Fig. 2-12).

Definition of a minimum-advance wavelet. A minimum-advance wavelet is one obtained from a minimum-delay wavelet by changing k to $-k$.

Thus, when we reverse the direction of time, a minimum-delay wavelet becomes a *minimum-advance wavelet.* Therefore, a minimum-advance sequence is the time reverse of a minimum-delay sequence. As a result, a minimum-advance sequence is anti-causal. Further, if we replace k by $-k$, then the phase-lag $-\theta(\omega)$ is transformed to $+\theta(\omega)$. Thus, a minimum-advance sequence is also a maximum-negative-phase (anti-causal) sequence. Hence, the minimum-delay sequence $m_k = \{1, r_1r_2\}$ can be made minimum-advance by folding the sequence $\{1, r_1r_2\}$ about the origin (see Fig. 2-13). Now, the Laplace z-transform of the folded sequence m_{-k} becomes $r_1r_2z^{-1} + 1 = m_{11}^{(2)}(z^{-1})$. This property leads us to the following observation:

$$\begin{aligned} m_{22}^{(2)}(z) &= z^2 m_{11}^{(2)}(z^{-1}) \\ m_{12}^{(2)}(z) &= z^2 m_{21}^{(2)}(z^{-1}) \end{aligned} \qquad [2.3\text{-}15]$$

If we define the polynomials in $\mathbf{M}^{(1)}$ by:

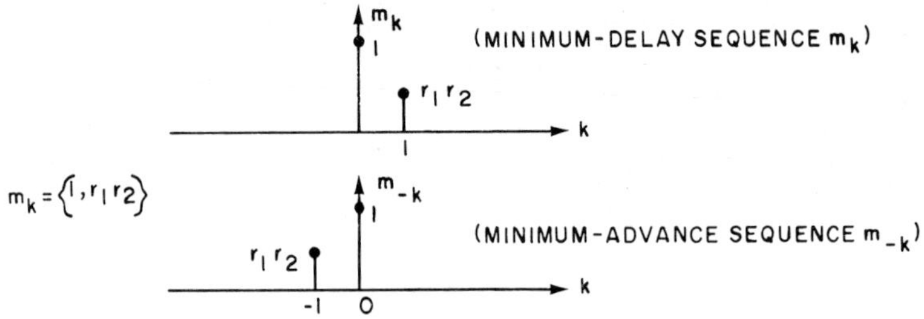

Fig. 2-13. Conversion of a minimum-delay sequence to a minimum-advance sequence by the process of folding.

$$\mathbf{M}^{(1)} = \begin{bmatrix} 1 & -r_1 z \\ -r_1 & z \end{bmatrix} = \begin{bmatrix} m_{11}^{(1)}(z) & z m_{21}^{(1)}(z^{-1}) \\ m_{21}^{(1)}(z) & z m_{11}^{(1)}(z^{-1}) \end{bmatrix} \qquad [2.3\text{-}16]$$

then we can rewrite equation [2.3-14] as:

$$\mathbf{M}^{(2)}\mathbf{M}^{(1)} = \begin{bmatrix} 1 & -r_2 z \\ -r_2 & z \end{bmatrix} \begin{bmatrix} m_{11}^{(1)}(z) & z m_{21}^{(1)}(z^{-1}) \\ m_{21}^{(1)}(z) & z m_{11}^{(1)}(z^{-1}) \end{bmatrix} = \begin{bmatrix} m_{11}^{(2)}(z) & z^2 m_{21}^{(2)}(z^{-1}) \\ m_{21}^{(2)}(z) & z^2 m_{11}^{(2)}(z^{-1}) \end{bmatrix} \qquad [2.3\text{-}17]$$

Performing the matrix multiplication indicated in [2.3-17] reveals the following recursion relations:

$$\begin{aligned} m_{11}^{(2)}(z) &= m_{11}^{(1)}(z) - r_2 z m_{21}^{(1)}(z) \\ m_{21}^{(2)}(z) &= -r_2 m_{11}^{(1)}(z) + z m_{21}^{(1)}(z) \end{aligned} \qquad [2.3\text{-}18]$$

with $m_{11}^{(1)}(z) = 1$ and $m_{21}^{(1)}(z) = -r_1$.

In general, we observe that:

$$\mathbf{M}^{(n)} \cdot \mathbf{M}^{(n-1)} \ldots \mathbf{M}^{(1)} = \begin{bmatrix} m_{11}^{(n)}(z) & z^n m_{21}^{(n)}(z^{-1}) \\ m_{21}^{(n)}(z) & z^n m_{11}^{(n)}(z^{-1}) \end{bmatrix} \qquad [2.3\text{-}19]$$

for $n = 1, 2, \ldots N$

The general recursion procedure for evaluation of these matrix products becomes:

$$\begin{aligned} m_{11}^{(n)}(z) &= m_{11}^{(n-1)}(z) - r_n z m_{21}^{(n-1)}(z) \\ m_{21}^{(n)}(z) &= -r_n m_{11}^{(n-1)}(z) + z m_{21}^{(n-1)}(z) \end{aligned} \qquad [2.3\text{-}20]$$

for $n = 2, 3, \ldots N$

with $m_{11}^{(1)}(z) = 1$ and $m_{21}^{(1)}(z) = -r_1$.

We shall now prove that the polynomials $m_{11}^{(n)}(z)$ $(n = 1, 2, \ldots N)$ are minimum-delay, i.e. possess no zeroes inside the unit circle. Our inductive proof is as follows:

For $n = 1$, the constant polynomial $m_{11}^{(1)}(z) = 1$ is clearly minimum-delay. Similarly, $m_{11}^{(2)}(z) = 1 + r_1 r_2 z$ is also minimum-delay, since $|r_1 r_2|$ is generally less than unity and the zeroes of $m_{11}^{(2)}(z)$ lie outside the unit circle. Let us now assume that $m_{11}^{(n-1)}(z)$ is minimum-delay. In order to show that $m_{11}^{(n)}(z)$ is minimum-delay, we use the theorem of Rouche. We shall give one version of Rouche's theorem which is useful for our purposes.

Rouche's theorem: If $P(z)$ and $Q(z)$ are analytic interior to the unit circle C, if they are continuous on C, $Q(z) \neq 0$ on C, and they satisfy the following condition on C:

$$|P(z)| < |Q(z)|, \quad z \in C$$

then the function $R(z) = P(z) + Q(z)$ has the same number of zeroes interior to C as does $Q(z)$.

If we denote the polynomials $Q(z)$ and $P(z)$ by:

$$\begin{aligned} Q(z) &= m_{11}^{(n-1)}(z) \\ P(z) &= -r_n z m_{21}^{(n-1)}(z) \end{aligned} \qquad [2.3\text{-}21]$$

then $R(z)$ is given by:

$$R(z) = m_{11}^{(n-1)}(z) - r_n z m_{21}^{(n-1)}(z) \qquad [2.3\text{-}22]$$

Comparison of [2.3-20] with [2.3-22] indicates that:

$$R(z) = m_{11}^{(n)}(z) \qquad [2.3\text{-}23]$$

Let us proceed to show that the inequality condition of Rouche's theorem is satisfied. Specifically, we will show that on C, i.e. $z = e^{-i\omega}$, we obtain:

$$|P(e^{-i\omega})| < |Q(e^{-i\omega})| \qquad [2.3\text{-}24]$$

or:

$$|r_n| \, |m_{21}^{(n-1)}(e^{-i\omega})| < |m_{11}^{(n-1)}(e^{-i\omega})|$$

We shall denote the determinant of the matrix product $\mathbf{M}^{(n-1)} \cdot \mathbf{M}^{(n-2)} \ldots \mathbf{M}^{(1)}$ by $\det[\mathbf{M}^{(n-1)} \cdot \mathbf{M}^{(n-2)} \ldots \mathbf{M}^{(1)}]$. Further:

$$\det[\mathbf{M}^{(n-1)} \cdot \mathbf{M}^{(n-2)} \ldots \mathbf{M}^{(1)}] = \det[\mathbf{M}^{(n-1)}] \det[\mathbf{M}^{(n-2)}] \ldots \det[\mathbf{M}^{(1)}] \qquad [2.3\text{-}25]$$

Now, $\det[\mathbf{M}^{(n)}]$, with $\mathbf{M}^{(n)}$ defined in [2.3-13], is given by

$$\det[\mathbf{M}^{(n)}] = z(1 - r_n^2) \qquad [2.3\text{-}26]$$

Using this fact, we can rewrite [2.3-25] as:

$$\det[\mathbf{M}^{(n-1)} \cdot \mathbf{M}^{(n-2)} \ldots \mathbf{M}^{(1)}] = z(1 - r_{n-1}^2) \cdot z(1 - r_{n-2}^2) \ldots z(1 - r_1^2) \qquad [2.3\text{-}27]$$

or:

$$\det[\mathbf{M}^{(n-1)} \cdot \mathbf{M}^{(n-2)} \ldots \mathbf{M}^{(1)}] = z^{n-1}(1-r_{n-1}^2)(1-r_{n-2}^2) \ldots (1-r_1^2) \quad [2.3\text{-}28]$$

From [2.3-19], with n replaced by $n-1$, we observe that:

$$\det[\mathbf{M}^{(n-1)} \cdot \mathbf{M}^{(n-2)} \ldots \mathbf{M}^{(1)}] = z^{n-1}[m_{11}^{(n-1)}(z)m_{11}^{(n-1)}(z^{-1}) - m_{21}^{(n-1)}(z)m_{21}^{(n-1)}(z^{-1})] \quad [2.3\text{-}29]$$

Equating [2.3-28] and [2.3-29], we obtain:

$$m_{11}^{(n-1)}(z)m_{11}^{(n-1)}(z^{-1}) - m_{21}^{(n-1)}(z)m_{21}^{(n-1)}(z^{-1}) = (1-r_{n-1}^2)(1-r_{n-2}^2) \ldots (1-r_1^2) \quad [2.3\text{-}30]$$

But since the squared value of a reflection coefficient is generally less than unity, then the factor $(1-r_n^2)$ is always positive, i.e. $(1-r_n^2) > 0$. Thus, the right-hand side of [2.3-30] is always positive and real, which allows us to write:

$$m_{11}^{(n-1)}(z)m_{11}^{(n-1)}(z^{-1}) - m_{21}^{(n-1)}(z)m_{21}^{(n-1)}(z^{-1}) > 0 \quad [2.3\text{-}31]$$

Evaluating [2.2-31] on the unit circle ($z = \mathrm{e}^{-i\omega}$), we get:

$$|m_{11}^{(n-1)}(\mathrm{e}^{-i\omega})|^2 - |m_{21}^{(n-1)}(\mathrm{e}^{-i\omega})|^2 > 0 \quad [2.3\text{-}32]$$

or:

$$|m_{11}^{(n-1)}(\mathrm{e}^{-i\omega})|^2 > |m_{21}^{(n-1)}(\mathrm{e}^{-i\omega})|^2 \quad [2.3\text{-}33]$$

Rewriting [2.3-33], we obtain:

$$\frac{|m_{21}^{(n-1)}(\mathrm{e}^{-i\omega})|}{|m_{11}^{(n-1)}(\mathrm{e}^{-i\omega})|} < 1, \quad |m_{11}^{(n-1)}(\mathrm{e}^{-i\omega})| \neq 0 \quad [2.3\text{-}34]$$

Now since $|r_n| \leqslant 1$, then we can write:

$$\frac{|r_n|\,|m_{21}^{(n-1)}(\mathrm{e}^{-i\omega})|}{|m_{11}^{(n-1)}(\mathrm{e}^{-i\omega})|} < 1 \quad [2.3\text{-}35]$$

or:

$$|r_n|\,|m_{21}^{(n-1)}(\mathrm{e}^{-i\omega})| < |m_{11}^{(n-1)}(\mathrm{e}^{-i\omega})| \quad [2.3\text{-}36]$$

which proves the assertion in [2.3-24]. Thus, the polynomial $R(z) = m_{11}^{(n)}(z)$ has the same number of zeroes interior to C as does the polynomial $Q(z) = m_{11}^{(n-1)}(z)$. Since $m_{11}^{(n-1)}(z)$ is minimum-delay, then $m_{11}^{(n)}(z)$ is also minimum-delay, i.e. both have no zeroes in C. This completes the inductive proof, and shows that $m_{11}^{(n)}(z)$ is minimum-delay for all $n = 1, 2, \ldots N$. Hence, the recursion in [2.3-20] preserves the minimum-delay property of $m_{11}^{(n)}(z)$.

Using [2.3-19], we can rewrite [2.3-12] as:

$$\begin{bmatrix} U^{N+1}(z) \\ D^{N+1}(z) \end{bmatrix} = \frac{z^{-N/2}}{t_1' t_2' \ldots t_N'} \begin{bmatrix} m_{11}^{(N)}(z) & z^N m_{21}^{(N)}(z^{-1}) \\ m_{21}^{(N)}(z) & z^N m_{11}^{(N)}(z^{-1}) \end{bmatrix} \begin{bmatrix} U^{(1)}(z) \\ D^{(1)}(z) \end{bmatrix} \quad [2.3\text{-}37]$$

with the elements of the square matrix defined by the recursion relations of [2.3-20]. From our problem formulation, we observe that $U^{N+1}(z) = 0$, since there is no upgoing waveform impinging on the bottom of interface N. Under this condition, we can extract the following equation from [2.3-37]:

$$0 = \frac{z^{-N/2}}{t_1' t_2' \ldots t_N'} [m_{11}^{(N)}(z) U^{(1)}(z) + z^N m_{21}^{(N)}(z^{-1}) D^{(1)}(z)] \quad [2.3\text{-}38]$$

or:

$$U^{(1)}(z) = \frac{-z^N m_{21}^{(N)}(z^{-1})}{m_{11}^{(N)}(z)} D^{(1)}(z) \quad [2.3\text{-}39]$$

But the polynomial $D^{(1)}(z)$ is expressed as:

$$D^{(1)}(z) = 1 + r_0' U^{(1)}(z) \quad [2.3\text{-}40]$$

where the term 1 is due to the unit impulse excitation at $k = 0$ by the choice of time origin as the time of the shot. Substitution of [2.3-40] in [2.3-39] gives:

$$U^{(1)}(z) = \frac{-z^N m_{21}^{(N)}(z^{-1})}{m_{11}^{(N)}(z) - r_0 z^N m_{21}^{(N)}(z^{-1})} \quad [2.3\text{-}41]$$

which is our desired *reflection response* for an N-layered system. We note that the denominator of [2.3-41] is minimum-delay (by Rouche's theorem), and therefore the reflection response is stable as one might physically expect. Since $U^{(1)}(z)$ is the Laplace z-transform of the sequence $u_0^{(1)}, u_1^{(1)}, u_2^{(1)}, \ldots u_k^{(1)}$, measured at the seismometer, it is effectively the impulse response of our system. Thus, the transfer function of the N-layered earth is $F^{(N)}(z) = U^{(1)}(z)$, and [2.3-41] becomes:

$$F^{(N)}(z) = \frac{-z^N m_{21}^{(N)}(z^{-1})}{m_{11}^{(N)}(z) - r_0 z^N m_{21}^{(N)}(z^{-1})}, \quad N \geqslant 1 \quad [2.3\text{-}42]$$

where $m_{11}^{(N)}(z) = m_{11}^{(N-1)}(z) - r_N z m_{21}^{(N-1)}(z)$

$m_{21}^{(N)}(z) = -r_N m_{11}^{(N-1)}(z) + z m_{21}^{(N-1)}(z), \quad N \geqslant 2$

$m_{11}^{(1)}(z) = 1, \quad m_{21}^{(1)}(z) = -r_1, \quad$ and $F^{(0)}(z) = 0$

Let us expand $F^{(N)}(z)$ for a few cases. If we let N denote the number of layers and $F^{(N)}(z)$ the Laplace z-transform of the reflection response for N layers, then:

N	$F^{(N)}(z)$
0	$F^{(0)}(z) = 0$
1	$F^{(1)}(z) = \dfrac{r_1 z}{1 + r_0 r_1 z}$
2	$F^{(2)}(z) = \dfrac{r_1 z + r_2 z^2}{1 + (r_0 r_1 + r_1 r_2)z + r_0 r_2 z^2}$
3	$F^{(3)}(z) = \dfrac{r_1 z + (r_2 + r_3 r_2 r_1)z^2 + r_3 z^3}{1 + (r_0 r_1 + r_1 r_2 + r_2 r_3)z + (r_0 r_2 + r_1 r_3 + r_0 r_1 r_2 r_3)z^2 + r_0 r_3 z^3}$

[2.3-43]

In addition to the reflected energy received at our seismometer, there is also energy transmitted into the air (layer zero) and into the other infinite half-space layer $N + 1$. Thus, the reflection response of an N-layered system could also be defined as the sequence $t'_0 u_0^{(1)}, t'_0 u_1^{(1)}, t'_0 u_2^{(1)}, \ldots t'_0 u_k^{(1)} \ldots$ if we choose to perform our measurement in air, i.e. the bottom of layer zero. We define the *transmission response* as the sequence $d_{N/2-1}^{(N+1)}, d_{N/2}^{(N+1)}, d_{N/2+1}^{(N+1)}, \ldots$ which is transmitted into the lower infinite half-space (layer $N + 1$). Now, due to our problem formulation, the Laplace z-transform of this sequence, denoted by $F_{\mathrm{T}}^{(N)}(z)$, is given by:

$$F_{\mathrm{T}}^{(N)}(z) = D^{(N+1)}(z), \quad N \geqslant 1 \qquad [2.3\text{-}44]$$

If we perform our measurement in air, then we can define $F_{\mathrm{R}}^{(N)}(z)$ as the Laplace z-transform of the reflection response $t'_0 u_0^{(1)}, t'_0 u_1^{(1)}, t'_0 u_2^{(1)}, \ldots t'_0 u_k^{(1)}, \ldots$ In symbols:

$$F_{\mathrm{R}}^{(N)}(z) = t'_0 U^{(1)}(z) = t'_0 F^{(N)}(z) \qquad [2.3\text{-}45]$$

Now, from [2.3-37], we can extract the equation:

$$D^{(N+1)}(z) = \frac{z^{-N/2}}{t'_1 t'_2 \ldots t'_N} [m_{21}^{(N)}(z) U^{(1)}(z) + z^N m_{11}^{(N)}(z^{-1}) D^{(1)}(z)] \qquad [2.3\text{-}46]$$

Substitution of [2.3-40] in [2.3-46] and recalling that $F^{(N)}(z) = U^{(1)}(z)$, we get:

$$D^{(N+1)}(z) = \frac{z^{-N/2}}{t'_1 t'_2 \ldots t'_N} [F^{(N)}(z)\{m_{21}^{(N)}(z) - r_0 z^N m_{11}^{(N)}(z^{-1})\} + z^N m_{11}^{(N)}(z^{-1})] \qquad [2.3\text{-}47]$$

Substitution of [2.3-42] in [2.3-47] yields:

$$D^{(N+1)}(z) = \frac{z^{-N/2}}{t'_1 t'_2 \ldots t'_N} \left[\frac{z^N m_{11}^{(N)}(z) m_{11}^{(N)}(z^{-1}) - z^N m_{21}^{(N)}(z) m_{21}^{(N)}(z^{-1})}{m_{11}^{(N)}(z) - r_0 z^N m_{21}^{(N)}(z^{-1})} \right] \qquad [2.3\text{-}48]$$

If we let $\det[\mathbf{M}^{(N)} \cdot \mathbf{M}^{(N-1)} \ldots \mathbf{M}^{(1)}]$ denote the determinant of the resultant matrix product $\mathbf{M}^{(N)} \cdot \mathbf{M}^{(N-1)} \ldots \mathbf{M}^{(1)}$, then inspection of [2.3-19] indicates that:

$$\det[\mathbf{M}^{(N)} \cdot \mathbf{M}^{(N-1)} \ldots \mathbf{M}^{(1)}] = z^N m_{11}^{(N)}(z) m_{11}^{(N)}(z^{-1}) - z^N m_{21}^{(N)}(z) m_{21}^{(N)}(z^{-1}) \quad [2.3\text{-}49]$$

which is identical to the numerator of [2.3-48]. Further:

$$\det[\mathbf{M}^{(N)} \cdot \mathbf{M}^{(N-1)} \ldots \mathbf{M}^{(1)}] = \det[\mathbf{M}^{(N)}] \det[\mathbf{M}^{(N-1)}] \ldots \det[\mathbf{M}^{(1)}] \quad [2.3\text{-}50]$$

Referring to the definition of $\mathbf{M}^{(n)}$ in [2.3-13], [2.3-50] can be rewritten as:

$$\det[\mathbf{M}^{(N)} \cdot \mathbf{M}^{(N-1)} \ldots \mathbf{M}^{(1)}] =$$

$$\det \begin{bmatrix} 1 & -r_N z \\ -r_N & z \end{bmatrix} \det \begin{bmatrix} 1 & -r_{N-1} z \\ -r_{N-1} & z \end{bmatrix} \ldots \det \begin{bmatrix} 1 & -r_1 z \\ -r_1 & z \end{bmatrix} \quad [2.3\text{-}51]$$

Expansion of [2.3-51] gives:

$$\det[\mathbf{M}^{(N)} \cdot \mathbf{M}^{(N-1)} \ldots \mathbf{M}^{(1)}] = z(1 - r_N^2) \cdot z(1 - r_{N-1}^2) \ldots z(1 - r_1^2) \quad [2.3\text{-}52]$$

Using the relation from [2.3-1] that $t_n t_n' = 1 - r_n^2$, [2.3-52] reduces to:

$$\det[\mathbf{M}^{(N)} \cdot \mathbf{M}^{(N-1)} \ldots \mathbf{M}^{(1)}] = z^N (t_1 t_2 \ldots t_N)(t_1' \cdot t_2' \ldots t_N') \quad [2.3\text{-}53]$$

Combining equations [2.3-49] and [2.3-53], we obtain:

$$z^N m_{11}^{(N)}(z) m_{11}^{(N)}(z^{-1}) - z^N m_{21}^{(N)}(z) m_{21}^{(N)}(z^{-1}) = z^N (t_1 t_2 \ldots t_N)(t_1' \cdot t_2' \ldots t_N') \quad [2.3\text{-}54]$$

Substitution of [2.3-54] into the numerator of [2.3-48] yields:

$$D^{(N+1)}(z) = \frac{z^{N/2}[t_1 t_2 \ldots t_N]}{m_{11}^{(N)}(z) - r_0 z^N m_{21}^{(N)}(z^{-1})} \quad [2.3\text{-}55]$$

Finally, substitution of [2.3-55] in [2.3-44] gives:

$$F_T^{(N)}(z) = \frac{z^{N/2}[t_1 t_2 \ldots t_N]}{m_{11}^{(N)}(z) - r_0 z^N m_{21}^{(N)}(z^{-1})}, \quad N \geqslant 1 \quad [2.3\text{-}56]$$

where the $z^{N/2}$ factor accounts for the delay time $N/2$ for direct transmission through N layers. Equations [2.3-42], [2.3-45], and [2.3-56], respectively, describe the transfer functions of the reflection responses and transmission response of the earth to a unit impulse. (Note that these equations all have the same denominator $m_{11}^{(N)}(z) - r_0 z^N m_{21}^{(N)}(z^{-1})$. By Rouche's theorem, this denominator is minimum-delay.) Further, we have now completed our convolution model by relating the parameters of the transfer function $F(z)$ to the reflection coefficients of the subsurface structure. Specifically, if we consider an N-layered earth, then $F^{(N)}(z)$ is the corresponding "reflection" transfer function which relates the source wavelet $\{s_k\}$ to the uncorrupted measured sequence $\{x_k\}$. Later, we will discuss the situation when $\{x_k\}$ is corrupted by measurement noise $\{n_k\}$, resulting in the noisy measurement $\{y_k\} = \{x_k\} + \{n_k\}$.

2.4. CHARACTERISTIC AND REFLECTION POLYNOMIALS

In general, the transfer function of any bounded linear time-invariant system can be expressed as the ratio of two polynomials in a complex variable. From the previous section, we see that the reflection trànsfer function $F^{(N)}(z)$ can be expressed as:

$$F^{(N)}(z) = \frac{E^{(N)}(z)}{A^{(N)}(z)} = \frac{\sum_{n=1}^{N} \epsilon_n^{(N)} z^n}{\sum_{n=0}^{N} a_n^{(N)} z^n} = \frac{\epsilon_1^{(N)} z + \epsilon_2^{(N)} z^2 + \ldots + \epsilon_N^{(N)} z^N}{a_0^{(N)} + a_1^{(N)} z + \ldots + a_N^{(N)} z^N} \qquad [2.4\text{-}1]$$

with $\epsilon_1^{(N)} = r_1$, $a_0^{(N)} = 1$, $\epsilon_N^{(N)} = r_N$, and $a_N^{(N)} = r_0 r_N$. Cast in this form, we see that $F^{(N)}(z)$ has the form of a standard recursive digital filter, whose output is dependent on present and past values of the input as well as present and past values of the output. Letting x_k represent the noise-free output of the filter and s_k the input, then [2.4-1] has the time-domain representation:

$$x_k + a_1^{(N)} x_{k-1} + a_2^{(N)} x_{k-2} + \ldots + a_N^{(N)} x_{k-N} = \epsilon_1^{(N)} s_{k-1} + \epsilon_2^{(N)} s_{k-2} + \ldots + \epsilon_N^{(N)} s_{k-N} \qquad [2.4\text{-}2]$$

where $a_N^{(N)} = r_0 r_N$ and $\epsilon_N^{(N)} = r_N$.

In the analysis of systems in general, stability is of paramount importance. Both linear and non-linear systems have associated analytical and graphical procedures from which one can determine the stability of the system. Such procedures might simply involve finding the pole locations in the complex plane while other methods involve more sophisticated complex variable analysis. For our purposes, we refer to the basic definition of stability, i.e. if all bounded inputs give rise to bounded outputs, then the system is stable. Now, as a physical fact, we know that the earth cannot produce an unbounded response to a bounded input. Thus, from purely physical grounds, we know that the sequence $\{x_k\}$ is always bounded and we conclude that the transfer function $F^{(N)}(z)$ represents a stable system.

The denominator of [2.4-1] is generally referred to as the characteristic polynomial of the system, which in this case is a polynomial in the complex variable z. If we let $A^{(N)}(z) = \Sigma_{n=0}^{N} a_n^{(N)} z^n$ be the characteristic polynomial of an N-layered system, then by equating $A^{(N)}(z)$ to zero we obtain the characteristic equation of an N-layered system given by:

$$a_N^{(N)} z^N + a_{N-1}^{(N)} z^{N-1} + \ldots + a_1^{(N)} z + 1 = 0 \qquad [2.4\text{-}3]$$

Now, the solution of [2.4-3] yields the roots or zeroes of $A^{(N)}(z)$ or the poles of $F^{(N)}(z)$. From phyiscal grounds, we stated that $F^{(N)}(z)$ is a stable system and therefore has no poles inside the unit circle. Thus, the Laplace z-transform $A^{(N)}(z)$ has no zeroes inside the unit circle and by definition is minimum-delay. In other words, the characteristic sequence $\{1, a_1^{(N)}, a_2^{(N)}, \ldots a_N^{(N)}\}$ is a minimum-delay sequence whose Laplace z-transform is the minimum-delay characteristic polynomial $A^{(N)}(z)$.

Equations [2.4-1] and [2.3-42] describe the layered earth model, consistent with the assumptions stated in section 2.3. Let us now examine this model in more mathematical detail. Our main goal is to show that certain mathematical approximations lead to simplified versions of the model, which are consistent with physical reasoning. These simplified models will be a valuable asset in understanding the physical theory of deconvolution discussed in Chapter 4. First, let us consider the case of the "small reflection coefficient" model.

Small reflection coefficient model

From the denominator of [2.3-42], we see that the characteristic polynomial $A^{(N)}(z)$ can be expressed as:

$$A^{(N)}(z) = m_{11}^{(N)}(z) - r_0 z^N m_{21}^{(N)}(z^{-1}), \quad N \geqslant 1 \qquad [2.4\text{-}4]$$

with $m_{11}^{(N)}(z)$ and $m_{21}^{(N)}(z)$ defined by the recursion relations given in [2.3-42]. Inspection of the denominators of [2.3-43] reveals that:

$$\begin{aligned} A^{(1)}(z) &= 1 + r_0 r_1 z \\ A^{(2)}(z) &= 1 + (r_0 r_1 + r_1 r_2)z + r_0 r_2 z^2 \qquad [2.4\text{-}5] \\ A^{(3)}(z) &= 1 + (r_0 r_1 + r_1 r_2 + r_2 r_3)z + (r_0 r_2 + r_1 r_3)z^2 + r_0 r_1 r_2 r_3 z^2 + r_0 r_3 z^3 \end{aligned}$$

Now, for the case of four layers ($N = 4$), we use [2.4-4] and the recursion in [2.3-42] to obtain:

$$\begin{aligned} A^{(4)}(z) &= 1 + (r_0 r_1 + r_1 r_2 + r_2 r_3 + r_3 r_4)z + (r_0 r_2 + r_1 r_3 + r_2 r_4)z^2 \\ &\quad + (r_4 r_3 r_2 r_1 + r_4 r_3 r_1 r_0 + r_3 r_2 r_1 r_0)z^2 + (r_0 r_3 + r_1 r_4)z^3 \\ &\quad + (r_4 r_2 r_1 r_0 + r_4 r_3 r_2 r_0)z + r_0 r_4 z^4 \end{aligned}$$

Examination of [2.4-5] persuades us to define the following serial correlation:

$$\phi_m^{(N)} = \sum_{n=0}^{N} r_n r_{n+m}\,; \quad m \geqslant 1, N \geqslant 1 \qquad [2.4\text{-}6]$$

with $\phi_0^{(N)} = 1$.

With this definition, the polynomials $A^{(1)}(z)$, $A^{(2)}(z)$, $A^{(3)}(z)$, and $A^{(4)}(z)$ in [2.4-5] become:

$$\begin{aligned} A^{(1)}(z) &= 1 + \phi_1^{(1)} z \\ A^{(2)}(z) &= 1 + \phi_1^{(2)} z + \phi_2^{(2)} z^2 \qquad [2.4\text{-}7] \\ A^{(3)}(z) &= 1 + \phi_1^{(3)} z + [\phi_2^{(3)} + r_0 r_1 r_2 r_3] z^2 + \phi_3^{(3)} z^3 \end{aligned}$$

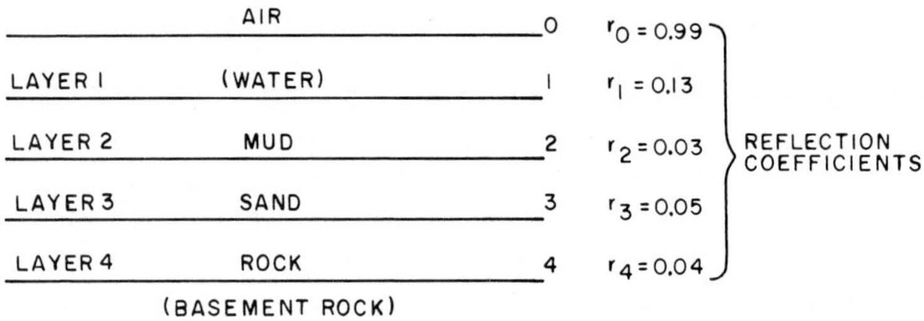

Fig. 2-14. A possible configuration of the first few layers encountered in marine geophysical exploration.

$$
\begin{aligned}
A^{(4)}(z) &= 1 + \phi_1^{(4)} z + [\phi_2^{(4)} + r_4 r_3 r_2 r_1 + r_4 r_3 r_1 r_0 + r_3 r_2 r_1 r_0] z^2 \\
&\quad + [\phi_3^{(4)} + r_4 r_3 r_2 r_0 + r_4 r_2 r_1 r_0] z^3 + \phi_4^{(4)} z^4
\end{aligned}
$$

Now, the magnitude of any reflection coefficient can never exceed unity, and in practice, these magnitudes (with the exception of r_0) are quite small, usually smaller than 0.1 or 0.05. For example, in marine geophysical exploration, one might encounter a situation as shown in Fig. 2-14. Practically speaking, significant impedance changes occur in the first few layers, with the air-water interface being the most significant. The values of $r_0 = 0.99$, $r_1 = 0.13$, $r_2 = 0.03$, $r_3 = 0.05$, and $r_4 = 0.04$ are typical of reflection coefficients encountered in practice. For these reflection coefficients, we see that:

$$
\begin{array}{lll}
\phi_1^{(4)} = 0.1361 & r_4 r_3 r_2 r_1 = & 0.0078 \times 10^{-3} \\
\phi_2^{(4)} = 0.0374 & r_4 r_3 r_1 r_0 = & 0.2574 \times 10^{-3} \\
\phi_3^{(4)} = 0.0547 & r_3 r_2 r_1 r_0 = & \underline{0.1930 \times 10^{-3}} \\
 & \text{sum} & 0.4582 \times 10^{-3} \\
\phi_4^{(4)} = 0.0396 & & \\
 & r_4 r_3 r_2 r_0 = & 0.0594 \times 10^{-3} \\
 & r_4 r_2 r_1 r_0 = & \underline{0.1544 \times 10^{-3}} \\
 & \text{sum} & 0.2138 \times 10^{-3}
\end{array}
$$

From the above values, we observe that:

$$\phi_2^{(4)} \gg r_4 r_3 r_2 r_1 + r_4 r_3 r_1 r_0 + r_3 r_2 r_1 r_0$$

$$\phi_3^{(4)} \gg r_4 r_3 r_2 r_0 + r_4 r_2 r_1 r_0$$

In general, $\phi_m^{(N)}$ is the sum of products of two reflection coefficients which is greater than the sum of products of four or more reflection coefficients. Thus, to a good approximation, the characteristic polynomial of an N-layered system may be given by:

$$A^{(N)}(z) = \sum_{l=0}^{N} a_l^{(N)} z^l \approx 1 + \phi_1^{(N)} z + \phi_2^{(N)} z^2 + \ldots + \phi_N^{(N)} z^N \qquad [2.4\text{-}8]$$

where the serial correlations are $\phi_m^{(N)} = \Sigma_{n=0}^{N} r_n r_{n+m}$ for $m \geqslant 1, N \geqslant 1$

We observe from [2.4-8] that the serial correlation $\{1, \phi_1^{(N)}, \phi_2^{(N)}, \ldots \phi_N^{(N)}\}$ has the form of a time autocorrelation sequence, except that $\phi_0^{(N)}$ is fixed at unity and not as $\Sigma_{n=0}^{N} r_n^2$ as it should be to be an autocorrelation function. The reflection coefficient sequence $\{r_0, r_1, \ldots r_N\}$ is an ordered sequence of numbers with the subscripts denoting the arrangement of these numbers in direct correspondence with the time of occurrence of a number. For example, the reflection coefficient r_7 at interface 7 corresponds to time $k = 7$ for the primary reflection for this interface. Thus, the sequence $\{r_0, r_1, \ldots r_N\}$ is ordered with respect to depth, since we defined our layered earth model with layer zero as the air and layer n deeper than layer $n-1$. The sequence of numbers $\{r_1, r_2, \ldots r_N\}$ can be thought of as the values of the primary reflections appearing on the seismogram. Fig. 2-15 depicts the primary reflections described above. We note that the actual amplitude value of the primary reflection from interface k at time k is given by $(t_1 t_2 \ldots t_{k-1})(t_1' t_2' \ldots t_{k-1}') r_k$ when a downgoing unit impulse is applied at the surface.

The term characteristic polynomial also has a physical interpretation. For example, since the sequence $a_0^{(N)}, a_1^{(N)}, \ldots a_N^{(N)}$ is related to the amount of correlation between reflection coefficients, then it "characterizes" a specific geographic location. If one could observe the characteristic sequence, then one would have a measure of the orderliness of the subsurface structure. In this context, the characteristic sequence truly characterizes the N-layered system.

Let us define the polynomial $E^{(N)}(z)$ by:

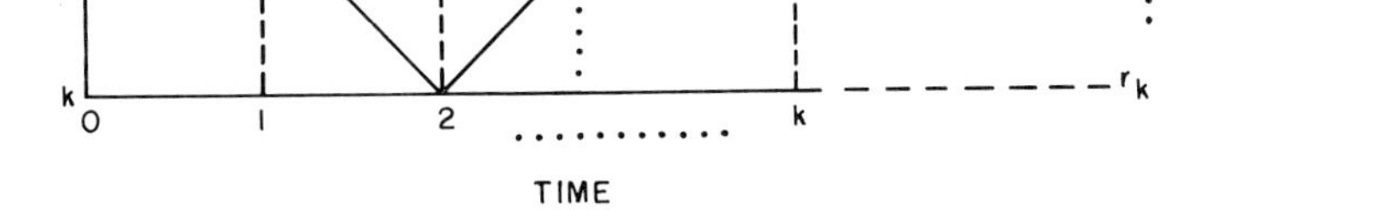

Fig. 2-15. The sequence $\{r_1, r_2, \ldots r_k, \ldots\}$ depicted as the amplitude values of the primary reflections occurring at time k from interface k.

$$E^{(N)}(z) = \sum_{n=1}^{N} \epsilon_n^{(N)} z^n \quad [2.4\text{-}9]$$

which is the numerator in [2.4-1]. Now from the numerator of [2.3-42], $E^{(N)}(z)$ can be expressed as:

$$E^{(N)}(z) = -z^N m_{21}^{(N)}(z^{-1}), \quad N \geqslant 1 \quad [2.4\text{-}10]$$

with $m_{21}^{(N)}(z^{-1})$ given by the recursion relations in [2.3-42]. Inspection of the numerators of [2.3-43] reveals that:

$$\begin{aligned} E^{(1)}(z) &= r_1 z \\ E^{(2)}(z) &= r_1 z + r_2 z^2 \\ E^{(3)}(z) &= r_1 z + (r_2 + r_3 r_2 r_1) z^2 + r_3 z^3 \end{aligned} \quad [2.4\text{-}11]$$

For the case of four layers ($N = 4$), we use [2.4-9] and the recursion of [2.3-42] to get:

$$E^{(4)}(z) = r_1 z + [r_2 + r_4 r_3 r_1 + r_3 r_2 r_1] z^2 + [r_3 + r_4 r_2 r_1 + r_4 r_3 r_2] z^3 + r_4 z^4$$

In the example above (refer to Fig. 2-14), the values of reflection coefficients were $r_1 = 0.13$, $r_2 = 0.03$, $r_3 = 0.05$, and $r_4 = 0.04$. For these values, we see that:

$r_4 r_3 r_1 =$	0.260×10^{-3}	$r_4 r_2 r_1 =$	0.156×10^{-3}
$r_3 r_2 r_1 =$	0.195×10^{-3}	$r_4 r_3 r_2 =$	0.060×10^{-3}
sum	0.455×10^{-3}	sum	0.216×10^{-3}

From the above values, we observe that:

$$r_2 \gg r_4 r_3 r_1 + r_3 r_2 r_1; \quad r_3 \gg r_4 r_2 r_1 + r_4 r_3 r_2$$

In considering many layers (N large), it will be true in general that we may neglect sums of products of three or more reflection coefficients in comparison to a single reflection coefficient. Thus, to a good approximation, the polynomial $E^{(N)}(z)$ can be expressed as:

$$E^{(N)}(z) = \sum_{l=1}^{N} \epsilon_l^{(N)} z^l \approx \sum_{l=1}^{N} r_l z^l, \quad N \geqslant 1 \quad [2.4\text{-}12]$$

where $r_1, r_2, \ldots r_N$ are the reflection coefficients.

Therefore, according to the small reflection coefficient approximation $\epsilon_n^{(N)} \approx r_n$, so the numerator of [2.4-1] is replaced by [2.4-12]. Thus, $E^{(N)}(z)$ has the interesting property of being (approximately) the Laplace z-transform of a sequence whose amplitudes are exactly the reflection coefficients. We shall refer to $E^{(N)}(z)$ as the *reflection polynomial* and the sequence

$\{r_1, r_2, \ldots r_N\}$ as the *reflection sequence*. However, unlike the characteristic polynomial $A^{(N)}(z)$, $E^{(N)}(z)$ has no reason to be minimum-delay, and in general, is mixed-delay, i.e. its Laplace z-transform contains zeroes (roots) both inside and outside the unit circle. The polynomial $E^{(N)}(z)$ contains the information that is needed in seismic exploration, namely the reflection coefficients at their respective time delays.

With the above "small reflection coefficient" approximations, the reflection transfer function $F^{(N)}(z)$ can be expressed as the ratio of the reflection polynomial $E^{(N)}(z)$ to the characterisitc polynomial $A^{(N)}(z)$. In symbols:

$$F^{(N)}(z) = \frac{E^{(N)}(z)}{A^{(N)}(z)} \approx \frac{r_1 z + r_2 z^2 + \ldots + r_N z^N}{1 + a_1^{(N)} z + a_2^{(N)} z^2 + \ldots + a_N^{(N)} z^N} \qquad [2.4\text{-}13]$$

Comparison of [2.4-13] with [1.5-14] shows that this approximation is equivalent to the Robinson seismic model for system reverberation (internal primary reflections), as discussed in section 1.5. Thus, the hypothetical source strengths $\{\epsilon_k\}$ in the Robinson model are actually the reflection coefficients to the stated approximation (i.e. the case of small reflection coefficients), and the difference equation operator $\{a_k\}$ in the model is equivalent to the characteristic sequence.

In addition to assuming that the magnitudes of the reflection coefficients are small, let us consider the so-called "random reflection coefficients" hypothesis proposed by Robinson (1954).

Random reflection coefficient hypothesis

In our discussion of statistical models in geophysics (Chapter 1), we noted that the subsurface layers of the earth were laid down in an unsystematic fashion. Oil exploration success shows that this statement is correct; it implies that various segments of the sequence $\{r_0, r_1, r_2, \ldots r_N\}$ are unsystematically related also. That is, various geologic depth intervals exhibit little correlation in their reflection coefficients. If the reflection coefficients are unsystematically related (in a geologic sense), then the serial correlation would account for this effect, i.e. the reflection coefficients would exhibit very small serial correlation. In this sense, we can approximate segments of the numerator polynomial $E^{(N)}(z)$ as having a unit spike as its autocorrelation.

Thus, we see that the reflection coefficients can be considered as uncorrelated numbers. However, the reflection coefficients are fixed (constant) values, that are determined by the matter with which the subsurface layers were created. In this context, how can the reflection coefficients be considered as random? First, the reflectivities of the geologic beds are not, in a sense, realizations of a random process. Specifically, the segment $\{r_k, r_{k+1}, r_{k+2}, \ldots r_{k+L}\}$ is not one realization of a random process, but the reflection coefficient r_k can be treated as a random variable. Our use of the term "random variable" does not imply that the variables which represent the reflectivities are those whose values are uncertain and can be determined at any given instance by a "chance" experiment. In other words, the seismic variables (reflection coefficients) are not random in the sense of the frequency interpretation of probability, because such variables are fixed by the geologic structure. Instead, the reflection coefficients ("random variables") are similar to, say, a variable which represents the billionth digit in the expansion of $\pi = 3.1415926 \ldots$, which, although unknown, is a definite fixed number. Thus, our interpretation of the word random would correspond to speaking of the probability distribution of the billionth digit in the expansion of π, and whether two successive digits in the expansion are uncorrelated. In the same manner, we can speak of the probability distribution of a deep reflection coefficient, and whether two successive reflection coefficients from deep interfaces are uncorrelated. However, the working geophysicist is concerned with exploration over large areas and analyzes many records. Any amount of data in large enough quantities takes on a statistical character, even if the individual piece of data is of a deterministic nature. In this sense, Robinson (1954) *treated* the reflection coefficients as random variables, which explains the so-called random reflection coefficient hypothesis.

Small and random (uncorrelated) reflection coefficient model

Under the assumption that the magnitudes of the reflection coefficients are much less than unity and under the random reflection hypothesis (for the entire geologic depth section) that the set of ordered reflection coefficients are uncorrelated ($\phi_1^{(N)} \approx \phi_2^{(N)} \approx \phi_3^{(N)} \ \ldots \ \phi_{N-1}^{(N)} \approx \phi_N^{(N)} \approx 0$), then $F^{(N)}(z)$ in [2.4-13] simplifies to:

$$F^{(N)}(z) \approx r_1 z + r_2 z^2 + \ldots + r_N z^N \qquad [2.4\text{-}14]$$

Note that [2.4-14] represents the Laplace z-transform of the reflection sequence $\{r_1, r_2, \ldots r_N\}$. Physically, this situation corresponds to observing only the primary reflections resulting from a downgoing unit impulse source applied at the surface (see Fig. 2-15), but with the interesting effect that the amplitudes of these primary reflections have been jacked up from $(t_1 t_2 \ldots t_k)(t_1' t_2 \ldots t_k')r_k$ to r_k. In other words, under the small and random reflection hypothesis, the effect of all the interbed multiples is only good, namely, to jack up the primaries on the seismogram to their proper strengths r_k and to make invisible the multiples. In all science, there is probably no better example of randomness in causing such a beautiful result. This result had practical value because it made possible essentially all the oil discovered in the period 1930—1960. During this period (before digital signal processing), only seismograms that fitted this model could be interpreted with good results, because the presence of strong multiples and reverberations made conventional seismic interpretation (i.e. before digital processing) impossible. Further discussion of these models and others will be given in Chapter 4.

Chapter 3

HOMOMORPHIC ANALYSIS AND SPECTRAL FACTORIZATION

3.1. HOMOMORPHISMS IN ENGINEERING AND SCIENCE

In Chapter 2 we related the constant parameters of the earth's reflection transfer function $F^{(N)}(z)$ to the reflection coefficients of the N layers comprising the subsurface structure. These results were dependent on our layered earth model and the underlying assumptions in the model construction. We defined the reflection polynomial given by $E^{(N)}(z) \approx r_1 z + r_2 z^2 + \ldots r_N z^N$ and observed that this is really the Laplace z-transform of the sequence $\{0, r_1, r_2, \ldots r_N\}$, which is our desired information. Thus, we have at our disposal a linear model of an N-layered earth and we now seek methods by which we can extract the reflection sequence $\{0, r_1, r_2, \ldots r_N\}$ from our observable (noise-free) data $\{x_k\}$. Let us proceed to develop these methods.

We first consider a few definitions. The first definition will regard the concept of a binary operation.

Definition 1. A binary operation, denoted by the symbol $\square$, on a set is a rule which assigns to each ordered pair of elements of the set some element of the set.

For example, consider the set of all real, bounded, causal sequences. If we denote the real, bounded, causal sequence $\{a_0, a_1, a_2, \ldots a_k, a_{k+1}, \ldots\}$ by $\{a_k\}$ and the real, bounded, causal sequence $\{b_0, b_1, \ldots b_k, b_{k+1}, \ldots\}$ by $\{b_k\}$, then (a_k, b_k) is an ordered pair of elements from the set of all real, bounded, causal sequences. Now $a_k \square b_k$ is the element of the set assigned to the ordered pair (a_k, b_k) by the binary operation $\square$. Consider the following examples:

Example 3.1-1. Let $\square$ denote addition, i.e. $\square = +$. In this case, $a_k \square b_k = \{a_k\} + \{b_k\} = \{a_0 + b_0, \quad a_1 + b_1, \ldots a_k + b_k, \quad a_{k+1} + b_{k+1}, \ldots\}$ is the element of the set assigned to the ordered pair (a_k, b_k) by the binary operation of addition. Note that if we had considered a different ordered pair, i.e. (b_k, a_k), we would have obtained the same result since $\{b_k\} + \{a_k\} = \{a_k\} + \{b_k\}$, which defines the commutative property of addition. However, if we had chosen the binary operation of subtraction, i.e. $\square = -$, then we see that the ordered pair (a_k, b_k) would yield a different result from the ordered pair (b_k, a_k). Hence, the concept of "ordered" pairs is an important part of the definition.

Example 3.1-2. Let $\square$ denote convolution, i.e. $\square = *$. In this case, $a_k \square b_k = a_k * b_k = \{\Sigma_{n=0}^{k} a_n b_{k-n}\}$ for $k = 0, 1, 2, \ldots$ Thus, $a_k * b_k$ is the element of the set of all real, bounded, causal sequences assigned to the ordered pair (a_k, b_k) by the binary operation of convolution. Since convolution is commutative, i.e. $a_k * b_k = b_k * a_k$, then we would have obtained the same results had we considered the ordered pair (b_k, a_k). Now that we understand the idea of binary operation, let us consider the definition of a homomorphism.

Definition 2. A mapping φ of a set S with binary operation $\square$ into a set S' with binary operation $\square'$ is a *homomorphism* if:

$$\varphi(a \square b) = \varphi(a) \square' \varphi(b)$$

for all elements a and b in S.

The algebraic concept of homomorphisms is basic in mathematics (Hille and Phillips, 1948) and has been applied to engineering systems by Oppenheim and Schafer (1975). If we replace the word set by group, field, ring, or algebra, we obtain the analogous definitions for a group homomorphism, field homomorphism, etc. Let us now consider some examples of homomorphisms.

Example 3.1-3. As in the previous examples, let us consider the set S of all real, bounded, causal sequences with the binary operation of addition, i.e. $\square = +$. Let us also consider the linear mapping ζ_L given by:

$$\zeta_L(\cdot) = \sum_{n=-\infty}^{\infty} (\cdot)\, z^n$$

which defines the Laplace (two-sided) z-transform. For two elements a_k and b_k in S, we observe that:

$$\zeta_L(a_k + b_k) = \zeta_L(a_k) + \zeta_L(b_k)$$

which defines the additive property of the Laplace z-transform, and in general, of any linear mapping. (This includes the Laplace and Fourier transforms and the linear operations of differentiation and integration.) However, the mapping ζ_L can be thought of as a mapping of the set S with the binary operation of addition into the set S', consisting of the Laplace z-transforms of all real, bounded, causal sequences with the binary operation of addition ($\square' = +$). In this context, ζ_L satisfies our definition 2 and is considered as a homomorphism. In general, we see that the additive property of any linear mapping is a homomorphism.

Example 3.1-4. Suppose we define the set S and the mapping ζ_L as above but we consider the binary operation of convolution on S. For two elements, a_k and b_k in S, we know that:

$$\zeta_L(a_k * b_k) = \zeta_L(a_k) \cdot \zeta_L(b_k) = A(z) \cdot B(z)$$

In this case ζ_L can be thought of as a mapping of the set S with the binary operation of convolution into the set S', consisting of the Laplace z-transforms of all real, causal, and bounded sequences with the binary operation of multiplication, i.e. $\square' = \cdot$. Thus, ζ_L satisfies our definition 2 and in this context is a homomorphism. Analogously, the continuous-time Laplace and Fourier transforms also map sets of real, bounded, causal time functions with the binary operation of convolution into sets of Laplace and Fourier transforms of such functions with the binary operation of multiplication, making the Laplace and Fourier transforms homomorphisms in this sense. However, one should realize that these linear mappings (transforms) per se are not homomorphisms. They are only considered homomorphisms in view of their operation on signals which have been combined through some binary operation and satisfy definition 2. In concluding this example, we add that the inverse mapping ζ_L^{-1}, defined by:

$$\zeta_L^{-1}(\cdot) = \frac{1}{2\pi i} \oint_C (\cdot) z^{-k-1}\, dz$$

with C denoting a clockwise circular contour of radius defined by the region of convergence of $(\cdot)$ in the complex z-plane, gives the following result when operating on the Laplace z-transforms $A(z)$ and $B(z)$:

$$\zeta_L^{-1}[A(z) + B(z)] = \zeta_L^{-1}[A(z)] + \zeta_L^{-1}[B(z)]$$

$$\zeta_L^{-1}[A(z) \cdot B(z)] = \zeta_L^{-1}[A(z)] * \zeta_L^{-1}[B(z)]$$

Thus, ζ_L^{-1} is also a homomorphism by definition 2 and in light of our discussion involving the sets S and S' above. Fig. 3-1 shows the Laplace z-transform as a homomorphism.

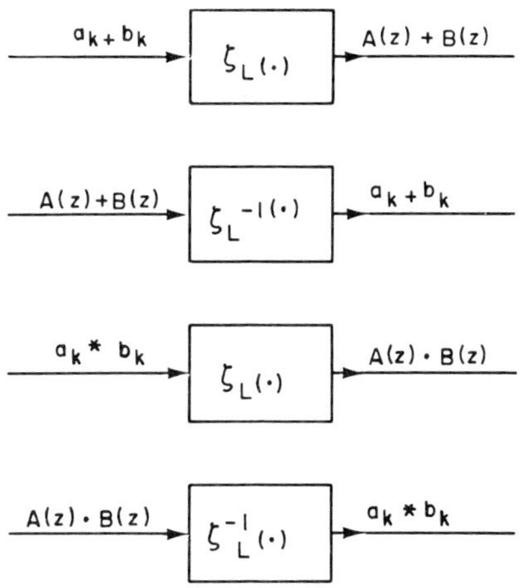

Fig. 3-1.The Laplace z-transforms $\zeta_L(\cdot)$ and its inverse $\zeta_L^{-1}(\cdot)$ depicted as homomorphisms.

Example 3.1-5. Let S be the set of all minimum-delay Laplace z-transforms with the binary operation of multiplication, i.e. $\square = \cdot$. Let S' be the set of all the analytic logarithms of these corresponding transforms with the binary operation of addition ($\square' = +$) and both sets defined for $|z| \leqslant 1$. Now the natural logarithm log (•) is a homomorphism under these conditions since:

$$\log [A(z) \cdot B(z)] \;=\; \log [A(z)] + \log [B(z)]$$

and definition 2 is satisfied. Such a mapping is useful in many fields of science, since it transforms multiplication into addition. In fact, the construction of the so-called Bode plots used in analyzing the frequency response of linear systems is based on such a homomorphism, although it is never actually viewed in this abstract manner. Suppose we consider the inverse mapping $e^{(\cdot)} \equiv \exp(\cdot)$ from S' to S. In symbols:

$$\exp \{\log [A(z)] + \log [B(z)]\} \;=\; \exp \{\log [A(z)]\} \cdot \exp \{\log [B(z)]\}$$

$$= A(z) \cdot B(z)$$

and exp (•) also represents a homomorphism under these conditions. Fig. 3-2 shows the analytic logarithm log (•) and its inverse exp (•) as homomorphisms.

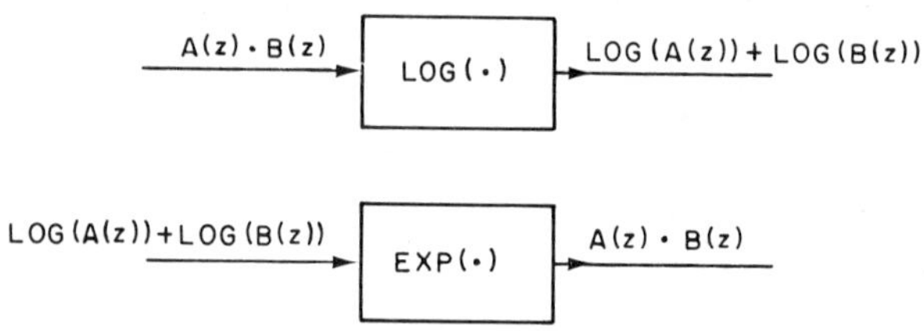

Fig. 3-2. The natural logarithm log (•) and its inverse exp (•) viewed as homomorphisms in the context of example 3.1-5.

Although the above examples discuss operations quite common to the signal processing engineer, they are still homomorphisms in the abstract sense. These homomorphisms are essential to the decomposition and factorization of signals and their spectra. In the next sections, we shall relate such homomorphisms to the general problem of spectral factorization.

3.2. SPECTRAL FACTORIZATION

Let us assume that we have at our disposal some arbitrary power (or energy) spectral density denoted by $\Phi(\omega)$, where ω is the continuous frequency variable defined over the closed interval $[-\pi, \pi]$ with units of radians per sample interval. Now $\Phi(\omega)$ is arbitrary in the sense that it does not have to be a rational function.

A power spectral density function $\Phi(\omega)$ must satisfy the following conditions in order to be factored so as to yield a causal system:

(1) $\Phi(\omega)$ is non-negative on the interval $-\pi \leqslant \omega \leqslant \pi$
(2) $\Phi(\omega)$ has finite area on the interval $-\pi \leqslant \omega \leqslant \pi$ [3.2-1]
(3) $\log[\Phi(\omega)]$ has finite area on the interval $-\pi \leqslant \omega \leqslant \pi$

Now the problem of spectral factorization is the problem of extracting a causal system $B(z)$ (i.e. the Laplace z-transform of a causal sequence $\{b_k\}$) whose amplitude spectrum is the square root of the power spectrum. In symbols:

$$|B(\omega)| = [\Phi(\omega)]^{\frac{1}{2}} \quad [3.2\text{-}2]$$

Szegö (1915) and Kolmogorov (1939) recognized that a solution to this problem could be obtained by making use of the classical method of determining a potential function with an assigned value for its real part on the unit circle. As is well known in potential theory (see Bateman, 1944), Schwarz's classical expression is:

$$F(z) = \frac{1}{2\pi}\int_{-\pi}^{\pi} \frac{1+z\,e^{i\omega'}}{1-z\,e^{i\omega'}}\, Re\,F(\omega')\,d\omega', \quad |z|<1 \quad [3.2\text{-}3]$$

with ω' as the integration variable and $z = r\,e^{-i\omega}$. This equation gives the function $F(z)$ whose real part on the unit circle is $Re\,F(\omega)$. The Schwarz expression follows from the famous result of Poisson on potential theory. Now $\log|B(\omega)|$ is the real part of the function $\log B(\omega)$. Further:

$$Re\log B(\omega) = \log|B(\omega)| = \tfrac{1}{2}\log[\Phi(\omega)] \quad [3.2\text{-}4]$$

Thus, the required solution is found by merely substituting $\frac{1}{2}\log\Phi(\omega)$ for $Re\,F(\omega)$ and substituting $\log B(z)$ for $F(z)$ in Schwarz's expression to obtain:

$$\log B(z) = \frac{1}{2\pi}\int_{-\pi}^{\pi} \frac{1+z\,e^{i\omega'}}{1-z\,e^{i\omega'}}\,\tfrac{1}{2}\log\Phi(\omega')\,d\omega', \quad |z|<1 \quad [3.2\text{-}5]$$

Thus, the Szegö-Kolmogorov solution to the spectral factorization problem is:

$$B(z) = \exp[\log B(z)]$$

$$= \exp\left[\frac{1}{4\pi}\int_{-\pi}^{\pi} \frac{1+z\,e^{i\omega'}}{1-z\,e^{i\omega'}}\log\Phi(\omega')\,d\omega'\right], \quad |z|<1 \quad [3.2\text{-}6]$$

Whereas Szegö and Kolmogorov were only concerned with the real part of the function $\log[B(z)]$ on the unit circle, Robinson (1954) was concerned with the imaginary part as well; namely both the real and the imaginary parts have physical meaning. Thus, it is important to consider the complex logarithm of the spectrum. Therefore, the spectrum of the causal system $B(z)$ is written in polar form as:

$$B(\omega) = |B(\omega)|e^{i\theta(\omega)} \qquad [3.2\text{-}7]$$

where the absolute value $|B(\omega)|$ is the amplitude (or magnitude) spectrum and the argument $\theta(\omega)$ is the phase spectrum. As a result, the complex logarithm of the spectrum is:

$$\log B(\omega) = \log |B(\omega)| + i\theta(\omega) \qquad [3.2\text{-}8]$$

where the real part $\log |B(\omega)|$ is the log-amplitude (or log-magnitude) spectrum and the imaginary part is the phase spectrum. Robinson then asked the question: What is the phase spectrum that results from Schwarz's expression, and he came to the conclusion that $-\theta(\omega)$ is none other than the minimum-negative-phase spectrum, or equivalently, $\theta(\omega)$ is that particular phase spectrum that renders the causal system $B(z)$ a minimum-delay system.

We shall now elaborate on these results. Let us go in the reverse direction and start with the assumption that the causal system $B(z)$ is minimum-delay. By minimum-delay property (10) given in section 2.1, we know that:

$$B(z) \neq 0 \quad \text{for } |z| < 1 \qquad [3.2\text{-}9]$$

The logarithmic function is well behaved as long as its argument is neither zero nor infinity. Since $B(z)$ is neither zero nor infinite within the unit circle, it follows that $\log B(z)$ is well behaved within the unit circle. Consequently, $\log B(z)$ has the power series representation (Maclaurin series):

$$\log B(z) = \sum_{k=0}^{\infty} \beta_k z^k, \quad |z| < 1 \qquad [3.2\text{-}10]$$

(As we will see in the next section, β_k is called the kepstrum.) The complex variable z represents a point within the unit circle (i.e. $|z| < 1$). However, we can consider the limit of $\log B(z)$ as the interior point z approaches a point on the circumference of the unit circle; that is, we consider the limit of $\log B(z)$ as $z \to e^{-i\omega}$. This limit is:

$$\log B(\omega) = \sum_{k=0}^{\infty} \beta_k \, e^{-i\omega k} \qquad [3.2\text{-}11]$$

Separating each part of equation [3.2-11] into real and imaginary parts, we obtain:

$$\log |B(\omega)| + i\theta(\omega) = \left[\beta_0 + Re\left\{\sum_{k=1}^{\infty} \beta_k \, e^{-i\omega k}\right\}\right] + i\left[Im\left\{\sum_{k=1}^{\infty} \beta_k \, e^{-i\omega k}\right\}\right] \qquad [3.2\text{-}12]$$

Let us consider the real part of equation [3.2-12], that is:

$$\log |B(\omega)| = \beta_0 + Re\left\{\sum_{k=1}^{\infty} \beta_k \, e^{-i\omega k}\right\} \qquad [3.2\text{-}13]$$

Now $\log |B(\omega)|$ is a periodic function of ω and has the Fourier series expansion:

$$\log |B(\omega)| = \sum_{k=-\infty}^{\infty} \beta_k' \, e^{-i\omega k} \qquad [3.2\text{-}14]$$

Hence, the $\{\beta_k'\}$ are the Fourier coefficients which can be computed as the Fourier transform:

$$\beta_k' = \frac{1}{2\pi}\int_{-\pi}^{\pi} \log |B(\omega)| \, e^{i\omega k} \, d\omega \quad \text{for } k = 0, \pm 1, \pm 2, \ldots \qquad [3.2\text{-}15]$$

But the real part of equation [3.2-13] can be written as:

$$\log |B(\omega)| = \beta_0 + \left[\sum_{k=1}^{\infty} \frac{\beta_k}{2} e^{-i\omega k}\right]^* + \sum_{k=1}^{\infty} \frac{\beta_k}{2} e^{-i\omega k} \qquad [3.2\text{-}16]$$

or:

$$\log |B(\omega)| = \beta_0 + \sum_{k=-\infty}^{-1} \frac{\beta_{-k}^*}{2} e^{-i\omega k} + \sum_{k=1}^{\infty} \frac{\beta_k}{2} e^{-i\omega k} \qquad [3.2\text{-}17]$$

Expanding equation [3.2-14] we obtain:

$$\log |B(\omega)| = \ldots \beta_{-2}' \, e^{i\omega 2} + \beta_{-1}' \, e^{i\omega} + \beta_0' + \beta_1' \, e^{-i\omega} + \beta_2' \, e^{-i\omega 2} + \ldots \qquad [3.2\text{-}18]$$

and by comparison of equations [3.2-17] and [3.2-18] we arrive at a relationship between the $\{\beta_k\}$ and $\{\beta_k'\}$ coefficients. Thus, given the amplitude spectrum $|B(\omega)|$, we can compute the $\{\beta_k'\}$ by equation [3.2-15] and thereby obtain $\{\beta_k\}$ by the formulas:

$$\begin{aligned} \beta_0 &= \beta_0' && \text{for } k = 0 \\ \beta_k &= 2\beta_k' && \text{for } k = 1, 2, 3, \ldots \\ \beta_k &= 0 && \text{for } k = -1, -2, -3, \ldots \end{aligned} \qquad [3.2\text{-}19]$$

The resulting $\{\beta_k\}$ are the coefficients of the power series of $\log B(z)$. Hence, we can find the required minimum-delay system $B(z)$ by:

$$B(z) = \exp\{\log B(z)\} \qquad [3.2\text{-}20]$$

which is:

$$\sum_{k=0}^{\infty} b_k z^k = \exp\left\{\sum_{k=0}^{\infty} \beta_k z^k\right\} \qquad [3.2\text{-}21]$$

A formula for obtaining the system coefficients $\{b_k\}$ from the known coefficients $\{\beta_k\}$ can be derived as follows. We take the derivative of equation [3.2-21]:

$$\frac{d}{dz}\sum_{k=0}^{\infty} b_k z^k = \exp\left\{\sum_{k=0}^{\infty} \beta_k z^k\right\} \frac{d}{dz}\sum_{k=0}^{\infty} \beta_k z^k \qquad [3.2\text{-}22]$$

and then divide equation [3.2-22] by [3.2-21] to obtain:

$$\frac{\dfrac{d}{dz}\sum_{k=0}^{\infty} b_k z^k}{\sum_{k=0}^{\infty} b_k z^k} = \frac{d}{dz}\sum_{k=0}^{\infty} \beta_k z^k \qquad [3.2\text{-}23]$$

Equation [3.2-23] becomes:

$$\sum_{k=1}^{\infty} k b_k z^{k-1} = \left(\sum_{k=0}^{\infty} b_k z^k\right)\left(\sum_{k=1}^{\infty} k\beta_k z^{k-1}\right) \qquad [3.2\text{-}24]$$

or:

$$\sum_{k=0}^{\infty} (k+1) b_{k+1} z^k = \left(\sum_{k=0}^{\infty} b_k z^k\right)\left(\sum_{k=0}^{\infty} (k+1)\beta_{k+1} z^k\right) \qquad [3.2\text{-}25]$$

which gives the recursive relation (see Oppenheim and Schafer, 1975):

$$(k+1)b_{k+1} = \sum_{n=0}^{k} b_n (k+1-n)\beta_{k+1-n}, \quad k = 0, 1, 2, \ldots \qquad [3.2\text{-}26]$$

Thus, we have obtained an implicit relation between $\{b_k\}$ and $\{\beta_k\}$ which can be used to compute $\{b_k\}$ from $\{\beta_k\}$ (or vice versa).

Let us now look at the imaginary part of equation [3.2-12], namely:

$$\theta(\omega) = Im\left\{\sum_{k=1}^{\infty} \beta_k \, e^{-i\omega k}\right\} \qquad [3.2\text{-}27]$$

which is:

$$\theta(\omega) = -i\left\{-\left[\sum_{k=1}^{\infty} \frac{\beta_k}{2} e^{-i\omega k}\right]^* + \sum_{k=1}^{\infty} \frac{\beta_k}{2} e^{-i\omega k}\right\} \qquad [3.2\text{-}28]$$

$$= i\sum_{k=-\infty}^{-1} \beta'_k \, e^{-i\omega k} - i\sum_{k=1}^{\infty} \beta'_k \, e^{-i\omega k}$$

Using equation [3.2-15], namely:

$$\beta_k' = \frac{1}{2\pi}\int_{-\pi}^{\pi} \log |B(\omega')| e^{i\omega' k}\, d\omega' \quad \text{for } k = 0, \pm 1, \pm 2, \ldots \tag{3.2-29}$$

we can rewrite equation [3.2-28] as:

$$\theta(\omega) = \frac{i}{2\pi}\int_{-\pi}^{\pi} \left[\sum_{k=-\infty}^{-1} e^{-i(\omega-\omega')k} - \sum_{k=1}^{\infty} e^{-i(\omega-\omega')k}\right] \log |B(\omega')| d\omega' \tag{3.2-30}$$

or:

$$\theta(\omega) = \frac{-1}{2\pi}\int_{-\pi}^{\pi} 2\left[\sum_{k=1}^{\infty} \sin(\omega-\omega')k\right] \log |B(\omega')| d\omega' \tag{3.2-31}$$

We note that:

$$\frac{\sin(\omega-\omega')}{1-2r\cos(\omega-\omega')+r^2} = \sum_{k=1}^{\infty} r^{k-1}\sin(\omega-\omega')k, \quad r<1 \tag{3.2-32}$$

Now, taking the limit of equation [3.2-32] as $r \to 1$, we get:

$$\begin{aligned}\lim_{r\to 1} \sum_{k=1}^{\infty} r^{k-1}\sin(\omega-\omega')k &= \frac{\sin(\omega-\omega')}{1-2\cos(\omega-\omega')+1} \\ &= \frac{\sin(\omega-\omega')}{2[1-\cos(\omega-\omega')]}\end{aligned} \tag{3.2-33}$$

With the aid of the trigonometric identities:

$$\sin^2\left(\frac{\omega-\omega'}{2}\right) = \frac{1-\cos(\omega-\omega')}{2}$$

$$\sin(\omega-\omega') = 2\sin\left(\frac{\omega-\omega'}{2}\right)\cos\left(\frac{\omega-\omega'}{2}\right)$$

$$\cot\left(\frac{\omega-\omega'}{2}\right) = \cos\left(\frac{\omega-\omega'}{2}\right)\Big/\sin\left(\frac{\omega-\omega'}{2}\right)$$

equation [3.2-33] becomes:

$$\lim_{r\to 1}\left\{\sum_{k=1}^{\infty} r^{k-1}\sin(\omega-\omega')k\right\} = \sum_{k=1}^{\infty} \sin(\omega-\omega')k = \tfrac{1}{2}\cot\left(\frac{\omega-\omega'}{2}\right) \tag{3.2-34}$$

Substitution of [3.2-34] in [3.2-31] yields:

$$\theta(\omega) = \frac{-1}{2\pi} P \int_{-\pi}^{\pi} \cot\left(\frac{\omega-\omega'}{2}\right) \log |B(\omega')| \, d\omega' \qquad [3.2\text{-}35]$$

where in equation [3.2-35] the symbol P denotes that the integral has its Cauchy principal value due to the singularity of cot $[(\omega - \omega')/2]$ at $\omega = \omega'$. (A rigorous account of the above limiting process and the derivation of equation [3.2-35] in the context of potential theory is given by Fatou (1906).) Thus, the final expression for the minimum-negative-phase spectrum $-\theta(\omega)$ in terms of the amplitude spectrum $|B(\omega)|$ is:

$$-\theta(\omega) = \frac{1}{2\pi} P \int_{-\pi}^{\pi} \cot\left(\frac{\omega-\omega'}{2}\right) \log |B(\omega')| \, d\omega' \qquad [3.2\text{-}36]$$

or in terms of the power spectrum $\Phi(\omega)$ is:

$$-\theta(\omega) = \frac{1}{4\pi} P \int_{-\pi}^{\pi} \cot\left(\frac{\omega-\omega'}{2}\right) \log [\Phi(\omega')] \, d\omega' \qquad [3.2\text{-}37]$$

We say that log $|B(\omega)|$ and $-\theta(\omega)$ are a Hilbert transform pair.

Finally, let us derive the Schwarz expression. We have:

$$\log [B(z)] = \sum_{k=0}^{\infty} \beta_k z^k = \beta_0' + 2 \sum_{k=1}^{\infty} \beta_k' z^k \qquad [3.2\text{-}38]$$

which is, upon using equation [3.2-29] for β_k':

$$\log B(z) = \frac{1}{2\pi} \int_{-\pi}^{\pi} [1 + 2 \sum_{k=1}^{\infty} e^{i\omega' k} z^k] \log |B(\omega')| d\omega' \qquad [3.2\text{-}39]$$

But:

$$1 + 2 \sum_{k=1}^{\infty} (e^{i\omega'} z)^k = \frac{1 + z\, e^{i\omega'}}{1 - z\, e^{i\omega'}}, \quad |z| < 1 \qquad [3.2\text{-}40]$$

Hence, we obtain the Schwarz expression:

$$\log B(z) = \frac{1}{2\pi} \int_{-\pi}^{\pi} \frac{1 + z\, e^{i\omega'}}{1 - z\, e^{i\omega'}} \log |B(\omega')| d\omega', \quad |z| < 1 \qquad [3.2\text{-}41]$$

which is the same as equation [3.2-5] if we substitute log $|B(\omega)| = \frac{1}{2} \log \Phi(\omega)$.

Specialization to the case of real signals

In our previous discussion, we considered the cases where $\{b_k\}$ was a complex signal. Let us now specialize our results for the case of real signals.

Equation [3.2-11] can be rewritten as:

$$\log B(\omega) = \left\{ \beta_0 + \sum_{k=1}^{\infty} \beta_k \cos \omega k \right\} + i \left\{ - \sum_{k=1}^{\infty} \beta_k \sin \omega k \right\} \qquad [3.2\text{-}42]$$

with β_0 and β_k as real numbers.

Substitution of [3.2-8] in [3.2-42] and equating real and imaginary parts gives:

$$\log |B(\omega)| = \beta_0 + \sum_{k=1}^{\infty} \beta_k \cos \omega k = \text{log-amplitude spectrum (even function of } \omega\text{: symmetric)} \qquad [3.2\text{-}43]$$

$$\theta(\omega) = - \sum_{k=1}^{\infty} \beta_k \sin \omega k = \text{phase spectrum (odd function of } \omega\text{; antisymmetric)}$$

An example of a log-amplitude and phase spectrum variation is given in Fig. 3-3.

We can multiply both sides of the first equation in [3.2-43] by $\cos \omega n$ and integrate over $[-\pi, \pi]$. Doing so yields:

$$\begin{aligned} \int_{-\pi}^{\pi} \cos \omega n \log |B(\omega)|\, d\omega &= \int_{-\pi}^{\pi} \beta_0 \cos \omega n\, d\omega \\ &+ \int_{-\pi}^{\pi} \left\{ \sum_{k=1}^{\infty} \beta_k \cos \omega k \cos \omega n \right\} d\omega, \quad n = 0, 1, 2, \ldots \end{aligned} \qquad [3.2\text{-}44]$$

Term-by-term integration of the infinite series in equation [3.2-44] gives the result:

$$\begin{aligned} \beta_0 &= \frac{1}{2\pi} \int_{-\pi}^{\pi} \log |B(\omega)|\, d\omega \\ \beta_k &= \frac{1}{\pi} \int_{-\pi}^{\pi} \cos \omega k \log |B(\omega)|\, d\omega, \quad k = 1, 2, 3, \ldots \end{aligned} \qquad [3.2\text{-}45]$$

where β_0 and β_k are real.

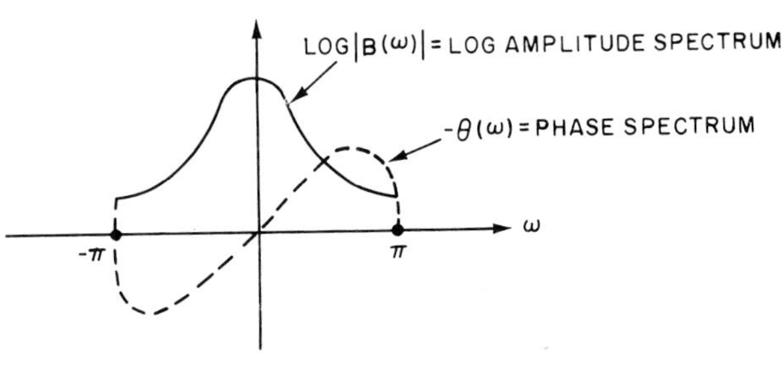

Fig. 3-3. Log-amplitude spectrum (even function of ω) and phase spectrum (odd function of ω).

Thus, given the real coefficients β_0 and $\{\beta_k\}$ from equation [3.2-45], we can use equation [3.2-26] to obtain the real signal $\{b_k\}$.

3.3. THE KEPSTRUM

The idea of the kepstrum appears in the classical work of Poisson (1823), Schwarz (1872), Szegö (1915), and Kolmogorov (1939), and has been applied to geophysical problems by Robinson (1954), Bogert et al. (1963), Bogert and Ossanna (1966), Oppenheim and Schafer (1975), and others. In this section we bring all this work together and develop new properties of the kepstrum in terms of homomorphisms, so as to bring out its essential features and symmetries. As a result, the kepstrum evolves as an important construct in the theory of signal analysis and time series, and takes its place among such established functions as the autocorrelation and spectrum. Finally, we show the usefulness of the kepstrum in geophysical applications, both from a theoretical and computational point of view.

In the case of a minimum-delay system, $B(z)$ and $\log B(z)$ are analytic functions for $|z| \leqslant 1$ and possess Taylor series expansions (causal Laplace z-transforms) given by:

$$B(z) = \sum_{k=0}^{\infty} b_k z^k, \quad |z| \leqslant 1$$

$$\log B(z) = \sum_{k=0}^{\infty} \beta_k z^k, \quad |z| \leqslant 1 \qquad [3.3\text{-}1]$$

with the $\{\beta_k\}$ given by equations [3.2-15] and [3.2-19] in the case of complex $\{b_k\}$ and by [3.2-45] in the case of real $\{b_k\}$. Recalling our discussion on homomorphisms, we see that:

$$\exp\{\log B(z) + \log 1\} = B(z) \cdot 1 \qquad [3.3\text{-}2]$$

describes the non-linear mapping $\exp\{\cdot\}$ as a homomorphism, with the set S being the set of all analytic logarithms with the binary operation of addition and the set S' being the set of all analytic Laplace z-transforms with binary operation of multiplication and $|z| \leqslant 1$ in both sets. Substitution of equation [3.3-1] in equation [3.3-2] yields the desired result:

$$\exp\{\beta_0 + \beta_1 z + \beta_2 z^2 + \ldots\} = b_0 + b_1 z + b_2 z^2 + \ldots \qquad [3.3\text{-}3]$$

We call this equation Kolmogorov's equation.

Putting $z = 0$ in [3.3-3] we observe that:

$$b_0 = e^{\beta_0} = \exp\left\{\frac{1}{2\pi}\int_{-\pi}^{\pi}\log|B(\omega)|\,d\omega\right\} \quad [3.3\text{-}4]$$

For some applications, it is convenient to scale the $\{b_k\}$ so that $b_0 = 1$; that is, we write:

$$B(z) = b_0[1 + \frac{b_1}{b_0}z + \frac{b_2}{b_0}z^2 + \ldots] \quad [3.3\text{-}5]$$

or:

$$B(z) = e^{\beta_0}\ [1 + \frac{b_1}{b_0}z + \frac{b_2}{b_0}z^2 + \ldots] \quad [3.3\text{-}6]$$

With this scaling, Kolmogorov's equation [3.3-3] is rewritten as:

$$\exp\{\beta_1 z + \beta_2 z^2 + \ldots\} = 1 + \frac{b_1}{b_0}z + \frac{b_2}{b_0}z^2 + \ldots \quad [3.3\text{-}7]$$

where we have the desired minimum-delay scaled sequence $\{1, b_1/b_0, b_2/b_0, \ldots\}$ in terms of the coefficients $\{\beta_k\}$, defined by equations [3.2-15] and [3.2-19] for the case of complex $\{b_k\}$ and by equation [3.2-45] for the case of real $\{b_k\}$. The sequence $\{0, \beta_1, \beta_2, \ldots\}$ is causal with a pure delay of one time unit (i.e. its term for $k = 0$ is zero). Now for the case of real $\{b_k\}$, from equation [3.2-43] we observe that $\{\beta_k\}$ represents the inverse Fourier cosine transform of the even function $\log|B(\omega)|$ and the inverse Fourier sine transform of the odd function $-\theta(\omega)$, which is the minimum-negative-phase characteristic. Thus, we have the interesting result that our desired minimum-delay sequence $\{1, b_1/b_0, b_2/b_0, \ldots\}$ is related to a causal sequence $\{\beta_k\}$ by the homomorphism in equation [3.3-2]. We shall define the power series expansion $\beta_1 z + \beta_2 z^2 + \beta_3 z^3 + \ldots$ appearing in the exponent of Kolmogorov's equation as the Kolmogorov Equation Power Series (KEPS). Further, since KEPS is the Laplace z-transform of the sequence $\{0, \beta_1, \beta_2, \ldots\} = \{\beta_k\}$, we can think of $\{\beta_k\}$ as being a time response. Thus, we shall define the sequence $\{0, \beta_1, \beta_2, \ldots\} = \{\beta_k\}$ as the Kolmogorov Equation Power Series Time Response and call this sequence the "kepstrum", where we have added the Latin singular ending "um" to denote one kepstrum and use the Latin plural ending "a" to denote more than one kepstrum, i.e. kepstra.

Thus, we see that the power spectral density (power spectrum) $\Phi(\omega)$ can be factored into two factors: (1) a minimum-delay factor $B(\omega)$ which contains no zeroes or poles within the unit circle in the complex z-plane and (2) a minimum-advance factor $B^*(\omega)$ which contains no zeroes and poles outside of the unit circle. (The superscript * denotes here the complex conjugate of a quantity, so $B^*(\omega)$ is the complex conjugate of $B(\omega)$.) Spectral factorization is basically concerned with determining the minimum-negative-phase characteristic $-\theta(\omega)$, since $|B(\omega)|$ is simply $[\Phi(\omega)]^{\frac{1}{2}}$. Hence, from knowledge of $\Phi(\omega)$, we can calculate the β_k coefficients (equations [3.2-15] and [3.2-19] for complex $\{b_k\}$ and equation [3.2-45] for real $\{b_k\}$), which we call the kepstrum, and from these coefficients we are able to determine $\theta(\omega)$. However, in the case of real signals, the minimum-negative-phase characteristic $-\theta(\omega)$ is essentially the Fourier sine transform of the kepstrum $\{\beta_k\}$, and we observe that the kepstrum and minimum-negative-phase characteristic $-\theta(\omega)$ are related in the same fashion as any bounded time function is related to its spectrum. In this sense the kepstrum is a time response, whose Fourier sine transform is the minimum-negative-phase spectrum $-\theta(\omega)$. Recalling our discussion on minimum-delay, we noted that one consequence of a minimum-negative-phase function is that if the magnitude of the frequency response $|B(\omega)|$ is specified for $-\pi \leqslant \omega \leqslant \pi$, then the corresponding phase spectrum $\theta(\omega)$ is also specified. We derived the unique relationship between $|B(\omega)|$ and $-\theta(\omega)$ in equation [3.2-36] such that given $|B(\omega)|$, we can obtain $-\theta(\omega)$. By similar methods we could derive the expression from which we could obtain $|B(\omega)|$ given $-\theta(\omega)$. With these facts in mind, we have an additional consideration in the concept of minimum-delay for we observe that given the kepstrum $\{\beta_k\}$, then we uniquely determine the amplitude and phase spectrum of a minimum-delay function. For example, given the kepstrum $\{\beta_k\}$, we take the Fourier sine transform of $\{\beta_k\}$ yielding $-\theta(\omega)$, and take the Fourier cosine transform of $\{\beta_k\}$ yielding the log-amplitude spectrum $\log|B(\omega)|$. Let us now relate our knowledge of homomorphisms to the above discussion.

Through the homomorphism in [3.3-2], the kepstrum is related to the minimum-delay scaled sequence $\{1, b_1/b_0, b_2/b_0, \ldots\}$. Further, the Fourier sine transform can also be viewed as a homomorphism which maps the kepstrum into the minimum-negative-phase spectrum $-\theta(\omega)$. Now suppose $\Phi(\omega)$ was a known rational function which was obtained by passing white noise with a power spectral density of unity through a linear system whose real impulse response $\{1, b_1/b_0, b_2/b_0, \ldots\}$ was unknown. In this sense, the transfer function $B(\omega)$ is related to $\Phi(\omega)$ by $\Phi(\omega) = |B(\omega)|^2$. Let us consider the problem of determining the impulse response of this system given $\Phi(\omega)$ in terms of homomorphisms. Our first step would be to take the inverse Fourier cosine transform of the even quantity $\log[\Phi(\omega)]^{\frac{1}{2}}$ by use of equation [3.2-45]. This would yield the real kepstrum $\{\beta_k\}$, where the inverse Fourier cosine transform can be considered as a homomorphism due to its additive property. Next, we would use the homomorphism in [3.3-2] to relate the kepstrum to the desired scaled impulse response $\{1, b_1/b_0, b_2/b_0, \ldots\}$. This procedure would yield equation [3.3-7]. This example illustrates an application of the spectral factorization method in the context of the kepstrum and related homomorphisms.

Let us now consider the kepstrum per se, i.e. we shall investigate its properties in terms of a signal processing operation, although we will always remember its relation to the spectral factorization method and minimum-negative-phase spectrum $-\theta(\omega)$.

Now equation [3.3-7] can also be used to calculate the kepstrum $\{\beta_k\}$ given knowledge of the minimum-delay sequence $\{1, b_1/b_0, b_2/b_0, \ldots\}$. This idea will prove useful in our discussion of decomposition methods.

In the following discussion, we assume knowledge of the minimum-delay sequence $\{1, b_1/b_0, b_2/b_0, \ldots\} = \{b'_k\}$ and desire to compute the kepstrum $\{0, \beta_1, \beta_2, \ldots\}$ of $\{b'_k\}$. Through the composition of various homomorphisms, we can arrive at the kepstrum by:

(1) Computing the Laplace z-transform of the sequence $\{b'_k\}$. The linear mapping ζ_L can be viewed as a homomorphism from the set of all real, minimum-delay* sequences with binary operation of convolution to the set of all minimum-delay Laplace z-transforms with binary operation of multiplication. In this context, we shall write the sequence $\{b'_k\}$ as $b'_k * \delta_k$, where the unit impulse δ_k is from the set of real, minimum-delay sequences. In symbols:

$$\zeta_L(b'_k * \delta_k) = \zeta_L(b'_k) \cdot \zeta_L(\delta_k) = B'(z) \cdot 1$$

where:

$$B'(z) = \sum_{k=0}^{\infty} b'_k z^k = 1 + b'_1 z + b'_2 z^2 + \ldots, \quad b'_0 = 1$$

and $B'(z) \neq 0$ for $|z| \leqslant 1$.

(2) Take the analytic logarithm of $B'(z) \cdot 1$. The non-linear mapping log can be viewed as a homomorphism from the set of all minimum-delay Laplace z-transforms with binary operation of multiplication to the set of all analytic logarithms with binary operation of addition. In symbols:

$$\log [B'(z) \cdot 1] = \log B'(z) + \log (1)$$

where $\log B'(z)$ is analytic for $|z| \leqslant 1$ and:

$$\begin{aligned}\log B'(z) &= \beta_1 z + \beta_2 z^2 + \beta_3 z^3 + \ldots = \text{KEPS} \\ &= z(\beta_1 + \beta_2 z + \beta_3 z^2 + \ldots)\end{aligned}$$

(3) Now by taking the inverse Laplace z-transform of $\log B'(z)$, we arrive at the kepstrum $\{0, \beta_1, \beta_2, \ldots\}$ of $\{b'_k\}$. The linear mapping ζ_L^{-1} can be viewed as a homomorphism from the set of analytic logarithms with binary operation of addition to the set of all real, causal sequences with binary operation of addition. In symbols:

$$\zeta_L^{-1} [\log B'(z) + \log (1)] = \zeta_L^{-1} [\log B'(z)] + \zeta_L^{-1} [\log (1)]$$

where $\zeta_L^{-1} [\log B'(z)] = \{0, \beta_1, \beta_2, \beta_3, \ldots\} = \{\beta_k\}$ is the kepstrum of the minimum-delay sequence $\{1, b_1/b_0, b_2/b_0, \ldots\}$.

---

* Minimum-delay sequences are causal sequences.

By expressing b'_k as $b'_k * \delta_k$ in the above procedure, we see that the convolution $b'_k * \delta_k$, from the set of all minimum-delay sequences with binary operation of convolution, was mapped into the addition $\{\beta_k\} + \{0\}$, from the set of all real, causal sequences with binary operation of addition. But recall that the sequence $\{1, b'_1, b'_2, \ldots\}$ and its associated kepstrum $\{0, \beta_1, \beta_2, \ldots\}$ are related to the minimum-delay factor $B(\omega)$ of the power spectral density $\Phi(\omega)$. Now suppose that more than one minimum-delay factor of $\Phi(\omega)$ is desired. That is, let us assume that $B(\omega)$ can be expressed as:

$$B(\omega) = A(\omega) \cdot C(\omega) = |A(\omega)|\,|C(\omega)|\, e^{i[\theta_A(\omega) + \theta_C(\omega)]} \qquad [3.3\text{-}8]$$

where $A(z)$ and $B(z)$ are both analytic functions for $|z| \leqslant 1$ and have no zeroes or poles inside or on the unit circle. Under this condition, $\Phi(\omega)$ can be considered to have two minimum-delay factors, $A(\omega)$ and $C(\omega)$, and can be expressed as:

$$\Phi(\omega) = |A(\omega)\, C(\omega)|^2 = A(\omega)\, C(\omega) \cdot A^*(\omega)\, C^*(\omega) \qquad [3.3\text{-}9]$$

Under these conditions, Kolmogorov's equation [3.3-3] can be expressed as:

$$\exp\{(\alpha_1 z + \alpha_2 z^2 + \ldots) + (\gamma_1 z + \gamma_2 z^2 + \ldots)\} \qquad [3.3\text{-}10]$$
$$= (1 + \frac{a_1}{a_0} z + \frac{a_2}{a_0} z^2 + \ldots)(1 + \frac{c_1}{c_0} z + \frac{c_2}{c_0} z^2 + \ldots)$$

where $A(z) = \sum_{k=0}^{\infty} a_k z^k, \quad |z| \leqslant 1$

$$C(z) = \sum_{k=0}^{\infty} c_k z^k, \quad |z| \leqslant 1$$

$$b_0 = a_0 c_0 = e^{(\alpha_0 + \gamma_0)} = e^{\beta_0} \quad \text{(see equation [3.3-4])}$$

$$\log A(z) = \sum_{k=0}^{\infty} \alpha_k z^k, \quad |z| \leqslant 1$$

$$\log C(z) = \sum_{k=0}^{\infty} \gamma_k z^k, \quad |z| \leqslant 1$$

Thus, we have manipulated the spectral-factorization results to include the factorization of $\Phi(\omega)$ into more than one minimum-delay factor. This process can be continued ad infinitum. For this case, the Kolmogorov equation power series (KEPS) becomes:

$$\text{KEPS} = [(\alpha_1 + \gamma_1)z + (\alpha_2 + \gamma_2)z^2 + \ldots] = [\beta_1 z + \beta_2 z^2 + \ldots] \qquad [3.3\text{-}11]$$

and we see that:

$$\{b'_k\} = a'_k * c'_k, \quad \{\beta_k\} = \alpha_k + \gamma_k \qquad [3.3\text{-}12]$$

where $\{b'_k\} = \{1, b_1/b_0, b_2/b_0, \ldots\}$ $\quad \{\beta_k\} = \{0, \beta_1, \beta_2, \beta_3, \ldots\}$

$\{a'_k\} = \{1, a_1/a_0, a_2/a_0, \ldots\}$ $\quad \{\alpha_k\} = \{0, \alpha_1, \alpha_2, \alpha_3, \ldots\}$

$\{c'_k\} = \{1, c_1/c_0, c_2/c_0, \ldots\}$ $\quad \{\gamma_k\} = \{0, \gamma_1, \gamma_2, \gamma_3, \ldots\}$

and $\{b'_k\}$, $\{a'_k\}$, and $\{c'_k\}$ are minimum-delay; $\{\beta_k\}$, $\{\alpha_k\}$, and $\{\gamma_k\}$ are causal. Now given the sequence $\{b'_k\} = a'_k * b'_k$, we can follow the exact same procedure as outlined above to obtain the kepstrum $\{\beta_k\}$. That is, we can perform the successive homomorphisms ζ_L, log, and ζ_L^{-1} on $a'_k * c'_k$ to obtain the kepstrum $\{\beta_k\} = \{\alpha_k\} + \{\gamma_k\} = \{0, \alpha_1 + \gamma_1 + \alpha_2 + \gamma_2, \ldots\}$. In doing so, we see that the convolution $a'_k * c'_k$, from the set of all real minimum-delay sequences with binary operation of convolution, is mapped into the addition $\{\alpha_k\} + \{\gamma_k\}$, from the set of all real, causal sequences with binary operation of addition. Further, $\{\alpha_k\}$ is the kepstrum of the sequence $\{a'_k\}$ and $\{\gamma_k\}$ is the kepstrum of the sequence $\{c'_k\}$. Let us define the ordered composition* of homomorphisms ζ_L, log, and ζ_L^{-1} by the mapping K, where K is the resulting homomorphism and:

$$K \equiv \zeta_L^{-1} \circ \log \circ \zeta_L \qquad [3.3\text{-}13]$$

We shall define K as the kepstral operator and, from our above discussion, we see that K has the following properties:

(a) $K(a'_k * c'_k) = K(a'_k) + K(c'_k) = \{\alpha_k\} + \{\gamma_k\}$

Recall that the multiplication of two Fourier transforms $A(\omega)$ and $C(\omega)$ results in the addition of their corresponding phase spectra $\theta_A(\omega)$ and $\theta_C(\omega)$. (See equation [3.3-8].) Since the addition $\{\alpha_k\} + \{\gamma_k\}$ is essentially the inverse Fourier sine transform of the total phase spectrum $\theta_A(\omega) + \theta_C(\omega)$, then the superposition of $\{\alpha_k\}$ and $\{\gamma_k\}$ represents a "phase vs. time" variation associated with spectrum $\theta_A(\omega) + \theta_C(\omega)$

(b) $K^{-1}(\alpha_k + \gamma_k) = K^{-1}(\alpha_k) * K^{-1}(\gamma_k)$

where $K^{-1} = \zeta_L^{-1} \circ \exp \circ \zeta_L$ and $K^{-1}(\alpha_k) = \{a'_k\}$, $K^{-1}(\gamma_k) = \{c'_k\}$

(c) $K(\delta_k) = \{0\} =$ null sequence

$K^{-1}(\{0\}) = \delta_k$ $\qquad$ [3.3-14]

---

* The symbol $\circ$ serves to separate the three transformations ζ_L, log, ζ_L^{-1}. The notation $\zeta_L^{-1} \circ \log \circ \zeta_L$ is read "the mapping ζ_L composed with the mapping log composed with the mapping ζ_L^{-1}".

Since $\{a'_k\}$ and $\{c'_k\}$ are elements from the set of all real, minimum-delay sequences with binary operation of convolution, and $\{\alpha_k\}$ and $\{\gamma_k\}$ are elements from the set of all real, causal, bounded sequences with binary operation of addition, then K is also a group homomorphism.

Definition 3: A group, denoted by $[G, \square]$, is a set G, together with a binary operation $\square$ on G, such that the following axioms are satisfied:

(1) The binary operation is associative.

(2) There is an element e in G such that $a \square e = e \square a = a$. The element e is called the identity element for the binary operation $\square$ on G.

(3) For each a in G, there is an element a^{-1} in G with the property that $a^{-1} \square a = a \square a^{-1} = e$. The element a^{-1} is called the inverse of a with respect to the binary operation $\square$.

Note: If the binary operation $\square$ is also commutative, then the group $[G, \square]$ is called an abelian group.

Consider the following example:

Example 3.3-1. As in the previous examples, we will consider the set G to be the set of all real, bounded causal sequences. Let $\square$ denote convolution, i.e. $\square = *$. Since $(a_k * b_k) * c_k = a_k * (b_k * c_k)$, then convolution is associative. The identity element e is the unit impulse δ_k, where $e = e_k = \delta_k = \{1, 0, 0, \ldots 0, 0, \ldots\}$ and the condition $a_k * e_k = e_k * a_k = \{a_k\}$ is satisfied. Now the inverse sequence $\{a_k^{-1}\}$ is found from the defining equation $a_k^{-1} * a_k = a_k * a_k^{-1} = \delta_k$. However, the null sequence $\{0, 0, \ldots 0, 0, \ldots\}$ does not have an inverse. Further, all the non-minimum-delay sequences contained in the set G do not have inverses, since these inverse sequences would be unbounded and not an element of G. Thus, the set of all real, bounded, causal sequences with convolution as the binary operation does not form a group. However, the set of all real, minimum-delay sequences with convolution as the binary operation does form a group, since the inverse sequences are also minimum-delay and contained in the defining set of all real, minimum-delay sequences. In concluding this example, we would like to point out that the set of all real, causal bounded sequences with addition as the binary operation does form a group. In this case, the null sequence is the identity element and the inverse sequence $\{a_k^{-1}\}$ of the sequence $\{a_k\}$ would be $\{a_k^{-1}\} = -\{a_k\}$.

Thus, the kepstral operator K is a mapping of the group $[G, *]$ into the group $[G', +]$, where G is the set of all real, minimum-delay sequences and G' is the set of all real, causal, bounded sequences. By comparing property (a) of [3.3-14] with definition 2, we have shown that K is also a group homomorphism. We note that the familiar linear operators of Laplace and Fourier map functions of time into functions of a complex frequency and frequency respectively, but the Hilbert transform is a linear mapping which maps functions of time into functions of time. Thus, our kepstral operator, although it is a non-linear operator, also maps time functions into time functions. Recall that the kepstrum is a time response.

Now that we understand some of the basic properties of the kepstral operator (group homomorphism) K, let us investigate the kepstrum $\{\beta_k\} = K(b'_k)$ corresponding to the real minimum-delay sequence $\{b'_k\}$. Also, let us not forget that the time response $\{\beta_k\}$ is an important quantity in the spectral factorization method, since the minimum-negative-phase characteristic $-\theta(\omega)$ is dependent upon $\{\beta_k\}$, which is found from knowledge of the power spectral density $\Phi(\omega)$. Given the sequence $\{1, b'_1, b'_2, \ldots\}$ with kepstrum $\{0, \beta_1, \beta_2, \beta_3, \ldots\}$, we shall define the following quantities:

Reverse sequence. Given the minimum-delay sequence $\{b'_k\}$, we form the reverse sequence $\{b'_{-k}\}$ by folding $\{b'_k\}$ about the origin. The resulting sequence $\{b'_{-k}\}$ is minimum-advance.

If $\{b'_k\} = \{1, b'_1, b'_2, \ldots\}$, then:

$$\zeta_L(b'_k) = B'(z) = \sum_{k=0}^{\infty} b'_k z^k$$

where $b'_0 = 1$ and $|z| \leqslant 1$.

Now $\{b'_{-k}\} = \{\ldots - b'_{-2}, b'_{-1}, 1, 0, 0, \ldots\}$ and:

$$\zeta_L(b'_{-k}) = B'(z^{-1}) = \sum_{k=0}^{\infty} b'_k z^{-k}$$

where $b'_0 = 1$ and $|z| \leqslant 1$.

Recall that $\log B'(z) = \beta_1 z + \beta_2 z^2 + \ldots$ is the causal KEPS.

Thus, $\log B'(z^{-1}) = \beta_1 z^{-1} + \beta_2 z^{-2} + \ldots$ is the anti-causal KEPS. We observe that given $K(b'_k) = \{\beta_k\}$, $K(b'_{-k})$ is found by folding $\{\beta_k\}$ about the origin, i.e. $K(b'_{-k}) = \{\beta_{-k}\}$.

$$\{b'_k\} = \{1, b'_1, b'_2, \ldots\} \longleftrightarrow K(b'_k) = \{0, \beta_1, \beta_2, \beta_3 \ldots\}$$

$$\{b'_{-k}\} = \{\ldots b'_{-2}, b'_{-1}, 1, 0, \ldots\} \longleftrightarrow K(b'_{-k}) = \{\ldots \beta_{-3}, \beta_{-2}, \beta_{-1}, 0, 0, 0\}$$

In words:

minimum-delay $\longleftrightarrow$ causal kepstrum

minimum-advance sequence $\longleftrightarrow$ anti-causal kepstrum

Inverse sequence. Recall that the set of all real, minimum-delay sequences with the binary operation of convolution form an abelian group. Thus, each element of the set has an inverse. Given the sequence $\{b'_k\}$, we define the inverse sequence $\{b'^{-1}_k\}$ by the relation:

$$b'_k * b'^{-1}_k = \delta_k$$

where the unit impulse δ_k is the identity element of the group. Thus, $\{b'^{-1}_k\}$ is given by:

$\{b_k'^{-1}\} = \zeta_L^{-1}\{[B'(z)]^{-1}\}$

Now $\log [B'(z)]^{-1} = -\log B'(z) = -\beta_1 z - \beta_2 z^2 - \beta_3 z^3 \ldots$
Thus, $K(b_k'^{-1}) = \{-\beta_1, -\beta_2, -\beta_3, \ldots\}$ = kepstrum of inverse sequence $\{b_k'^{-1}\}$. We observe that given the kepstrum $\{\beta_k\}$ of the sequence $\{b_k'\}$, the kepstrum $K(b_k'^{-1})$ is found by a reflection of $\{\beta_k\}$ about the positive-time axis, i.e. $K(b_k'^{-1}) = -\{\beta_k\}$.

In words:

minimum-delay sequence ⟷ causal kepstrum

inverse of minimum-delay sequence ⟷ negative of causal kepstrum

Reverse-inverse sequence. We define the sequence $\{b_{-k}'^{-1}\}$ as the inverse of the reverse sequence. Based on the previous two properties, the KEPS for this sequence is given by:

$$\log [B'(z^{-1})]^{-1} = -\log [B'(z^{-1})] = -\beta_1 z^{-1} - \beta_2 z^{-2} - \beta_3 z^{-3} - \ldots$$

$$K(b_{-k}'^{-1}) = \{\ldots -\beta_3, -\beta_2, -\beta_1, 0, 0, 0, \ldots\}$$

We observe that given $K(b_k')$, $K(b_{-k}'^{-1})$ is found by folding $K(b_k')$ about the origin then performing a reflection about the negative-time axis. Alternatively, given $K(b_k')$, $K(b_{-k}'^{-1})$ is also found by performing a reflection about the positive-time axis giving $K(b_k'^{-1})$, then folding $K(b_k'^{-1})$ about the origin resulting in the desired kepstrum $K(b_{-k}'^{-1})$, i.e. $K(b_{-k}'^{-1}) = -\{\beta_{-k}\}$.

In words:

minimum-delay sequence ⟷ causal kepstrum

reverse of inverse of minimum-delay sequence ⟷ reverse of negative of causal kepstrum

(inverse of minimum-advance sequence ⟷ negative of anti-causal kepstrum)

We have previously obtained an implicit relation between $\{b_k\}$ and $\{\beta_k\}$ which can be used to compute $\{b_k\}$ from $\{\beta_k\}$ (or vice versa) (see equation [3.2-26]). Let us proceed to develop an explicit relation between $\{b_k\}$ and $\{\beta_k\}$ based on our present knowledge of the kepstrum. Replacing b_k in equation [3.2-25] with b_k', we get:

$$\sum_{k=0}^{\infty} (k+1)\, b_{k+1}' z^k = \sum_{k=0}^{\infty} b_k' z^k \sum_{k=0}^{\infty} (k+1)\, \beta_{k+1} z^k \qquad [3.3\text{-}15]$$

where $b_0' = 1$.

Rewriting equation [3.3-15] we obtain:

$$\frac{1}{B'(z)} \cdot \sum_{k=0}^{\infty} (k+1)\, b_{k+1}' z^k = \sum_{k=0}^{\infty} (k+1)\, \beta_{k+1} z^k \qquad [3.3\text{-}16]$$

where $B'(z) = \Sigma_{k=0}^{\infty} b'_k z^k$. But we have previously shown that the Laplace z-transform of the inverse sequence $\{b_k'^{-1}\}$ is $[B'(z)]^{-1} = 1/B'(z)$. Using this fact, equation [3.3-16] can be expressed as:

$$\zeta_L(b_k'^{-1}) \cdot \zeta_L[(k+1)b'_{k+1}] = \zeta_L[(k+1)\beta_{k+1}] \qquad [3.3\text{-}17]$$

The inverse Laplace z-transform of equation [3.3-17] gives:

$$\beta_{k+1} = \frac{1}{k+1} \cdot [b_k'^{-1} * (k+1)b'_{k+1}], \quad k = 0, 1, 2, \ldots \qquad [3.3\text{-}18]$$

or:

$$K(b'_k) = \{\beta_{k+1}\} = \sum_{n=0}^{k} \left(\frac{n+1}{k+1}\right) b'_{n+1} b_{k-n}'^{-1}, \quad k = 0, 1, 2, \ldots \qquad [3.3\text{-}19]$$

which explicitly gives the kepstrum $\{\beta_k\}$ in terms of the minimum-delay scaled sequence $\{b'_k\}$.

Let us now consider the following examples:

Example 3.3-2. Recall that the impulse response of a single-layered earth, which was discussed in section 2.2, is given by:

$$X^{(R)}(z) = \frac{r_1 z}{1 - r'_0 r_1 z} \qquad |z| < \frac{1}{|r'_0 r_1|} \qquad [1]$$

(refer to equation [2.2-9] and Fig. 2-5).

Let us remove the pure delay from this response by considering the sequence $\{a_k\}$ whose Laplace z-transform is:

$$A(z) = z^{-1} X^{(R)}(z) = \frac{r_1}{1 - r'_0 r_1 z}, \qquad |z| < \frac{1}{|r'_0 r_1|} \qquad [2]$$

Thus, by advancing the sequence $\{x_k^{(R)}\}$ by one time unit, we have removed the pure delay (linear phase component) associated with $\{x_k^{(R)}\}$ and have defined the minimum-delay sequence $\{a_k\} \equiv \{x_{k+1}^{(R)}\}$, which has no zeroes or poles inside the circle, since $|r'_0 r_1|$ is generally less than one. The inverse Laplace z-transform of [2] yields:

$$\{a_k\} = \{r_1, -r_0 r_1^2, r_0^2 r_1^3, -r_0^3 r_1^4, r_0^4 r_1^5, \ldots \qquad [3]$$

Let us consider the values $r_0 = 1$ and $r_1 = 0.8$. Now scaling $\{a_k\}$ such that the leading coefficient is 1.0, we obtain:

$$\{a'_k\} = \{1.0, -0.8, 0.64, -0.51, 0.41, -0.33, \ldots\} \qquad [4]$$

where $a'_k = (1/r_1)a_k$ = scaled sequence. Now the kepstrum of $\{a'_k\}$, denoted by the time sequence $\{\alpha_k\} = \{0, \alpha_1, \alpha_2, \ldots\}$, can be found by using the recursion formula [3.2-26] with α_k replacing β_k and a'_k replacing b_k. Doing so, we obtain the result:

$$\begin{aligned}
\alpha_1 &= a'_1 = -0.8 \\
\alpha_2 &= a'_2 - \tfrac{1}{2}\alpha_1 a'_1 = 0.64 - \tfrac{1}{2}\cdot(-0.8)\cdot(-0.8) = 0.32 \\
\alpha_3 &= a'_3 - \tfrac{1}{3}\alpha_1 a'_2 - \tfrac{2}{3}\alpha_2 a'_1 = -0.17 \\
\alpha_4 &= a'_4 - \tfrac{1}{4}\alpha_1 a'_3 - \tfrac{2}{4}\alpha_2 a'_2 - \tfrac{3}{4}\alpha_3 a'_1 = 0.11 \\
\alpha_5 &= a'_5 - \tfrac{1}{5}\alpha_1 a'_4 - \tfrac{2}{5}\alpha_2 a'_3 - \tfrac{3}{5}\alpha_3 a'_2 - \tfrac{4}{5}\alpha_4 a'_1 = -0.06 \\
&\cdot \\
&\cdot
\end{aligned} \qquad [5]$$

Thus, we see that the kepstrum $\alpha_k = K(a'_k)$ of a'_k is given by the sequence $K(a'_k) = \{0, -0.8, 0.32, -0.17, 0.11, -0.06, \ldots\}$. Further, the reverse sequence $\{a'_{-k}\}$ is given by:

$$\{a'_{-k}\} = \{\ldots -0.33, 0.41, -0.51, 0.64, -0.8, 1.0, 0.0, 0.0, \ldots\} \qquad [6]$$

which results from folding the sequence $\{a'_k\}$ about the origin. Now we showed that the kepstrum $K(a'_{-k})$ can be obtained by folding $\{\alpha_k\} = K(a'_k)$ about the origin. Thus, $K(a'_{-k}) = \{\alpha_{-k}\}$ = kepstrum of $\{a'_{-k}\}$ is given by:

$$K(a'_{-k}) = \{\ldots -0.06, 0.11, -0.17, 0.32, -0.8, 0.0, 0.0, \ldots\} \qquad [7]$$

Let us proceed to obtain the inverse sequence $\{a_k'^{-1}\}$ defined by the relation:

$$a'_k * a_k'^{-1} = \delta_k; \qquad [8]$$

Now we know that $\{a_k'^{-1}\}$ exists, since we are considering the group $[G, *]$ with G being the set of all real, minimum-delay sequences. The Laplace z-transform of [8] gives:

$$A'(z)\cdot[A'(z)]^{-1} = 1 \qquad [9]$$

where $A'(z) = (1/r_1)A(z)$ = the scaled Laplace z-transform, i.e. the Laplace z-transform of $\{a'_k\}$. Now inspection of [2] shows that $A'(z)$ is given by:

$$A'(z) = \frac{1}{1 + r_0 r_1 z} = \frac{1}{1 + 0.8z}, \quad |z| < \frac{1}{0.8} \qquad [10]$$

Now $\zeta_L(a_k'^{-1}) = [A'(z)]^{-1}$ is given by:

$$\zeta_L(a_k'^{-1}) = [A'(z)]^{-1} = 1 + 0.8z \qquad [11]$$

Taking the inverse Laplace z-transform of [11] shows that the inverse sequence $\{a_k'^{-1}\}$ is given by:

$$\{a_k'^{-1}\} = \{1.0, 0.8, 0.0, 0.0, \ldots\} \qquad [12]$$

which is minimum-delay and consists of only two non-zero values. Now we showed that the kepstrum $K(a_k'^{-1})$ can be obtained from $K(a_k')$ by a reflection of $K(a_k')$ about the positive-time axis. Thus, we showed that $K(a_k'^{-1}) = -\{\alpha_k\}$. Hence:

$$K(a_k'^{-1}) = \{0, 0.8, -0.32, 0.17, -0.11, 0.06, \ldots\} \quad [13]$$

The reverse-inverse sequence $\{a_{-k}'^{-1}\}$ is found by folding the inverse sequence $\{a_k'^{-1}\}$ about the origin. Doing this yields:

$$\{a_{-k}'^{-1}\} = \{\ldots 0.0, 0.0, 0.0, 0.8, 1.0, 0.0, 0.0, \ldots\} \quad [14]$$

which is an anti-causal, minimum-advance sequence. The kepstrum $K(a_{-k}^{-1})$ can be obtained by folding $K(a_k')$ about the origin then performing a reflection about the negative-time axis. That is, we showed that $K(a_{-k}'^{-1}) = -\{\alpha_{-k}\}$. Hence:

$$K(a_{-k}'^{-1}) = \{\ldots 0.06, -0.11, 0.17, -0.32, 0.8, 0.0, 0.0, 0.0, \ldots\} \quad [15]$$

Instead of using the recursion formula [3.2-26] to obtain the kepstrum $\{\alpha_k\}$, we could have proceeded according to the definition of the kepstral operator K. That is $K(a_k') = \zeta_L^{-1} \circ \log \circ \zeta_L(a_k')$ is given by:

$$K(a_k') = \zeta_L^{-1}\,\{\log A'(z)\} = \zeta_L^{-1}\left\{\log\left(\frac{1}{1+0.8z}\right)\right\} = \zeta_L^{-1}\,\{-\log(1+0.8z)\} \quad [16]$$

But log $(1 + az)$ has the power series expansion (Oppenheim and Schafer, 1975):

$$\log(1+az) = \frac{az}{1} - \frac{a^2}{2}z^2 + \frac{a^3}{3}z^3 - \ldots \quad [17]$$

or:

$$\log(1+az) = \sum_{k=1}^{\infty} \frac{(-1)^{k+1}}{k} a^k z^k, \quad |z| < \frac{1}{|a|} \quad [18]$$

which is the Maclaurin series expansion and includes the unit circle in its region of convergence. For $a = 0.8$, we would have arrived at the result:

$$K(a_k') = \zeta_L^{-1}\,\{-\log(1+0.8z)\} = \{0, -0.8, 0.32, -0.17, \ldots\} \quad [19]$$

as obtained earlier for $K(a_k')$. The sequences $\{a_k'\}$, $\{a_{-k}'\}$, $\{a_k'^{-1}\}$, $\{a_{-k}'^{-1}\}$ and the corresponding kepstra $K(a_k')$, $K(a_{-k}')$, $K(a_k'^{-1})$, $K(a_{-k}'^{-1})$ are shown in Fig. 3-4. Inspection of Fig. 3-4a, b, e and f reveals that the kepstrum of either a minimum-delay or minimum-advance sequence, i.e. either $\{a_k'\}$ or $\{a_{-k}'\}$, attentuates more rapidly in time than the original time sequence $\{a_k'\}$ or $\{a_{-k}'\}$. This is explained by considering the power series expansion of log $(1 + az)$ in [18] and noting that the rapid attenuation of the kepstrum is due to the $1/k$ term in the series expansion. Also, inspection of Fig. 3-4c, d, g and h indicates that the kepstrum of either a minimum-delay or minimum-advance finite sequence, i.e. $\{a_k'^{-1}\}$ or $\{a_{-k}'^{-1}\}$, is still an infinite-length sequence.

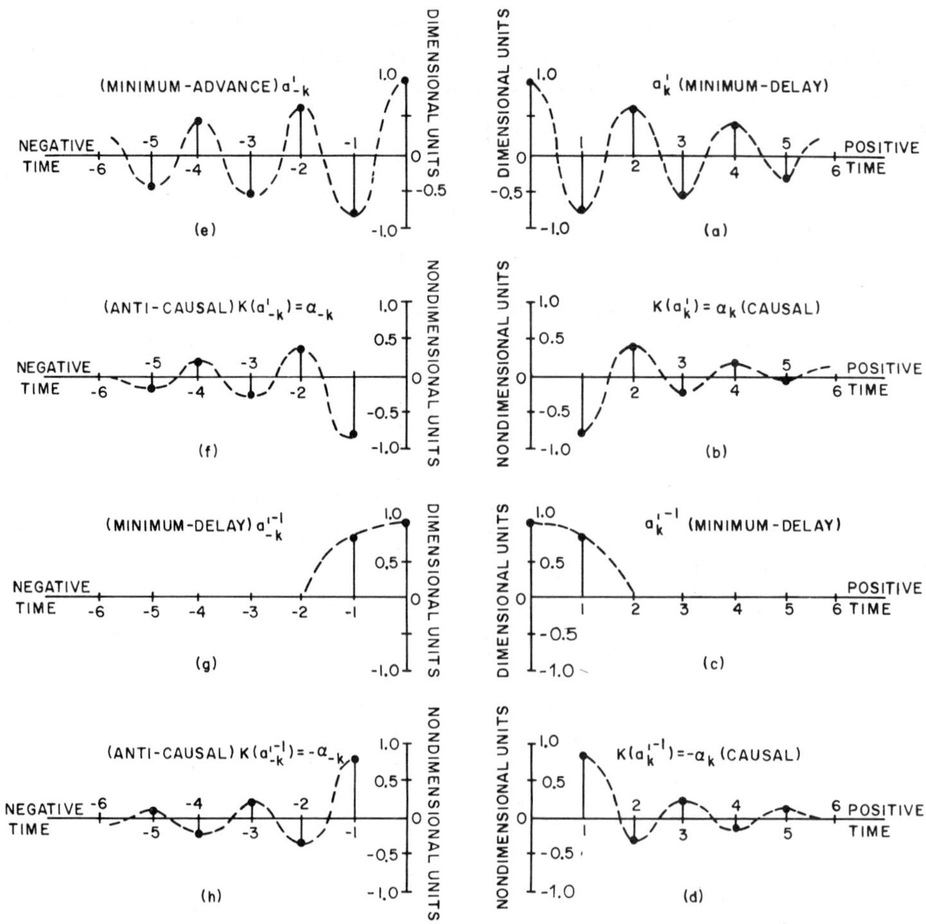

Fig. 3-4. Scaled sequence $\{a'_k\}$ of example 3.3-2, the reverse, inverse, and reverse-inverse of $\{a'_k\}$, and the corresponding kepstra for reflection coefficients $r_0 = 1.0$ and $r_1 = 0.8$.

This feature is also explained by considering the series expansion in [18] and noting that the logarithmic homomorphism in the kepstral operator will sometimes produce such infinite series expansions. In example 3.3-4, we shall prove that the kepstra of sequences with rational Laplace z-transforms are infinite-length sequences, but not all kepstra are infinite in length.

Example 3.3-3. At this point, the kepstrum appears to be nothing more than an abstract concept, which incorporates a group homomorphism K and related mathematical properties and operations. However, we realize that the kepstrum has meaning in terms of factoring the power spectrum $\Phi(\omega)$, but how does it relate to time series which result from exciting the earth with impulsive sources and how can we effectively use the kepstrum to help the geophysicist identify the subsurface structure, i.e. to extract the reflection sequence $\{0, r_1, r_2, \ldots r_N\}$ from an N-layered earth? In example 3.3-2, we considered the single-layered earth model of section 2.2 and removed the pure delay associated with the impulse response $\{x_k^{(\mathrm{R})}\}$ by defining the sequence $\{a_k\} = \{x_{k+1}^{(\mathrm{R})}\}$. We then scaled $\{a_k\}$ such that the first term in the sequence $\{a_k'\}$ was unity and we now have at our disposal the kepstrum $K(a_k') = \alpha_k$, which we obtained by the recursion formula [3.2-26] or by direct application of the kepstral operator K (see equation [3.3-13]). As a practical consideration, let us recover the known source wavelet δ_k given the scaled time series $\{a_k'\} = \{1, a_1', a_2', \ldots\}$ by use of the kepstrum. Now if we pass $\{a_k'\}$ through a linear filter with impulse response $\{a_k'^{-1}\}$, then we can recover the source wavelet δ_k, since $a_k' * a_k'^{-1} = \delta_k$. Thus, the problem is to construct the filter with impulse response $\{a_k'^{-1}\}$. This filter is sometimes referred to as an inverse filter or a deconvolution operator. The classical solution to this problem is to take the Laplace z-transform of the time-series $\{a_k'\}$ giving $A'(z)$; find the inverse of $A'(z)$ giving $[A'(z)]^{-1}$; then determine the inverse Laplace z-transform of $1/A'(z)$ giving $\{a_k'^{-1}\}$. Let us now solve this same problem in terms of the kepstrum. Recall that the kepstrum of $\{a_k'\}$ is given by $K(a_k' * \delta_k) = K(a_k') + K(\delta_k) = \alpha_k + \{0\}$, where $K(\delta_k) = \{0\}$ = the null sequence. Now we also showed that $K^{-1}(-\alpha_k) = \{a_k^{-1}\}$. Thus, in terms of the kepstrum solution we would find the kepstrum $K(a_k') = \{\alpha_k\}$ of the scaled time series $\{a_k'\}$; then compute the inverse of $\{\alpha_k\}$ by performing the reflection $-\{\alpha_k\}$; then apply the inverse kepstral operator $K^{-1}(-\alpha_k)$ giving $\{a_k'^{-1}\}$. If we carefully examine the two approaches, we shall see that they are *identical* in terms of abstract mathematical operations. Let us write out the exact operations in each case in order to clarify this point.

Classical approach

(1) Given the set G of real, minimum-delay sequences with binary operation of convolution, we consider the group homomorphism ζ_{L} where:

$$\zeta_{\mathrm{L}}(a_k' * \delta_k) = \zeta_{\mathrm{L}}(a_k') \cdot \zeta_{\mathrm{L}}(\delta_k) = A'(z) \cdot 1$$

and ζ_{L} is a group homomorphism since it maps the convolutional group $[G, *]$ into the multiplicative group $[G', \cdot]$, where G' is the set of all minimum-delay Laplace z-transforms with binary operation of multiplication. In the engineering sense, we have merely computed the Laplace z-transform of $\{a_k'\}$.

(2) Now since ζ_{L} has mapped the convolutional group into the multiplicative group, any mathematical operations we perform on the ordered pair $(A'(z), 1)$ are in terms of the binary operation of multiplication. Now by definition, the inverse of a multiplicative group is given by the defining relation:

$$A'(z) \cdot [A'(z)]^{-1} = 1$$

where the identity element in this group is unity. In the engineering sense, we have computed the inverse of $A'(z)$, i.e. $1/A'(z)$.

(3) Given the ordered pair $(1/A'(z), 1)$, we now consider the inverse group homomorphism ζ_L^{-1} which maps $[G', \cdot]$ into $[G, *]$. In symbols:

$$\zeta_L^{-1}\left[\frac{1}{A'(z)}\cdot 1\right] = \zeta_L^{-1}\left\{\frac{1}{A'(z)}\right\} * \zeta_L^{-1}\{1\} = a_k'^{-1} * \delta_k$$

Now we observe the important property that ζ_L maps the identity element of $[G, *]$, i.e. δ_k, into the identity element of $[G', \cdot]$, i.e. 1.0. Similarly, ζ_L^{-1} maps the inverse element in $[G', \cdot]$, in this case $[A'(z)]^{-1} = 1/A'(z)$, into the inverse element in $[G, *]$, which is $\{a_k'^{-1}\}$ in this case. In the engineering sense, we have simply computed the inverse Laplace z-transform of $1/A'(z)$.

Thus, in the abstract sense, we have been using homomorphisms in engineering continuously. In fact, we see that for the groups considered, the Laplace z-transform is essentially a group homomorphism.

Kepstrum approach

(1) Given the group $[G, *]$, we now consider a different group homomorphism called the kepstral operator K. In this case:

$$K(a_k' * \delta_k) = K(a_k') + K(\delta_k) = \{\alpha_k\} + \{0\}$$

where the group homomorphism K maps the convolutional group $[G, *]$ into the additive group $[G', +]$, and G' is the set of all real, causal kepstra with binary operation of addition. This step is analogous to computing the Laplace z-transform of $\{a_k'\}$ in the classical approach.

(2) Now since K has mapped the convolutional group into the additive group, any mathematical operations we perform on the ordered pair $(\alpha_k, \{0\})$ are in terms of the binary operation of addition. Now by definition, the inverse of an additive group is defined by the relation:

$$\{\alpha_k\} + \{\alpha_k^{-1}\} = \{0\}$$

where the identity element in this group is the null sequence $\{0\}$ and we see that $\{\alpha_k^{-1}\} = -\{\alpha_k\}$. This step is analogous to computing the inverse $[A'(z)]^{-1}$ in the z "complex" domain, where in this case we compute an analogous inverse in the kepstral domain, which is essentially a "real" time domain.

(3) Given the ordered pair $(-\alpha_k, \{0\})$, we now consider the inverse kepstral operator K^{-1} which maps $[G', +]$ into $[G, *]$. Hence:

$$K^{-1}(-\alpha_k + \{0\}) = K^{-1}(-\alpha_k) * K^{-1}(\{0\}) = a_k'^{-1} * \delta_k$$

Again we see that the important property of a group homomorphism in which the inverse element in $[G', +]$ is mapped into the inverse element in $[G, *]$ and the identity element in $[G', +]$ is mapped into the identity element in $[G, *]$. This step is analogous to computing the inverse Laplace z-transform of $1/A'(z)$.

Regardless of which approach is used, the gist of this example is as follows:

(a) In the classical approach of inverse filtering or deconvolution, we synthesize the inverse filter $\{a_k'^{-1}\}$ in the complex z-plane using the binary operation of multiplication.

(b) In the kepstrum approach to inverse filtering or deconvolution, we synthesize the inverse filter $\{a_k'^{-1}\}$ in a "real" time domain using the binary operation of addition. Moreover, in both instances we assumed knowledge of the source wavelet $\{s_k\}$, which was the minimum-delay unit impulse δ_k. In practice, our source is usually mixed-delay-advance, i.e. it contains zeroes both inside and outside the unit circle, and the transfer function $F^{(N)}(z)$ is generally mixed-delay-advance also. Later, we shall consider these cases in the context of synthesizing inverse filters and will discuss the usefulness of both the classical and kepstral approaches.

Example 3.3-4. An important class of sequences are those which have rational Laplace z-transforms. In particular, we showed that the transfer function of an N-layered earth could be approximated by the rational function $F^{(N)}(z)$ where:

$$F^{(N)}(z) = \frac{r_1 z + r_2 z^2 + r_3 z^3 + \ldots r_N z^N}{1 + \phi_1^{(N)} z + \phi_2^{(N)} z^2 + \ldots + \phi_N^{(N)} z^N}, \quad N \geqslant 1 \tag{1}$$

where $r_0, r_1, r_2, \ldots r_N$ are small real numbers (that cluster around zero) which represent the reflection coefficients of the subsurface layers and the coefficients $\phi_m^{(N)}$ are the serial correlations given by:

$$\phi_m^{(N)} = \sum_{n=0}^{N} r_n r_{n+m}, \quad m \geqslant 1, N \geqslant 1 \tag{2}$$

Now [1] can be rewritten as:

$$F^{(N)}(z) = \frac{r_1 z \left[1 + \frac{r_2}{r_1} z + \frac{r_3}{r_1} z^2 + \ldots + \frac{r_N}{r_1} z^{N-1}\right]}{1 + \phi_1^{(N)} z + \phi_2^{(N)} z^2 + \ldots + \phi_N^{(N)} z^N} \tag{3}$$

Let us investigate the kepstrum $\{\alpha_k^{(N)}\}$ of the scaled impulse response $a_k'^{(N)}$ of this N-layered system, where we shall define the scaled impulse response $\{a_k'^{(N)}\} = \{r_1^{-1} f_{k+1}^{(N)}\}$ to have an amplitude of unity at time $k = 0$. Thus, letting $A'^{(N)}(z) = (r_1 z)^{-1} F^{(N)}(z)$ and factoring [3] we get:

$$A'^{(N)}(z) =$$

$$\frac{A_0 \prod_{i=1}^{N} (1 + a_i z) \prod_{i=1}^{N} (1 + b_i z)(1 + b_i^* z) \prod_{i=1}^{N} (1 + c_i z^{-1}) \prod_{i=1}^{N} (1 + h_i z^{-1})(1 + h_i^* z^{-1})}{\prod_{i=1}^{N} (1 + m_i z) \prod_{i=1}^{N} (1 + n_i z)(1 + n_i^* z) \prod_{i=1}^{N} (1 + s_i z^{-1}) \prod_{i=1}^{N} (1 + t_i z^{-1})(1 + t_i^* z^{-1})} \tag{4}$$

$$\text{where } A_0 = \frac{\prod_{i=1}^{N} s_i \prod_{i=1}^{N} |t_i|^2}{\prod_{i=1}^{N} c_i \prod_{i=1}^{N} |h_i|^2}$$

and $|a_i| < 1$, $|b_i| < 1$, $|c_i| < 1$, $|h_i| < 1$, $|m_i| < 1$, $|n_i| < 1$, $|s_i| < 1$, $|t_i| < 1$; with a_i, c_i, m_i, and s_i as real numbers and b_i, h_i, n_i, and t_i as complex numbers.

For mathematical simplicity, we have factored [3] into the form in [4]. Now we realize that $A'^{(N)}(z)$ can have only N zeroes and N poles, and in fact only $N-1$ finite zeroes and N finite poles. However, by setting the appropriate coefficients to zero, i.e. some of the $a_i = 0$, some of the $b_i = 0$, etc., we can arrive at the true factorization of $A'^{N}(z)$. Nevertheless, the form of [4] will simplify our manipulations in calculating the kepstrum of $a_k'^{(N)}$.

Now the zeroes and poles of $A'^{(N)}(z)$ need not necessarily be simple; furthermore, we have shown them to be either real or complex numbers. However, since the reflection coefficients are always real numbers in our analysis, any complex poles or zeroes must always occur in conjugate pairs. This factorization is accounted for in [4].

The factors $(1 + a_i z)$ and $(1 + b_i z)(1 + b_i^* z)$ represent the real and complex conjugate minimum-delay zeroes of $A'^{(N)}(z)$, i.e. zeroes outside the unit circle, and the factors $(1 + m_i z)$ and $(1 + n_i z)(1 + n_i^* z)$ characterize the minimum-delay poles of $A'^{(N)}(z)$, i.e. the poles outside the unit circle. Similarly, the minimum-advance zeroes, i.e. zeroes inside the unit circle, are characterized by the real factors $(1 + c_i z^{-1})$ and the complex conjugate factors $(1 + h_i z^{-1})(1 + h_i^* z^{-1})$. The poles located inside the unit circle are represented by the factors $(1 + s_i z^{-1})$ and $(1 + t_i z^{-1})(1 + t_i^* z^{-1})$. But from physical grounds, we know that the denominator of $A'^{(N)}(z)$, i.e. the characteristic polynomial, will always be minimum-delay. Hence, practically speaking, we need not concern ourselves with poles contained inside the unit circle or the factors $(1 + s_i z^{-1})$ or $(1 + t_i z^{-1})(1 + t_i^* z^{-1})$. Previously, we showed that all the minimum-delay sequences with Laplace z-transform $(1 + a_i z)$ have kepstra of the form:

$$\zeta_L^{-1} \{\log [1 + a_i z]\} = \frac{(-1)^{k+1} a_i^k}{k}, \quad \text{for } k > 0, \; a_i \text{ real} \qquad [5]$$

Using [5] as our basic building block, we can derive the kepstrum of the scaled impulse response $\{a_k'^{(N)}\}$. Now the reverse sequence corresponding to [5] has the Laplace z-transform $1 + a_i z^{-1}$ with $|a_i| < 1$. Recall that the kepstrum of this folded minimum-advance sequence is found by folding the kepstrum of the original minimum-delay sequence. From this property, we conclude that the kepstrum associated with $1 + c_i z^{-1}$, denoted by γ_k, has the form:

$$\gamma_k = \zeta_L^{-1}\{\log[1 + c_i z]\}\Big|_{k=-k} = \frac{(-1)^k c_i^{-k}}{k}, \quad \text{for } k < 0, \; c_i \text{ real} \qquad [6]$$

The inverse sequence corresponding to [5] has the Laplace z-transform $1/(1 + a_i z)$ with $|a_i| < 1$. Recall that the kepstrum of an inverse sequence is found by reflecting the kepstrum of the original minimum-delay sequence (equation [5]) about the positive-time axis. From this property, we conclude that the kepstrum associated with $1/(1 + m_i z)$, denoted by μ_k, is given by:

$$\mu_k = -\zeta_L^{-1}\{\log[1 + m_i z]\} = \frac{-(-1)^{k+1} m_i^k}{k}, \quad \text{for } k > 0, \; m_i \text{ real} \qquad [7]$$

The reverse-inverse sequence associated with equation [5] has the Laplace z-transform $1/(1 + a_i z^{-1})$ with $|a_i| < 1$. Recall that the kepstrum of the reverse-inverse sequence can be found by folding the kepstrum of the original minimum-delay sequence (equation [5]) about the origin then performing a reflection about the negative-time axis. From this property, we conclude that the kepstrum associated with the factor $1/(1 + s_i z^{-1})$, denoted by σ_k, is given by:

$$\sigma_k = -\zeta_L^{-1}\{\log[1 + s_i z]\}\Big|_{k=-k} = \frac{-(-1)^k s_i^{-k}}{k}, \quad \text{for } k < 0, \; s_i \text{ real} \qquad [8]$$

Let us now investigate the kepstrum of a sequence which contains complex conjugate poles or zeroes. Consider the minimum-delay sequence with Laplace z-transform $(1 + b_i z)(1 + b_i^* z)$. Proceeding directly from the definition of the kepstral operator K, we see that the kepstrum of this sequence, denoted by $\{\beta_k\}$, can be represented by:

$$\beta_k = \frac{(-1)^{k+1} b_i^k}{k} + \frac{(-1)^{k+1}(b_i^*)^k}{k}, \quad \text{for } k > 0 \qquad [9]$$

But b_i is a complex number which may be expressed as:

$$b_i = |b_i| \exp(i\phi_{b_i}), \quad b_i^* = |b_i| \exp(-i\phi_{b_i}) \qquad [10]$$

Hence, [9] may be rewritten as:

$$\beta_k = \frac{(-1)^{k+1}}{k} |b_i|^k \left[\exp(ik\phi_{b_i}) + \exp(-ik\phi_{b_i})\right] \qquad [11]$$

Using the fact that $\cos k\phi_{b_i} = \frac{1}{2}[\exp(ik\phi_{b_i}) + \exp(-ik\phi_{b_i})]$

$$\beta_k = \frac{2(-1)^{k+1}}{k} |b_i|^k \cos k\phi_{b_i}, \quad \text{for } k > 0 \qquad [12]$$

Now if we fold equation [12] about the origin, we obtain the kepstrum of the minimum-advance sequence with Laplace z-transform $(1 + b_i z^{-1}) \times (1 + b_i^* z^{-1})$. We conclude that factors of the form $(1 + h_i z^{-1})(1 + h_i^* z^{-1})$ have the kepstrum η_k given by:

$$\eta_k = \frac{2(-1)^k}{k} |h_i|^{-k} \cos k\phi_{b_i}, \quad \text{for } k < 0 \qquad [13]$$

By using the properties of the inverse sequence and inverse-reverse sequence, we can show that the factors $1/(1 + n_i z)(1 + n_i^* z)$ have kepstrum ν_k given by:

$$\nu_k = \frac{-2(-1)^{k+1}}{k} |n_i|^k \cos k\phi_{n_i}, \quad \text{for } k > 0 \qquad [14]$$

and factors of the form $1/(1 + t_i z^{-1})(1 + t_i^* z^{-1})$ have kepstrum τ_k where:

$$\tau_k = \frac{-2(-1)^k}{k} |t_i|^{-k} \cos k\phi_{t_i}, \quad \text{for } k < 0 \qquad [15]$$

Thus, the kepstrum of the scaled impulse response $\{a_k'^{(N)}\}$ can now be found by adding the kepstra associated with each of the factors in [4]. (Recall the property that the kepstrum of the convolution of two sequences is the addition of their individual kepstra.) Using our results in equations [5], [6], [7], [8], [12], [13], [14], and [15], and considering the product of all factors in equation [4] as the result of convolutions of minimum-delay sequences, reverse sequences, and inverse-reverse sequences, we arrive at the kepstrum of $\{a_k'^{(N)}\}$, denoted by $\{\alpha_k^{(N)}\}$ and expressed as:

$$\{\alpha_k^{(N)}\} = \sum_{i=1}^{N} \frac{(-1)^k}{k} \left[c_i^{-k} + 2|h_i|^{-k} \cos k\phi_{h_i} \right] - \sum_{i=1}^{N} \frac{(-1)^k}{k} \left[s_i^{-k} + 2|t_i|^{-k} \cos k\phi_{t_i} \right], \quad \text{for } k < 0$$

$$\{\alpha_k^{(N)}\} = \log A_0, \quad \text{for } k = 0 \text{ and } A_0 > 0 \qquad [16]$$

$$\{\alpha_k^{(N)}\} = \sum_{i=1}^{N} \frac{(-1)^{k+1}}{k} \left[a_i^k + 2|b_i|^k \cos k\phi_{b_i} \right] - \sum_{i=1}^{N} \frac{(-1)^{k+1}}{k} \left[m_i^k + 2|n_i|^k \cos k\phi_{n_i} \right], \quad \text{for } k > 0$$

where we have assumed that the sign of A_0 is positive. Equation [16] is the most general form for the kepstrum of a sequence with rational Laplace z-transform, and for such sequences the kepstrum is an infinite-length sequence. For $k > 0$, we see that the kepstrum constants a_i, b_i, m_i, and n_i are associated with the minimum-delay portion of $A'^{(N)}(z)$, and for $k < 0$, the kepstrum constants c_i, h_i, s_i, and t_i are related to the minimum-advance portion of the rational function $A'^{(N)}(z)$. In fact, $-a_i^{-1}$, $-b_i^{-1}$, and $-b_i^{*-1}$ are the zeroes outside the unit circle and $-m_i^{-1}$, $-n_i^{-1}$, and $-n_i^{*-1}$ are the poles outside the unit circle. Thus, we see the relationship between the spectral factorization method and the kepstrum. The time response of the KEPS for $k > 0$ is shaped by the minimum-delay poles and zeroes. This time response corresponds to the β_k coefficients (equation [3.2-45] for real signals), which contains damped cosine terms, with the rate of damping and frequency of oscillation controlled by the magnitudes and arguments of the minimum-delay poles and zeroes. Thus, the kepstrum can serve as an indicator which tells us whether a given sequence is minimum-advance, mixed-delay-advance, or minimum-delay. If we observe no kepstrum for $k > 0$ but have a response for $k < 0$, we have a minimum-advance sequence. Similarly, no contribution for $k < 0$ but a response for $k > 0$ indicates a minimum-delay sequence. Also, we see that the "kepstral" time and the familiar "real" time are not the same. In kepstral time, we can always observe the past and the future at the same time, which of course does not correspond to our concept of real time. Thus, we should keep in mind that the time response attached to the acronym kepstrum is not real time.

In concluding this section, we would like to point out that not all kepstra are infinite-length sequences. For example, consider the sequence $\{1, 1, 1/2, 1/6, 1/24, 1/120, \ldots\}$ which is essentially $\{1/k!\}$ for $k = 0, 1, 2, \ldots$ The kepstrum of this sequence is δ_{k-1}, which is a unit impulse at $k = 1$. However, the Laplace z-transform of $1/k!$ is e^z which is certainly not a rational function. We can expect that the impulse responses of linear time-invariant systems have infinite-length kepstra but must be aware that maybe certain source wavelets are not characterized by rational Laplace z-transforms. If this is so, we can expect that the kepstra of certain source wavelets might be finite-length sequences, for example, like the kepstrum of $1/k!$. In any event, we hope that the concepts of homomorphisms and spectral factorization have made the kepstrum a meaningful signal processing idea. We wish to show the engineer that he uses homomorphisms continuously and that the abstractness of these mappings have practical applications, for example, in factoring the power spectrum of a process, or mapping convolution into addition, etc. Regardless of which viewpoint we take, it is most important that we remember the roots of the term kepstrum, i.e. we are actually factoring the spectrum of the observed signal. In the next chapter, we shall relate the kepstrum and its spectral factorization features to the idea of deconvolution.

Chapter 4

DECONVOLUTION

4.1. PREDICTIVE DECONVOLUTION

In the previous chapters we have discussed the evolution of geophysical models, statistical considerations in interpreting models, and the importance of understanding the governing physical process which the model describes. As a review, we first considered the reflection seismic method, which was modeled by a convolutional process, with the source wavelet $\{s_k\}$ and the input and noise-free reflection response $\{x_k\}$ as the output of a linear time-invariant system with impulse response $\{f_k\}$. We noted that similar models exist in other areas of science, for example, radar and sonar, where one considers the superposition of amplitude-scaled and time-delayed replicas. However, the exploration geophysicist is interested in the earth's subsurface structure, and in particular the reflection coefficients. In our convolution model (refer to Fig. 1-1), these reflection coefficients were not observable, i.e. there was not an explicit dependence between the observable geophysical time series and the desired reflection coefficients. This led us to the development of a layered earth model, from which evolved a transfer function of the earth $F^{(N)}(z)$ in the form of a standard recursive digital filter. Through physical considerations and approximations, we arrived at a useful transfer function that explicitly contained the reflection coefficients as its parameters (see equation [2.4-13]). This transfer function, equation [2.4-13], is valid when the magnitude of any reflection coefficient is small, and in practice $|r_n| < 0.1$. Moreover, the numerator of $F^{(N)}(z)$ is precisely the Laplace z-transform of the reflection coefficient sequence $\{0, r_1, r_2, \ldots r_N\}$ (equation [2.4-12]) and the denominator of $F^{(N)}(z)$ is expressed as the Laplace z-transform of a serial correlation function of the reflection coefficients (equation [2.4-8]). Equation [2.4-13] provides the mathematical justification for the Robinson seismic model of section 1.5. Thus, from good physical insight gained from our layered earth model, we shall now attempt to extract the reflection sequence $\{0, r_1, r_2, \ldots r_N\}$ from the observable noise-free time series = observed seismic record (reflection response) $\{x_k'\}$. In discussing various methods for accomplishing this task, we shall always jointly consider the physical models and their relation to a particular mathematical technique. However, good physical insight from our model will force us to consider mathematical techniques which are appropriate to the physical problem and not allow us to get lost in mathematical methods per se. We shall now consider the deconvolution problem in this context.

Definition of deconvolution. Given an observed time series which was generated by the convolution of multiple signal components, deconvolution is the process of decomposing the observed time series into its constituent members.

We note that although the convolutional process involves linear transformations, deconvolution is not restricted to the class of linear transformations, i.e. there are non-linear transformations which realize the process of deconvolution. Before proceeding any further, let us briefly review the theory of convolution.

When two Laurent series:

$$A(z) = \sum_{k=-\infty}^{\infty} a_k z^k, \quad B(z) = \sum_{k=-\infty}^{\infty} b_k z^k$$

are multiplied together, a new series of the same type results, and is expressed as:

$$C(z) = A(z)B(z) = \sum_{k=-\infty}^{\infty} c_k z^k$$

Now the new coefficients c_k of the series $C(z)$ are related to the coefficients a_k and b_k by:

$$c_k = \sum_{k=-\infty}^{\infty} a_{k-n} b_n, \quad k = 0, \pm 1, \pm 2, \ldots \qquad [4.1\text{-}1]$$

The sequence $\{c_k\}_{-\infty}^{\infty}$ is called the *convolution* of the sequences $\{a_k\}_{-\infty}^{\infty}$ and $\{b_k\}_{-\infty}^{\infty}$ We arrive at the continuous analogue of this operation when we multiply two bilateral Laplace integrals:

$$A(s) = \int_{-\infty}^{\infty} a(t)\, e^{-st}\, dt, \quad B(s) = \int_{-\infty}^{\infty} b(t)\, e^{-st}\, dt$$

to obtain:

$$C(s) = A(s)\, B(s) = \int_{-\infty}^{\infty} c(t)\, e^{-st}\, dt$$

where:

$$c(t) = \int_{-\infty}^{\infty} a(t-\tau)\, b(\tau)\, d\tau$$

This combination of functions occurs so frequently in the study of linear systems that it may be regarded as one of the most fundamental operations of analysis. It is most fortunate that the reflection seismic process may be reasonably approximated by a convolution model. Moreover, time series representations of other physical processes (speech) or economic processes (price and quantity prediction, business cycle forecasting) are often simplified by a linear characterization of the process in question. We now give a formal description of a linear system.

Definition of a linear system. A system or process is simply a prescribed relationship between two quantities generally called the system input $\{s_k\}$ and the system output $\{x_k\}$ such that:

$$s_k \longrightarrow x_k$$

where the above notation ($\longrightarrow$) means that an input $\{s_k\}$ gives rise to an output $\{x_k\}$. Now a system is defined as being *linear* if and only if the system satisfies two properties:

(1) *Additive property* (superposition principle);

If $s_k^{(1)} \longrightarrow x_k^{(1)}$ and $s_k^{(2)} \longrightarrow x_k^{(2)}$
then $s_k^{(1)} + s_k^{(2)} \longrightarrow x_k^{(1)} + x_k^{(2)}$

(2) *Multiplicative property:*

If $s_k \longrightarrow x_k$ then $cs_k \longrightarrow cx_k$

where c is in general a complex constant. A system which does not satisfy both the additive and multiplicative properties is termed a non-linear system. A system is also called time-invariant or shift-invariant if the relationship between $\{s_k\}$ and $\{x_k\}$ is independent of time, i.e.:

if $s_k \longrightarrow x_k$ then $s_{k-n} \longrightarrow x_{k-n}$

Given the definitions of convolution and a linear shift-invariant system, let us see how convolution evolves in studying linear shift-invariant systems. Now, by definition, the impulse response $\{f_k\}$ of a linear shift-invariant system is the response to a unit impulse δ_k. Hence:

$$\delta_k \longrightarrow f_k$$

Because the system is time-invariant, we can write:

$$\delta_{k-n} \longrightarrow f_{k-n}$$

Using the multiplicative property of linear systems, we can add that:

$$s_n \delta_{k-n} \longrightarrow s_n f_{k-n}$$

From the additive property (superposition principle) of linear systems, we can further claim that:

$$\sum_{n=-\infty}^{\infty} s_n \delta_{k-n} \longrightarrow \sum_{n=-\infty}^{\infty} s_n f_{k-n}$$

But we know that $s_k = \Sigma_{n=-\infty}^{\infty} s_n \delta_{k-n}$ and:

$s_k \longrightarrow x_k$

Hence, we arrive at the result:

$$x_k = \sum_{n=-\infty}^{\infty} s_n f_{k-n}$$

or by the change of index $m = k - n$, we get:

$$x_k = \sum_{m=-\infty}^{\infty} f_m s_{k-m} \qquad [4.1\text{-}2]$$

which defines the convolution of the sequences $\{f_k\}$ and $\{s_k\}$. Thus, given the input $\{s_k\}$ to a linear shift-invariant system with impulse response $\{f_k\}$, the output of this sytem $\{x_k\}$ may be obtained by the convolution in [4.1-2]. With knowledge of convolution, linear shift-invariant systems, and the concept of deconvolution, let us examine the method of predictive deconvolution.

Before going directly to the mathematical details involved in the method of predictive convolution, let us consider the historical evolution of this concept, i.e. how the problem of the elimination of water reverberations and other types of multiple reflections led to its development. Now, as noted in Chapter 1, the seismic methods developed in the 1930's and early 1940's were not sophisticated enough to explore successfully many of the potential oil-producing regions of the world. In particular, the existing seismic methods were not very successful in exploring offshore areas because of the water reverberations. This water reverberation problem was first recognized on the basis of the ringing nature (i.e. apparent periodicity) on a series of seismograms obtained from the Persian Gulf and from Lake Maracaibo in Venezuela during the 1940's.

Now the seismic interpreter is interested in extracting a desired signal, consisting of the "deep" reflected events due to subsurface layers potentially containing hydrocarbons, from a recording which contains a great deal of interference (undesired multiple reflections) and possibly measurement noise. Marine seismic records are the response of the water layer and subsurface layers to a marine seismic source, as recorded by the hydrophones (transducers). The marine seismic source, which may be an air-gun blast or 15—25 lb of high-velocity dynamite detonated 4—6 ft below the surface, produces a disturbance which may (on the seismic time scale) be regarded essentially as a positive pressure impulse (Cole, 1948). After interactions with the water and subsurface layers, the hydrophones (pressure transducers in marine work) convert the received pressure variations to voltage, which is used to generate the observed seismic record or seismogram.

In marine seismic operations, the water-air interface is a flat, strong reflector. In our notation, the air-water reflection coefficient is r_0, so the water-air reflection coefficient is $-r_0$. For a plane wave incident on a water-air boundary, a positive excess pressure in the incident wave is reflected as a negative excess pressure, i.e. a condensation is reflected as a rarefaction. Thus, for all practical purposes, the reflection coefficient at the water-air interface is approximately $-r_0 = -1$, or in other words the air-water reflection coefficient is $r_0 = 1$. In many areas, the water-bottom interface is also a strong reflector. Hence, there exists an energy trap: a non-attenuating medium (water layer) bounded by two strong reflecting interfaces. A pulse generated in the water layer, or entering the water layer from below, will be successively reflected between the two interfaces, with a time interval dependent on the velocity of propagation in water and water depth, and an amplitude decay dependent on the reflection coefficient of the water bottom in question. As a result, desired reflections from deep subsurface layers are obscured by these water reverberations.

K.E. Burg et al. (1951) analyzed the water layer reverberation problem by waveguide theory, attempting to explain the appearance of sinusoidal variations on the seismic traces. Wadsworth et al. (1953) used adaptive linear digital filters governed by a time-gating procedure to process the reverberant seismic data. They also recognized that any statistical approaches to this problem would require the consideration of non-stationary processes. As an approximate method of treating the non-stationary phenomenon represented by a seismogram, they divided the record into time intervals which were short enough such that the observed process was approximately stationary over each interval. They then determined linear operators for each interval, which were optimum in the least-squares sense, for eliminating the reverberatory components of the seismic trace. The work of Robinson (1954) involved the development of statistical models for geophysical signal processing and the introduction of the method of predictive deconvolution for the elimination of unwanted seismic reverberations and other types of multiple reflections from the seismic data. Because of the non-stationary character of the observed data, the predictive deconvolution method in practice is made data adaptive by a time-gating procedure. Later, Backus (1959) treated the effect of the water layer as a linear filtering operation, and investigated the use of inverse filtering techniques for the water reverberation elimination. The basic difference between the Backus inverse filtering approach and the Robinson "statistical" approach was that Backus used the deterministic frequency domain concepts of circuit theory in his mathematical formulation, while Robinson hypothesized that the desired "deep" reflections could be treated as a random uncorrelated sequence and applied the method of least squares as a means of determining the linear operators, required for his predictive deconvolution method, from the seismic data.

To date, the predictive deconvolution process has found extensive application in the digital elimination of multiple reflections from seismic time series data. Let us now view this method with respect to our layered earth model, and in particular we shall show how the method of predictive deconvolution can be used to estimate the subsurface structure in prospecting for petroleum and natural gas.

There are many different approaches to the deconvolution of seismic traces. Each approach has certain merits; in the overall analysis we do not want to be restricted to any given approach, but we want to be able to make use of a variety of different ones. For any deconvolution process we must make use of a seismic model which takes into account both known and unknown factors.

The method of predictive deconvolution is based on the Robinson "statistical" seismic model, which we treated in detail by physical reasoning in section 1.5, and for which we gave the mathematical justification in Chapter 2. As we have seen, there are two cases of this model, namely the case of internal primary reflections and the case of external primary reflections.

The case of internal primary reflections may be described as the case of total system reverberation, that is, all the multiple reflections and refractions within the layered system are taken into account to within the stated approximations. As we have seen on p. 40, this model is characterized by the following properties:

(1) The model is a convolutional model represented by

$$x_k = \epsilon_k * w_k$$

with $\{\epsilon_k\}$ = reflection coefficients of the N reflecting interfaces of the entire system; $\{w_k\}$ = composite wavelet = $b_k * s_k$; $\{b_k\}$ = system reverberation wavelet; $\{s_k\}$ = source wavelet; and $\{x_k\}$ = observed seismic record (time series).

(*Note:* At this point, we have dropped the primed superscript on x_k'; that is, x_k' of section 1.5 becomes simply x_k from now on.)

(2) The composite wavelet $\{w_k\}$ is minimum-delay. This follows from the physical fact that $\{b_k\}$ is minimum-delay and $\{s_k\}$ can be approximately minimum-delay by properly designed field procedures.

(3) The reflection coefficients $\{\epsilon_n\} = \{r_n\}$ are small in magnitude, a fact generally observed in seismic prospecting.

(4) The reflection coefficients $\{\epsilon_1, \epsilon_2, \ldots \epsilon_N\}$ are uncorrelated over various intervals $[k, k + L]$ of the seismic record, a fact generally observed in seismic prospecting.

The case of external primary reflections may be described as the case of reflections from deep layers subject to reverberations due to surface layers; that is, no multiple reflections and refractions are taken into account within the so-called intermediate free space and the other multiple reflections and refractions are taken into account to within the stated approximations. As we have seen on pp. 43—44, this model is characterized by the following properties:

(1) The model is a convolutional model represented by:

$$x_k = \epsilon_k * w_k$$

with $\{\epsilon_k\}$ = reflection coefficient sequence of the deep layer reflecting interfaces; $\{w_k\}$ = composite wavelet = $b_k * b_k * s_k$; $\{b_k\}$ = surface-layer reverberation wavelet; $\{s_k\}$ = source wavelet; and $\{x_k\}$ = observed seismic record (time series).

(*Note:* We have dropped the primed superscript on x'_k of section 1.5, which we now write simply as x_k.)

(2) The composite wavelet $\{w_k\}$ is minimum-delay. This follows from the physical fact that $\{b_k\}$ is minimum-delay and $\{s_k\}$ can be minimum-delay by properly designed field procedures.

(3) The reflection coefficients $\{\epsilon_n\} = \{r_n\}$ are small in magnitude, a fact generally observed in seismic prospecting.

(4) The reflection coefficients $\{\epsilon_1, \epsilon_2, \ldots \epsilon_N\}$ are uncorrelated over various intervals $[k, k + L]$ of the seismic record, a fact generally observed in seismic prospecting.

On comparing the two cases, we see that their essential features, from the point of view of deconvolution, can be incorporated in one overall statistical model, namely: the Robinson statistical seismic model (for deconvolution) which is characterized by the following properties:

(1) The model is a convolutional model represented by:

$$x_k = \epsilon_k * w_k = w_0 \epsilon_k + w_1 \epsilon_{k-1} + w_2 \epsilon_{k-2} + \ldots$$

with $\{\epsilon_k\}$ = reflection coefficient sequence; $\{w_k\}$ = composite wavelet; and $\{x_k\}$ = observed seismic record (time series).

(2) The composite wavelet $\{w_k\}$ is minimum-delay.* Also, we shall assume that $w_0 = 1$.

(3) The reflection coefficients $\{\epsilon_n\} = \{r_n\}$ are small in magnitude.

(4) The reflection coefficients $\{\epsilon_1, \epsilon_2, \ldots \epsilon_N\}$ are uncorrelated over various intervals $[k, k + L]$.

In practice, we time gate the seismic trace so that the time interval of the gate corresponds to the depth interval in which we are interested. As a matter of convenience, the time series within the gate can be relabeled as $\{x_1, x_2, \ldots x_L\}$, denoting L time samples within the gate, and the reflection sequence for the corresponding depth interval can be relabeled as $\{\epsilon_1, \epsilon_2, \ldots \epsilon_L\}$. Under this condition, ϵ_1 is the reflection coefficient for the first interface of the depth section in question, ϵ_2 for the second interface of the desired depth section, etc., and x_1 occurs at the time of arrival of the direct (primary) reflection from the first interface of the desired depth section, x_2 occurs at the time of arrival of the primary reflection from the second interface of the depth section, etc.

The problem of predictive deconvolution can now be described as follows. Given the observed seismic record $\{x_k\}$ within a certain gate, determine the

---

* In section 4.2, we will use the more explicit notation $w_k^{(0)}$ for the minimum-delay wavelet.

corresponding reflection coefficient sequence $\{\epsilon_k\}$ to within a scale factor. Although the geophysicist is interested in mapping the subsurface structure, it is the *contrast* between events (reflections) due to different reflecting interfaces that is important. Because of the reverberations (i.e. multiple reflections), the direct (primary) reflections are masked and are therefore often hidden on the observed seismic record $\{x_k\}$.

As we have seen in section 1.5, the ideal deconvolution operator is $\{w_k^{-1}\}$, which we shall now denote as $\{q_k\}$, i.e. $\{q_k\} \equiv \{w_k^{-1}\}$. Thus, $\{q_k\}$ is the inverse of the composite wavelet $\{w_k\}$ and is defined as

$$q_k * w_k = \delta_k = \begin{cases} 1, & k = 0 \\ 0, & k \neq 0 \end{cases}$$

Applying this deconvolution operator to the seismic trace $\{x_k\}$ we get:

$$q_k * x_k = q_k * w_k * \epsilon_k = \epsilon_k$$

If the deconvolution operation were ideal, the masking effect of the multiple reflections would be entirely removed and the deconvolved seismic record would be simply $\{\epsilon_k\}$. As a result, the ideal deconvolved record would delineate each primary reflected event as to its time k and as to its strength ϵ_k. By use of a seismic velocity function, the arrival times (k) of these events can be converted to the depths of the corresponding reflecting horizons, from which a depth map can be constructed.

Hence, we seek a linear deconvolution operator that in the ideal case would be $\{q_k\}$. Such an operator can be obtained by the method of linear prediction, which we now proceed to explain.

Part 1: Linear prediction based on the infinite past; autocorrelation known

Suppose that we have at our disposal the sample values $\{\ldots x_{k-3}, x_{k-2}, x_{k-1}, x_k\}$. If these values were samples of a purely deterministic process, we would intuitively expect to be able to predict future values with little or no prediction error, i.e. the process would be highly predictable. However, suppose that the sequence of values $\{\ldots x_{k-3}, x_{k-2}, x_{k-1}, x_k\}$ are samples from a zero mean stationary random process denoted by $\{x_k\}$ and that the autocorrelation function of the process $\{x_k\}$, namely:

$$R_m \equiv E\{x_k x_{k-m}\} = R_{-m} \qquad [4.1\text{-}3]$$

is known for $m = 0, \pm1, \pm2, \ldots$ The symbol E denotes the expectation operator. We consider the case of real samples, but generalization to complex processes can be readily made, if desired. Let us now attempt to linearly predict the value of x_k given the infinite past $\{\ldots x_{k-3}, x_{k-2}, x_{k-1}\}$. If we denote our one-step estimate of x_k by $\hat{x}_k$, then the one-step predicted value of x_k, based on the infinite past, is:

$$\hat{x}_k \equiv -\sum_{n=1}^{\infty} a_n x_{k-n} \qquad [4.1\text{-}4]$$

The operator $\{-a_1, -a_2, -a_3, \ldots\}$ is called the *prediction operator* for prediction distance one, i.e. the one-step predicted value of x_k. The values of the predictive filter coefficients $\{-a_n\}_1^\infty$ are chosen according to the least-squares formulation, i.e. the prediction error e_k has minimum mean-square value or minimum variance, since $\{x_k\}$ is assumed zero mean. The one-step *prediction error* e_k is:

$$e_k \equiv x_k - \hat{x}_k = \sum_{n=0}^{\infty} a_n x_{k-n} \quad (\text{where } a_0 \equiv 1) \qquad [4.1\text{-}5]$$

Note that the coefficient a_0 is defined as equal to one. The operator $\{a_0, a_1, a_2, \ldots\}$ is called the *prediction-error operator* for prediction distance one. We shall denote the *prediction-error sequence* by $\{e_k\}$. From equation [4.1-5], the prediction error mean-square value (or variance) is:

$$\Upsilon_0 \equiv E\{e_k^2\} = \sum_{n=0}^{\infty} \sum_{m=0}^{\infty} a_n a_m R_{m-n} \qquad [4.1\text{-}6]$$

For a minimum* of equation [4.1-6], one computes:

$$\frac{\partial \Upsilon_0}{\partial a_l} = 2 \sum_{n=0}^{\infty} a_n R_{l-n}, \quad l \geqslant 1 \qquad [4.1\text{-}7]$$

and sets the result equal to zero, obtaining the values of the predictive filter coefficients $\{-a_n\}_1^\infty$ as the solution of the set of equations, called the *normal equations*, given by:

$$\sum_{n=0}^{\infty} a_n R_{l-n} = 0, \quad l \geqslant 1 \; (\text{where } a_0 \equiv 1) \qquad [4.1\text{-}8]$$

Let us investigate the autocorrelation Υ_j of the prediction-error sequence $\{e_k\}$. If the autocorrelation Υ_j is defined by $\Upsilon_j \equiv E\{e_k e_{k-j}\} = \Upsilon_{-j}$, then:

$$\Upsilon_j \equiv E\{e_k e_{k-j}\} = \sum_{n=0}^{\infty} \sum_{m=0}^{\infty} a_n a_m R_{m+j-n} \qquad [4.1\text{-}9]$$

or:

$$\Upsilon_j = \sum_{m=0}^{\infty} a_m \sum_{n=0}^{\infty} a_n R_{m+j-n} \qquad [4.1\text{-}10]$$

---

* It can be shown that the Hessian matrix of the function Υ_0 is positive definite, as required for a minimum.

Now the innermost sum on n in equation [4.1-10] is zero for $m + j \geqslant 1$, as dictated by equation [4.1-8]. If $j \geqslant 1$ in equation [4.1-10], then $j + m \geqslant 1$ since $m \geqslant 0$ in the outermost sum in equation [4.1-10]. Therefore, $\Upsilon_j = 0$ for $j \geqslant 1$ and because the autocorrelation Υ_j is a symmetric function of j (i.e. $\Upsilon_j = \Upsilon_{-j}$), it follows that:

$$\Upsilon_j = 0 \quad \text{for } j \neq 0 \qquad [4.1\text{-}11]$$

Thus, the prediction-error sequence $\{e_k\}$ is uncorrelated and therefore possesses a flat (white) power spectral density. Hence, the linear prediction-error filter characterized by the coefficients $\{a_n\}_0^\infty$ (where $a_0 \equiv 1$) is a whitening filter. The prediction-error sequence $\{e_k\}$ is also called the innovation process. The prediction-error operator $\{a_n\}_0^\infty$ (where $a_0 \equiv 1$) is depicted in Fig. 4-1. Now the zero-lag autocorrelation Υ_0 is the average power of the prediction-error sequence $\{e_k\}$; that is, the prediction-error variance given by equation [4.1-6] can be written as:

$$\Upsilon_0 \equiv E\{e_k^2\} = \sum_{m=0}^{\infty} a_m \sum_{n=0}^{\infty} a_n R_{m-n} \qquad [4.1\text{-}12]$$

Substitution of equation [4.1-8] in equation [4.1-12] gives the following useful expression for the prediction-error variance:

$$\Upsilon_0 = a_0 \sum_{n=0}^{\infty} a_n R_{-n} = \sum_{n=0}^{\infty} a_n R_{-n} = R_0 + \sum_{n=1}^{\infty} a_n R_n \qquad [4.1\text{-}13]$$

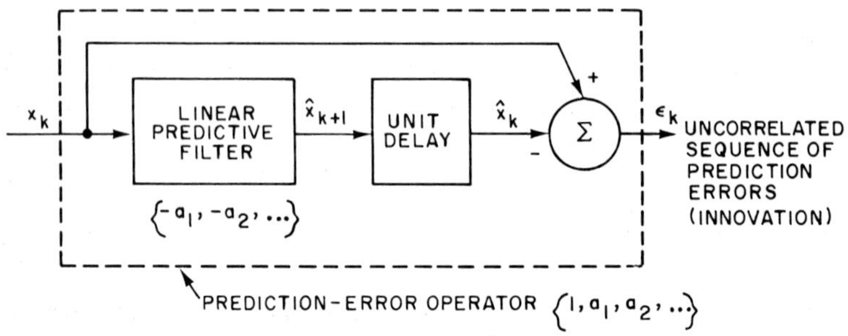

Fig. 4-1. Schematic diagram of the prediction-error operator $\{a_n\}_0^\infty$ (where $a_0 \equiv 1$).

Thus, for the case of linear prediction based on the infinite past and known autocorrelation, we see that the prediction-error sequence $\{e_k\}$ is purely random and has a unit spike autocorrelation function with zero-lag value given by equation [4.1-13].

Let us now solve the normal equations [4.1-8]. The normal equations hold only for $l \geqslant 1$, that is:

$$\sum_{n=0}^{\infty} a_n R_{l-n} = 0, \quad l \geqslant 1 \text{ (where } a_0 \equiv 1) \qquad [4.1\text{-}14]$$

If we let $\{h_l\}$ be a sequence such that:

h_l is unspecified (as yet) for $l \leqslant -1$
$h_0 = \Upsilon_0$ = prediction-error variance (by equation [4.1-13]) [4.1-15]
$h_l = 0$ for $l > 0$

then the normal equations may be written as:

$$\sum_{n=0}^{\infty} a_n R_{l-n} = h_l \text{ for all integers } l \qquad [4.1\text{-}16]$$

Let us define the Laplace z-transforms:

$$A(z) = \sum_{n=0}^{\infty} a_n z^n$$

$$\Phi(z) = \sum_{n=-\infty}^{\infty} R_n z^n \qquad [4.1\text{-}17]$$

$$H(z) = \sum_{n=-\infty}^{0} h_n z^n$$

Note that $H(z)$ contains no positive powers of z. Using the Laplace z-transforms in equation [4.1-17], the normal equations [4.1-16] can be rewritten as:

$$A(z)\,\Phi(z) = H(z) \qquad [4.1\text{-}18]$$

Recalling our discussion in Chapter 3 on spectral factorization, $\Phi(z)$ may be factored as:

$$\Phi(z) = \sigma^2 W(z)\, W^*(z) \qquad [4.1\text{-}19]$$

where the filter $W(z)$ is a minimum-delay filter. Here σ^2 is a positive constant, which we introduce so that we may require the leading coefficient w_0 be equal to one. Further:

$$W(z) = \sum_{n=0}^{\infty} w_n z^n \quad (\text{where } w_0 \equiv 1) \qquad [4.1\text{-}20]$$

Since we are dealing with real-valued coefficients, i.e. $\{w_n\}$ a sequence of real numbers, then:

$$W^*(z) = \sum_{n=0}^{\infty} w_n^* z^{-n} = \sum_{n=0}^{\infty} w_n z^{-n} = W(z^{-1}) \qquad [4.1\text{-}21]$$

That is:

$$\Phi(z) = \sigma^2 W(z)\, W(z^{-1}) \qquad [4.1\text{-}22]$$

where $W(z)$ contains no negative powers of z and $W(z^{-1})$ contains no positive powers of z. Substitution of equation [4.1-22] in [4.1-18] gives:

$$A(z)[\sigma^2 W(z)\, W(z^{-1})] = H(z) \qquad [4.1\text{-}23]$$

Since $W(z)$ is a minimum-delay Laplace z-transform, it follows that $W(z^{-1})$ is minimum-advance. Further, the inverse $1/W(z^{-1})$ represents an anti-causal stable sequence, i.e. $1/W(z^{-1})$ has a stable expansion in non-positive powers of z. Thus, we can rewrite equation [4.1-23] as:

$$\sigma^2 A(z)\, W(z) = H(z)/W(z^{-1}) \qquad [4.1\text{-}24]$$

Now the left-hand side of equation [4.1-24] has a stable expansion in non-negative powers of z, whereas the right-hand side of this equation has a stable expansion in non-positive powers of z. However, the only way this situation can occur is if each side of equation [4.1-24] is a constant. Thus:

$$\begin{aligned} \sigma^2 A(z)\, W(z) &= \text{constant} = C_0 \\ H(z)/W(z^{-1}) &= \text{constant} = C_0 \end{aligned} \qquad [4.1\text{-}25]$$

But the constant terms in the expansions $A(z)$ and $W(z)$ were respectively defined as $a_0 \equiv 1$ and $w_0 \equiv 1$. Hence, from the first equation in [4.1-25] we see that:

$$C_0 = \sigma^2 a_0 w_0 = \sigma^2 \qquad [4.1\text{-}26]$$

Also, the constant term in the expansion $H(z)$ was defined as the prediction-error variance Υ_0. But if the constant term in $W(z^{-1})$ is $w_0 \equiv 1$, then the constant term in $1/W(z^{-1})$ is also one. Thus, from the second equation of [4.1-25] we see that:

$$C_0 = \Upsilon_0 \cdot 1 \qquad [4.1\text{-}27]$$

Comparison of equations [4.1-26] and [4.1-27] indicates that:

$$\sigma^2 = \Upsilon_0 = \text{prediction-error variance} \qquad [4.1\text{-}28]$$

From the first equation in [4.1-25] we obtain:

$$\sigma^2 A(z)\, W(z) = \sigma^2 \qquad [4.1\text{-}29]$$

so that the solution of the normal equations is:

$$A(z) = 1/W(z) \qquad [4.1\text{-}30]$$

Thus, the required prediction-error operator is the inverse of the minimum-delay filter $W(z)$ which appears in the factorization of the power spectrum of the observed time series. Since the inverse of a minimum-delay function is also minimum-delay, it follows that $1/W(z)$ or $A(z)$ is minimum-delay, i.e. *the prediction-error operator $A(z)$ is minimum-delay.*

From the second equation in [4.1-25] we get:

$$H(z)/W(z^{-1}) = C_0 = \Upsilon_0 \qquad [4.1\text{-}31]$$

or:

$$H(z) = \Upsilon_0 W(z^{-1}) \qquad [4.1\text{-}32]$$

Hence, we see that $H(z)$ is determined by the minimum-advance factor of the spectral factorization method.

Part 2: Linear prediction based on the finite past; autocorrelation function known

Let us now consider a linear one-step prediction of x_k based on the past L values $\{x_{k-L}, \ldots x_{k-2}, x_{k-1}\}$. Knowledge of the autocorrelation function R_m for $|m| \leqslant L$ is available and the process $\{x_k\}$ is again assumed to be a zero mean stationary process. Now instead of equation [4.1-4], the predicted value of x_k is given by:

$$\hat{x}_k = -\sum_{n=1}^{L} a_n x_{k-n} \qquad [4.1\text{-}33]$$

The prediction error is:

$$e_k = x_k - \hat{x}_k = \sum_{n=0}^{L} a_n x_{k-n} \quad (\text{where } a_0 \equiv 1) \qquad [4.1\text{-}34]$$

with mean-square value (or variance):

$$\Upsilon_0 = E\{e_k^2\} = \sum_{n=0}^{L} \sum_{m=0}^{L} a_n a_m R_{m-n} \qquad [4.1\text{-}35]$$

For a minimum of equation [4.1-35], one computes:

$$\frac{\partial \Upsilon_0}{\partial a_l} = 2 \sum_{n=0}^{L} a_n R_{l-n}, \quad 1 \leqslant l \leqslant L \ (\text{where } a_0 \equiv 1) \qquad [4.1\text{-}36]$$

and sets the result in equation [4.1-36] equal to zero, obtaining the values of the prediction-filter coefficients $\{-a_1, -a_2, \ldots -a_L\}$ as the solution of the set of equations:

$$-\sum_{n=1}^{L} a_n R_{l-n} = R_l, \quad 1 \leqslant l \leqslant L \qquad [4.1\text{-}37]$$

Expansion of equation [4.1-37] gives:

$$
\begin{aligned}
-a_1R_0 - a_2R_{-1} - \ldots - a_LR_{1-L} &= R_1 \\
-a_1R_1 - a_2R_0 - \ldots - a_LR_{2-L} &= R_2 \\
&\cdot \\
&\cdot \\
-a_1R_{L-1} - a_2R_{L-2} - \ldots - a_LR_0 &= R_L
\end{aligned}
\qquad [4.1\text{-}38]
$$

The set of equations in [4.1-38] are known, in least-squares terminology, as the normal equations. In particular, the above method of generating equation [4.1-38] is referred to as the autocorrelation method. Equation [4.1-38] constitutes a set of L equations with L unknowns and can be solved for $\{-a_1, -a_2, \ldots -a_L\}$ by several direct methods such as the Gaussian elimination method or Crout reduction method. These general methods require the order of $L^3/3$ operations (multiplications and divisions) and L^2 storage locations. Further reduction in storage and computation time is possible in solving equation [4.1-38] because of its special form. Equation [4.1-38] can be expanded in matrix form as $\mathbf{R}a = g$ where:

$$
\underset{L \times L}{\begin{bmatrix} R_0 & R_{-1} & \ldots & R_{1-L} \\ R_1 & R_0 & \ldots & R_{2-L} \\ \vdots & \vdots & & \vdots \\ R_{L-1} & R_{L-2} & \ldots & R_0 \end{bmatrix}}
\underset{L \times 1}{\begin{bmatrix} -a_1 \\ -a_2 \\ \vdots \\ -a_L \end{bmatrix}}
=
\underset{L \times 1}{\begin{bmatrix} R_1 \\ R_2 \\ \vdots \\ R_L \end{bmatrix}}
= g \qquad [4.1\text{-}39]
$$

and $\mathbf{R}$ is an $L \times L$ *Toeplitz* and non-negative definite matrix. The following definitions are in order.

(1) *Toeplitz matrix.* The elements of the matrix ($\mathbf{R}$) along any northwest-southeast diagonal are identical.

(2) *Non-negative definite matrix.* The leading principal minors of the matrix ($\mathbf{R}$) are defined as Δ_i, $i = 1, 2, \ldots L$ where:

$$
\Delta_1 = R_0, \quad \Delta_2 = \begin{vmatrix} R_0 & R_{-1} \\ R_1 & R_0 \end{vmatrix}, \quad \Delta_3 = \begin{vmatrix} R_0 & R_{-1} & R_{-2} \\ R_1 & R_0 & R_{-1} \\ R_2 & R_1 & R_0 \end{vmatrix}, \quad \ldots \Delta_L = \det[\mathbf{R}]
$$

Note that $\det[\mathbf{R}]$ denotes the determinant of the matrix $\mathbf{R}$. A necessary and sufficient set of conditions that $\mathbf{R}$ be non-negative definite is:

$$\Delta_1 > 0, \quad \Delta_2 > 0, \quad \Delta_3 > 0, \quad \ldots \Delta_L > 0$$

or, in other words, all the principal minors of $\mathbf{R}$ are positive.

A set of equations such as equation [4.1-39] can be solved by the classical Toeplitz recursion occurring in the theory of polynomials orthogonal on the unit circle. Levinson (1947) and Durbin (1960) gave an algorithm for this recursion, whereas Wiggins and Robinson (1965) improved this recursive method by showing that the dot product used to compute the variance at each step in the Levinson-Durbin algorithm can be replaced by a recursive equation which increases both speed and accuracy. The general Toeplitz method assumes the $L \times 1$ column vector g does not necessarily contain the same elements found in the autocorrelation matrix **R**. However, because in equation [4.1-39] the column vector g comprises the same elements as **R**, the auxiliary Toeplitz method can be used. The auxiliary Toeplitz method requires $2L$ storage locations and the order of L^2 operations (multiplications and divisions) and is essentially twice as fast as the general Toeplitz method.

Now the prediction-error sequence $\{e_k\}$, also referred to as the residuals, has the autocorrelation Υ_j given by:

$$\Upsilon_j = E\{e_k e_{k-j}\} = \sum_{n=0}^{L} \sum_{m=0}^{L} a_n a_m R_{m+j-n} \qquad [4.1\text{-}40]$$

which is:

$$\Upsilon_j = \sum_{m=0}^{L} a_m \sum_{n=0}^{L} a_n R_{m+j-n} \qquad [4.1\text{-}41]$$

Now the inner sum in equation [4.1-41] is zero only for $1 \leqslant m + j \leqslant L$, as dictated by equation [4.1-37]. Therefore, unlike the case of linear prediction based on the infinite past, the autocorrelation $\Upsilon_j \neq 0$ for $j \geqslant 1$, and as a result, the prediction-error sequence does not necessarily possess uncorrelated elements e_k.

Let us now investigate the implications of the assumption that $\Upsilon_j = 0$ for $j \geqslant 1$. The assumption $\Upsilon_1 = 0$ requires that R_{L+1} satisfies:

$$R_{L+1} = -\sum_{n=1}^{L} a_n R_{L+1-n} \qquad [4.1\text{-}42]$$

The quantity R_{L+1} in equation [4.1-42] can be regarded as an approximation (prediction) of the autocorrelation at lag $L + 1$ based on the known autocorrelation values $\{R_1, R_2, \ldots R_L\}$. Therefore, assuming $\Upsilon_1 = 0$ is equivalent to assuming that the unknown (true) autocorrelation value R_{L+1} can be exactly predicted from the. past known values $\{R_1, R_2, \ldots R_L\}$. In addition to assuming $\Upsilon_1 = 0$, let us assume that $\Upsilon_2 = 0$. These assumptions require that R_{L+2} satisfies:

$$R_{L+2} = -\sum_{n=1}^{L} a_n R_{L+2-n} \qquad [4.1\text{-}43]$$

Proceeding in the same fashion, the assumption that $\Upsilon_1 = 0$, $\Upsilon_2 = 0$, $\Upsilon_3 = 0$, . . . requires:

$$R_{L+k} = -\sum_{n=1}^{L} a_n R_{L+k-n} \quad \text{for } k = 1, 2, 3, \ldots \qquad [4.1\text{-}44]$$

Equation [4.1-44] can be viewed as a recursion relation (starting with known values $R_1, R_2, \ldots R_L$) and can be considered to be an extrapolation of the known autocorrelation values into regions $R_{L+1}, R_{L+2}, R_{L+3}, \ldots$ where they are unknown.

Similar to the case of linear prediction based on the infinite past, the average power of $\{e_k\}$ is given by the variance:

$$\Upsilon_0 = E\{e_k^2\} = \sum_{n=0}^{L} a_n R_n = R_0 + \sum_{n=1}^{L} a_n R_n \qquad [4.1\text{-}45]$$

which follows from equation [4.1-41] with $j = 0$.

Hence, for the case of linear prediction based on the finite past (L samples) with known autocorrelation R_m for $|m| \leqslant L$, we see that the prediction-error sequence $\{e_k\}$ is not necessarily an uncorrelated sequence, i.e. does not necessarily have a unit spike autocorrelation function. One can still use the procedure indicated in Fig. 4-1 for the case of finite data records, but based on the preceding discussion, we must bear in mind that the prediction-error sequence $\{e_k\}$ is not precisely an uncorrelated sequence.

Part 3: Linear prediction based on the finite past; autocorrelation function unknown

In this situation, the autocorrelation values $\{R_m\}$ are unknown, and the only information available about the random sequence $\{x_k\}$ is a finite set of $L+1$ samples $\{x_{k-L}, \ldots x_{k-2}, x_{k-1}, x_k\}$. As before, we could form the one-step prediction $\hat{x}_k$ where:

$$\hat{x}_k = -\sum_{n=1}^{L} a_n x_{k-n} \qquad [4.1\text{-}46]$$

and the prediction error:

$$e_k = x_k - \hat{x}_k = \sum_{n=0}^{L} a_n x_{k-n} \quad (\text{where } a_0 \equiv 1) \qquad [4.1\text{-}47]$$

but we cannot minimize or utilize any ensemble averages involving the magnitude-squared error e_k^2 as was done in the previous discussions since there is only a finite segment of one member function (realization) of the process $\{x_k\}$ available. One approach to determining the prediction coefficients $\{-a_1, -a_2, \ldots -a_L\}$ is to estimate the autocorrelation values R_m from the available data and use these estimated values ($\hat{R}_m$) in the normal equations [4.1-38] or [4.1-39].

Given a finite set of data defined by the sequence $\{x_1, x_2, \ldots x_L\}$, a biased estimate of the autocorrelation sequence $\{R_m\}$ is given by:

$$\hat{R}_m = \frac{1}{L} \sum_{n=1}^{L-|m|} x_n x_{n-m}, \quad |m| < L \qquad [4.1\text{-}48]$$

where the bias B of the estimate in equation [4.1-48] is defined by:

$$B = \text{BIAS} \equiv R_m - E\{\hat{R}_m\} = \frac{|m|}{L} R_m \qquad [4.1\text{-}49]$$

Although equation [4.1-48] is a biased estimator of R_m, it is asymptotically unbiased, i.e. $B \to 0$ and $L \to \infty$. The variance of $\{\hat{R}_m\}$ in equation [4.1-48] is defined by:

$$\text{var}(\hat{R}_m) \equiv E\{(\hat{R}_m - E\{\hat{R}_m\})^2\} \qquad [4.1\text{-}50]$$

and following Box and Jenkins (1970) can be approximated by:

$$\text{var}(\hat{R}_m) \approx \frac{1}{L} \sum_{n=-\infty}^{\infty} [R_n^2 + R_{n+m} R_{n-m}] \qquad [4.1\text{-}51]$$

It is important to note that as m approaches the record length L, the bias B in equation [4.1-49] becomes appreciably large. Since the bias in this case is as large as the function that is being estimated, equation [4.1-48] does not appear to be a reasonable estimate of $\{R_m\}$ when m is on the order of L. However, the variance in equation [4.1-51] is decreased as the record length L increases and is not severely affected by m. The estimator $\hat{R}_m$ in equation [4.1-48] is consistent, due to the fact that B and $\text{var}(\hat{R}_m)$ both tend toward zero as the number of observations L increases.

An alternative estimator for the autocorrelation sequence $\{R_m\}$ is given by:

$$\hat{R}_m = \frac{1}{L - |m|} \sum_{n=1}^{L-|m|} x_n x_{n-m}, \quad |m| < L \qquad [4.1\text{-}52]$$

In this case, the bias $B = 0$ and the estimate $\hat{R}_m$ in equation [4.1-52] is termed an unbiased estimate of R_m. The variance of equation [4.1-52], as defined in equation [4.1-50], can be approximated (see Box and Jenkins, 1970) by:

$$\text{var}(\hat{R}_m) \approx \frac{1}{[L - |m|]^2} \sum_{n=-\infty}^{\infty} [R_n^2 + R_{n+m} R_{n-m}] \qquad [4.1\text{-}53]$$

Although equation [4.1-52] is an unbiased estimate of $\{R_m\}$, the variance in equation [4.1-53] is severely affected by m. As the value of m approaches L, the variance in equation [4.1-53] becomes undesirably large. Consequently, equation [4.1-52] does not provide a useful estimate for $\{R_m\}$ when m is on the order of L.

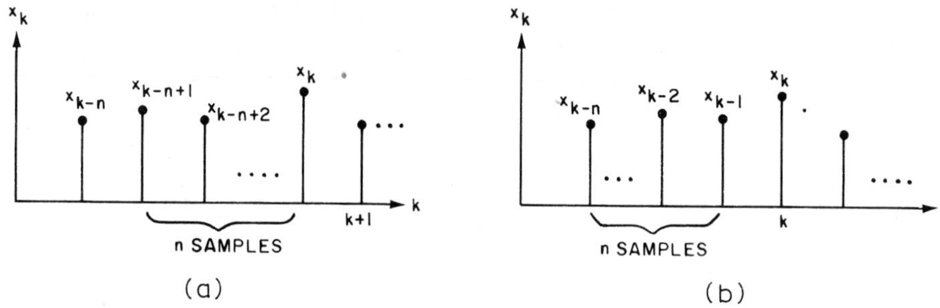

Fig. 4.2. (a) Backward prediction based on n future samples ($\hat{x}_{k-n}^{(b)} = \Sigma_{m=1}^{n} a_m^{(n)} x_{k-n+m}$). (b) Forward prediction based on n past samples ($\hat{x}_k^{(f)} = \Sigma_{m=1}^{n} a_m^{(n)} x_{k-m}$).

Jenkins and Watts (1968) conjecture that in many cases, the mean-squared error for the biased estimator (equation [4.1-48]) is less than that for the unbiased estimator (equation [4.1-52]). This statement, if valid, provides a rationale for using equation [4.1-48] over equation [4.1-52]. However, both estimators are asymptotically unbiased, so one can generally expect to improve the estimate of $\{R_m\}$ by considering larger data records, i.e. increasing the number of samples L.

An alternate approach to linear prediction for the case of finite-data records and unknown autocorrelation is the method of utilizing both forward and backward predictions in a symmetric manner in the Toeplitz recursion discussed in Chapter 5. The key to this symmetric Toeplitz algorithm, given by J.P. Burg (1968), lies in estimating the variance as the arithmetic mean of the forward prediction-error variance and the backward prediction-error variance. In the following paragraphs we give a brief account of this method.

The *forward* prediction of the value x_k based on n "past" values $x_{k-n}, \ldots x_{k-2}, x_{k-1}$ is denoted by $\hat{x}_k^{(f)}$ and defined by:

$$\hat{x}_k^{(f)} \equiv \sum_{m=1}^{n} a_m^{(n)} x_{k-m} \qquad [4.1\text{-}54]$$

The *backward* prediction* of the value x_{k-n} based on n "future" values $x_{k-n+1}, x_{k-n+2}, \ldots x_k$ is denoted by $\hat{x}_{k-n}^{(b)}$ and defined by:

$$\hat{x}_{k-n}^{(b)} \equiv \sum_{m=1}^{n} a_m^{(n)} x_{k-n+m} \qquad [4.1\text{-}55]$$

Fig. 4-2 depicts the forward and backward prediction process.

Now the nth-order forward residual $f_k^{(n)}$ and nth-order backward residual $b_k^{(n)}$ for $n \geqslant 1$ are defined according to:

---

* The term "hindsight" as well as the term "retrospection" are sometimes used instead of the term "backward prediction".

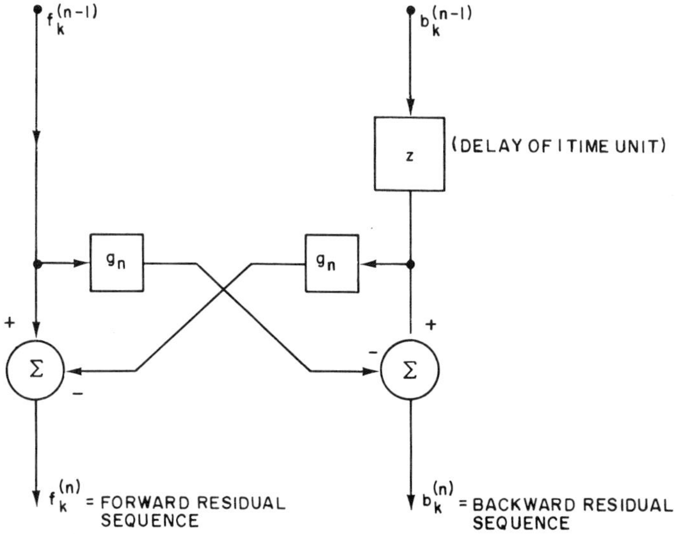

Fig. 4-3. Schematic diagram of the generation of the forward residual sequence and backward residual sequence.

$$\left.\begin{aligned} f_k^{(n)} &\equiv f_k^{(n-1)} - g_n b_{k-1}^{(n-1)} \\ b_k^{(n)} &\equiv b_{k-1}^{(n-1)} - g_n f_k^{(n-1)} \end{aligned}\right\} \quad \text{for } n+1 \leqslant k \leqslant L \qquad [4.1\text{-}56]$$

with the initialization $f_k^{(0)} \equiv x_k$, $b_k^{(0)} \equiv x_k$ for $1 \leqslant k \leqslant L$, where L is the total number of available data points.

A schematic diagram of equation [4.1-56] is provided in Fig. 4-3. But the forward residual sequence $f_k^{(n)}$ and backward residual sequence $b_k^{(n)}$ can be interpreted as one-step forward and backward prediction errors respectively. That is, if we define:

$$\begin{aligned} f_k^{(n)} &\equiv x_k - \sum_{m=1}^{n} a_m^{(n)} x_{k-m} = \text{forward prediction error} \\ b_k^{(n)} &\equiv x_{k-n} - \sum_{m=1}^{n} a_m^{(n)} x_{k-n+m} = \text{backward prediction error} \end{aligned} \qquad [4.1\text{-}57]$$

we find that Fig. 4-3 results, with g_n replaced by $a_n^{(n)}$. In this context, we can define the forward prediction-error sample variance by:

$$\Upsilon_0^{(f)} \equiv \frac{1}{L-n} \sum_{k=n+1}^{L} [f_k^{(n)}]^2 \qquad [4.1\text{-}58]$$

and the backward prediction-error sample variance by:

$$\Upsilon_0^{(b)} \equiv \frac{1}{L-n} \sum_{k=n+1}^{L} [b_k^{(n)}]^2 \qquad [4.1\text{-}59]$$

Under the assumption of stationarity of the process $\{x_k\}$, the theoretical forward prediction- and backward prediction-error variances are equal. Thus, we may combine the two sample variances by the standard statistical technique of forming their arithmetic mean, as given by:

$$\Upsilon_0^{(n)} \equiv \frac{\Upsilon_0^{(f)} + \Upsilon_0^{(b)}}{2} \qquad [4.1\text{-}60]$$

or:

$$\Upsilon_0^{(n)} = \frac{1}{2(L-n)} \sum_{k=n+1}^{L} \{[f_k^{(n)}]^2 + [b_k^{(n)}]^2\} \qquad [4.1\text{-}61]$$

Substitution of equation [4.1-56] in equation [4.1-61] gives:

$$\Upsilon_0^{(n)} = \frac{1}{2(L-n)} \sum_{k=n+1}^{L} \{[f_k^{(n-1)} - g_n b_{k-1}^{(n-1)}]^2 + [b_{k-1}^{(n-1)} - g_n f_k^{(n-1)}]^2\}, \qquad [4.1\text{-}62]$$

for $n \geqslant 1$

Now the coefficient g_n that minimizes the average sample variance $\Upsilon_0^{(n)}$ at the nth stage is the sample partial autocorrelation coefficient given by:

$$g_n = a_n^{(n)} = \frac{2 \sum_{k=n+1}^{L} f_k^{(n-1)} b_{k-1}^{(n-1)}}{\sum_{k=n+1}^{L} \{[f_k^{(n-1)}]^2 + [b_{k-1}^{(n-1)}]^2\}}, \quad n \geqslant 1 \qquad [4.1\text{-}63]$$

The usual definition of a sample partial autocorrelation is the ratio of the sample covariance to the geometric mean of the sample variances. However, in the case here, the two variances have the same theoretical value, so their arithmetic mean is used instead of their geometric mean. After we compute $a_n^{(n)}$ from equation [4.1-63], we may use the computed $g_n = a_n^{(n)}$ in equation [4.1-56] in order to obtain the residuals $f_k^{(n)}$ and $b_k^{(n)}$ from the previous residuals.

At this point we have only calculated the last linear predictive filter coefficient $a_n^{(n)}$. However, the key feature of the Toeplitz recursion (see Robinson, 1967) is the recognition that if the last nth-order coefficient $a_n^{(n)}$ can be evaluated, the remaining nth-order linear predictive filter coefficients $a_m^{(n)}$, $1 \leqslant m \leqslant n-1$, can be evaluated from $(n-1)$th-order filter coefficients. Thus, given $a_n^{(n)}$, we can find the remaining filter coefficients by the Toeplitz recursion relations:

$$a_m^{(n)} = a_m^{(n-1)} - a_n^{(n)} a_{n-m}^{(n-1)}, \quad m = 1, 2, \ldots n-1 \quad \text{(no terms if } n = 1) \qquad [4.1\text{-}64]$$

Hence, given a finite data record of a zero mean stationary process $\{x_k\}$ with unknown autocorrelation, we can find a linear predictive filter by the Burg method. The procedure is summarized below:

(1) *Initialization:*

$$f_k^{(0)} = x_k, \quad b_k^{(0)} = x_k, \quad 1 \leqslant k \leqslant L$$

where L is the total number of available data points.

(2) Calculation of the sample partial autocorrelation coefficient $g_n = a_n^{(n)}$ (equation [4.1-63]), via the forward and backward sequences (residuals) defined in equation [4.1-56], at the nth stage.

(3) Utilization of the Toeplitz recursion relations (equation [4.1-64]) at the nth stage.

Further, if the lattice structure of Fig. 4-3 is carried to the stage where further values of the coefficient g_n are very close to zero, then the forward and backward prediction errors are approximately uncorrelated.

Let us now relate the results from our layered earth model to the theory of linear prediction. The only information known to us is the observed seismic record $\{x_k\}$, since $\{\epsilon_k\}$ and $\{w_k\}$ are both unknown quantities. (Recall our requirement that the leading term in the sequence $\{w_k\}$ be one, i.e. $w_0 \equiv 1$. However, we note that such an assumption does not influence final seismic results.) Hence, we need some additional information in order to solve the problem. Now the reflection coefficient sequence $\{\epsilon_k\}$ is uncorrelated and can be *treated* as a random sequence with a white (flat) power spectrum. This follows from our discussion on statistical models (section 1.3) where we argued that the earth was unsystematically layered over geologic time and that we should not expect the sequence of reflection coefficients to have any systematic relation. We shall now make use of this physical property which we described as the random reflection coefficient hypothesis. Therefore, let us *treat* the reflection coefficient sequence $\{\epsilon_1, \epsilon_2, \ldots \epsilon_N\} = \{r_1, r_2, \ldots r_N\}$ over the interval in question as a random sequence, although we realize that the reflection coefficients are basically constant but unknown parameters.

Let us now turn to the theory of linear prediction, and in particular, the case of linear prediction based on the finite past with unknown autocorrelation. Regardless of which computational procedure we employ, e.g. solution of the normal equations formed by estimating the autocorrelation of the sequence $\{x_k\}$ via equations [4.1-48] or [4.1-52] or solution by the Burg algorithm, we can expect that the prediction errors $\{e_k\}$ will possess an approximately flat power spectrum, i.e. the sequence $\{e_k\}$ will be approximately uncorrelated. Further, the prediction-error operator $\{1, a_1, a_2, \ldots a_N\}$ will be basically a whitening filter. Now if the reflection coefficient sequence

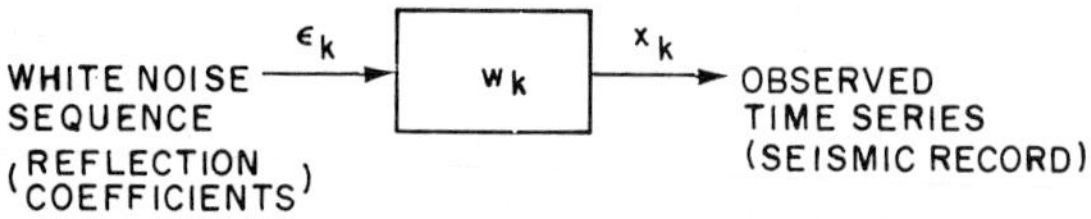

Fig. 4-4. The output $\{x_k\}$ of an N-layered earth modeled as a linear time-invariant system with impulse response $\{w_k\}$ driven by white noise $\{\epsilon_k\}$.

has an approximately white power spectrum, then the only source of non-white frequency components in $\{x_k\}$ is due to the composite wavelet $\{w_k\}$. Hence, the prediction-error operator removes the effect of the composite wavelet $\{w_k\}$ and yields an approximately uncorrelated sequence $\{e_k\}$. This uncorrelated sequence is the required approximation to the reflection coefficient sequence $\{\epsilon_k\}$. Let us elaborate. We can view the observed seismic record $\{x_k\}$ as the output of a linear time-invariant system with impulse response $\{w_k\}$ driven by a zero mean white noise sequence $\{\epsilon_k\}$ (see Fig. 4-4).

The representation in Fig. 4-4 is an appropriate mathematical realization of the seismic process which is useful for our purposes. With reference to Fig. 4-4, the power spectral density $\Phi(\omega)$ of the observed time series $\{x_k\} = w_k * \epsilon_k$ is:

$$\Phi(\omega) = |W(\omega)|^2 \sigma^2 \tag{4.1-65}$$

where $\Phi(\omega)$ = power spectrum of $\{x_k\}$; $|W(\omega)|^2$ = squared magnitude spectrum of the composite wavelet transfer function; and $\sigma^2 = E\{\epsilon_k^2\}$ = constant = flat power spectrum of $\{\epsilon_k\}$. We see that the power spectrum $\Phi(\omega)$ of the observed seismic record $\{x_k\}$ is shaped by the squared magnitude spectrum $|W(\omega)|^2$ of the composite wavelet $\{w_k\}$. But we have shown that the prediction-error operator is effectively the whitening filter $A(\omega) = 1/W(\omega)$. Thus, the output $\{e_k\}$ of the prediction-error operator will have an approximately flat power spectrum. A schematic diagram for the prediction-error operator is shown in Fig. 4-5. With reference to Fig. 4-5, the power spectral density of the output $\{e_k\}$ of the prediction-error operator is:

$$\text{output power spectrum} = |A(\omega)|^2\, \Phi(\omega) \tag{4.1-66}$$

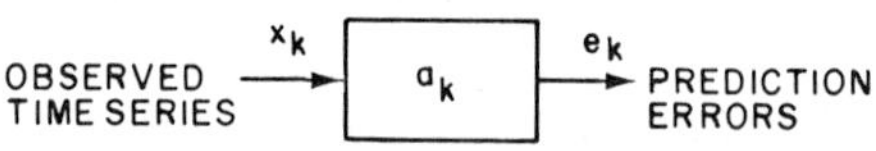

Fig. 4-5. The prediction-error operator $\{a_k\}$ (where $a_0 \equiv 1$) viewed as a linear time-invariant system.

Since $A(\omega) = 1/W(\omega)$, we have:

$$\text{output power spectrum} = \frac{1}{|W(\omega)|^2} \cdot \Phi(\omega) = \frac{1}{|W(\omega)|^2} \cdot |W(\omega)|^2 \sigma^2 = \sigma^2 \tag{4.1-67}$$

which is a constant; the output power spectrum is σ^2 for $-\pi \leqslant \omega \leqslant \pi$. From linear system theory, the autocorrelation R_m of $\{x_k\} = w_k * \epsilon_k$ (the seismic trace) can be expressed as:

$$R_m = \psi_m * \Upsilon_m \tag{4.1-68}$$

where ψ_m is the serial correlation $\psi_m = \sum_{k=0}^{\infty} w_k w_{k-m}$ and $\Upsilon_m = E\{\epsilon_k \epsilon_{k-m}\} = \sigma^2 \delta_m$ due to the random reflection coefficient hypothesis. Hence, equation [4.1-68] is:

$$R_m = \sigma^2 \psi_m \tag{4.1-69}$$

and we see that the shape of the autocorrelation R_m of the trace $\{x_k\}$ is the same as the shape of the serial correlation ψ_m of the composite wavelet $\{w_k\}$.

Equation [4.1-67] indicates that the prediction error $\{e_k\}$ is the desired uncorrelated reflection coefficient sequence $\{\epsilon_k\}$. For layered systems, therefore, we see that the prediction errors $\{e_k\}$ provide estimates of the reflection coefficients $\{\epsilon_k\}$ to within a constant scale factor. Thus, the prediction errors $\{e_k\}$ comprise the required deconvolved seismic record.

We see that the linear prediction method, i.e. the prediction-error operator technique, provides a linear approach to deconvolve the composite wavelet $\{w_k\}$ from our observed seismic record $\{x_k\}$. Due to the mathematical properties of linear prediction, we refer to this method as *predictive deconvolution*, originally formulated by Robinson (1954). To within a scale factor, the prediction errors, i.e. the prediction-error sequence $\{e_k\}$, are approximately the reflection coefficients. For the case when the source pulse $\{s_k\}$ can be approximated by a unit spike δ_k, the prediction-error operator is approximately the sequence of serial correlations $\{1, \phi_1^{(N)}, \phi_2^{(N)}, \ldots \phi_N^{(N)}\}$.

From detailed physical analysis of our layered earth model, we have shown the feasibility of estimating the reflection coefficients $\{\epsilon_k\}$ from a non-minimum delay seismic record $\{x_k\}$ via the theory of linear prediction and under the stated assumptions. It is important to note that our use of statistical averages, expectation operations, and assumptions of stationarity were strictly mathematical, for we showed that the seismic process is not a random process, i.e. it is not controlled by probabilistic laws. However, the mathematical techniques used in the study of linear prediction and stochastic processes offered a means of solving the deconvolution problem. Thus, we have *treated* the reflection coefficients as random variables and the observed time series $\{x_k\}$ as a random sequence, but our physical knowledge of the seismic process tells us that the process is of a deterministic nature and is generated by a physically fixed system made up of the earth's strata. Hence, it is important that we distinguish between the "random" assumptions of the mathematical technique and the deterministic nature of the physical model.

4.2. PREDICTIVE DECONVOLUTION TO ELIMINATE MULTIPLE REFLECTIONS

Deconvolution has proved to be an effective way to increase the resolving power of the seismograph. It is possible to process seismic data successfully without knowledge of many of the underlying geophysical and mathematical concepts. However, if one is interested in extracting pertinent information out of seismic data at a reasonable cost, and if one is concerned about the difference between a satisfactory deconvolution method and one which is better than satisfactory, then basic concepts become important and are worth learning about.

There are two basic approaches to seismic processing, namely the deterministic approach and the statistical approach. The deterministic approach is concerned with the construction of mathematical and physical models of the stratified earth in order better to understand seismic wave propagation. These models have no random elements; they are completely deterministic. The statistical approach to seismic processing is concerned with the building of seismic models which involve random elements. For example, in our statistical model the reflection coefficients are an uncorrelated "white" sequence. A major justification for using the statistical approach in seismology is due to the fact that large amounts of data must be processed. Any data in large enough quantities take on a statistical character, even if each individual piece of data is of a deterministic nature.

The model required for the application of predictive deconvolution is a statistical model. The model depends upon two basic hypotheses, namely, (1) the random reflection coefficient hypothesis that the reflection coefficient sequence associated with the subsurface interfaces can be represented as a random uncorrelated series, and (2) the deterministic hypothesis that the composite wavelet, consisting of the reverberatory and source wavelets, is minimum-delay. There are various ways of checking a model to see if it conforms with the physical situation. With the advent of routine digital field recording and extensive digital computer processing of seismic data in the last 15 years, it has been possible to test this statistical model on a large-scale basis with high-quality data. Because a seismic trace is made up of many overlapping wavelets, it is usually not possible to obtain direct measurements of the random reflection coefficient sequence or the wavelet shape. Hence, the random nature of the reflection coefficients and the minimum-delay nature of the reverberating waveform must be verified indirectly. This indirect verification has been conducted by applying the method of predictive deconvolution to seismic data.

After the seismic survey is completed, oil or gas wells are drilled on the basis of the survey. At that point, the results of deconvolution and other seismic processing methods are directly compared with the geology of the subsurface layers. The method of predictive deconvolution has been successfully used in the past 15 years in deconvolving virtually all seismic field records, both marine and land records. The general success of the method shows that the random reflection coefficient sequence hypothesis and the minimum-delay hypothesis associated with the composite wavelet are valid under a wide range of field situations.

The multiple reflection problem, or simply the multiple problem, in seismic operations may be described as follows. Certain interfaces are strong reflectors. As a result, seismic energy is successively reflected between these interfaces, thereby producing strong multiple reflections that appear on the seismic record. Consequently, reflections from deep horizons below these strong reflecting interfaces are obscured. In such cases, the elimination of these multiple reflections must be accomplished before satisfactory interpretations can be made from the seismic record.

In the previous section, we described the method of predictive deconvolution for the case of prediction distance equal to one, that is, unit-step deconvolution. This type of predictive decomposition is also called spike deconvolution because the deconvolution operator is one which is designed to convert the composite wavelet $\{w_k\}$ into a unit spike δ_k. If $\{q_k\} \equiv \{w_k^{-1}\}$ is the deconvolution operator, then $q_k * w_k = \delta_k$ defines spike deconvolution. Also, in geophysical terminology, the process of spike deconvolution "compresses" the composite wavelet into a spike. As we have seen, for spike deconvolution the observed seismic record $\{x_k\}$ is modeled by:

$$x_k = w_k * \epsilon_k \qquad [4.2\text{-}1]$$

where the reflection coefficient sequence $\{\epsilon_k\}$ is uncorrelated and the composite wavelet $\{w_k\}$ is minimum-delay. The ideal deconvolution operator is $\{q_k\}$ and the deconvolved record is

$$q_k * x_k = \epsilon_k \qquad [4.2\text{-}2]$$

The deconvolution operator $\{q_k\} \equiv \{w_k^{-1}\}$ may be regarded either as (1) a prediction-error operator with prediction distance equal to one time unit, or as (2) a "spike" operator that reduces the composite wavelet $\{w_k\}$ to a unit spike δ_k, that is:

$$w_k^{-1} * w_k = \delta_k = \begin{cases} 1 & \text{if } k = 0 \\ 0 & \text{if } k \neq 0 \end{cases} \qquad [4.2\text{-}3]$$

In this section we treat the prediction-error operator in the case of prediction distance α greater than unity, and show that it is an effective deconvolution operator for the elimination of multiple reflections from the seismogram. Before we concern ourselves with the method of determining this operator from actual field seismograms, it is useful to explore the physical reason for the success of α-step deconvolution in terms of a simple example. The prediction-error operator with various prediction distances is universally used by the petroleum exploration industry worldwide as an effective method of dealing with the multiple reflection problem. One of the first applications of this method was in the elimination of long-period multiple reflections from seismograms taken in the North Sea and in Alaska in the mid- and late-1960's, the final result being the discovery of the great British, Norwegian, and Alaskan oil and gas fields.

Simple example of α-step deconvolution

Let us now give a simple example of predictive deconvolution with prediction distance α in order to illustrate some of the mathematical principles involved. Let us consider the seismic model given by:

$$x_k = w_k * \epsilon_k \qquad [4.2\text{-}4]$$

where $\{\epsilon_k\}$ is the reflection coefficient sequence and $\{w_k\} = b_k * s_k$ is the composite wavelet. In order to keep the mathematics easy, we assume the simplest possible type of reverberation, namely, the reverberations produced by the characteristic polynomial:

$$A(z) = 1 + az^{\alpha} \qquad [4.2\text{-}5]$$

with $\alpha(\alpha > 1)$ denoting the two-way travel time in the layer and a a physical parameter characterizing the structure of the layer. Hence, the Laplace z-transform $B(z)$ of the reverberation wavelet $\{b_k\}$ is:

$$B(z) = \frac{1}{A(z)} = \frac{1}{1 + az^{\alpha}} = 1 - az^{\alpha} + a^2 z^{2\alpha} - a^3 z^{3\alpha} + \ldots \qquad [4.2\text{-}6]$$

so the reverberation wavelet $\{b_k\}$ is:

	↓primary	↓first multiple		↓second multiple		↓third multiple	
b_k = reverberation wavelet amplitude	1, 0, . . .	0, $-a$,	0, . . .	0, a^2,	0, . . .	0, $-a^3$,	0
k = time	0, 1, . . . $\alpha - 1$,	α, $\alpha + 1$, . . .		$2\alpha - 1$, 2α, $2\alpha + 1$, . . .		$3\alpha - 1$,	3α, $3\alpha + 1$

[4.2-7]

The value α is called the period of the multiple reflections, or simply the multiple period. Let us make *another special assumption* in order to keep our model simple; namely, let us assume that the source wavelet $\{s_k\}$ is made up of only two terms. That is:

$$\{s_k\} = \{s_0, s_1\} \qquad [4.2\text{-}8]$$

so its Laplace z-transform $S(z)$ is:

$$S(z) = s_0 + s_1 z \qquad [4.2\text{-}9]$$

Thus, the composite wavelet $\{w_k\} = s_k * b_k$ has the Laplace z-transform $W(z)$ given by:

$$W(z) = S(z)\,B(z) = (s_0 + s_1 z)\,B(z)$$
$$= s_0 + s_1 z - a s_0 z^{\alpha} - a s_1 z^{\alpha+1} + a^2 s_0 z^{2\alpha} + a^2 s_1 z^{2\alpha+1} + \ldots \qquad [4.2\text{-}10]$$

We shall assume $\alpha > 2$, so in the time domain the composite wavelet $\{w_k\}$ is:

	burst	quiet	burst	quiet	burst
w_k = composite wavelet amplitude	$s_0, s_1,$	$0, \ldots \quad 0,$	$-as_0, -as_1,$	$0, \ldots \quad 0,$	$a^2 s_0, \quad a^2 s_1, \ldots$
k = time	$0, \; 1,$	$2, \ldots \alpha - 1,$	$\alpha, \alpha+1,$	$\alpha+2, \ldots 2\alpha-1,$	$2\alpha, 2\alpha+1, \ldots$

[4.2-11]

This composite wavelet may be described as one exhibiting successive bursts of energy with quiet intervals or gaps between the bursts. The seismogram $\{x_k\}$ is the convolution of the reflection coefficient sequence $\{\epsilon_k\}$ with the composite wavelet $\{w_k\}$ or:

$$x_k = \sum_{n=0}^{k} \epsilon_n w_{k-n} = \sum_{n=0}^{k} w_n \epsilon_{k-n}$$
$$= w_0 \epsilon_k + w_1 \epsilon_{k-1} + w_2 \epsilon_{k-2} + \ldots \qquad [4.2\text{-}12]$$

Substitution of equation [4.2-11] in the last expression of equation [4.2-12] gives:

$$x_k = s_0 \epsilon_k + s_1 \epsilon_{k-1} - a s_0 \epsilon_{k-\alpha} - a s_1 \epsilon_{k-\alpha-1} + a^2 s_0 \epsilon_{k-2\alpha} + a^2 s_1 \epsilon_{k-2\alpha-1} - \ldots \qquad [4.2\text{-}13]$$

From equation [4.2-13] we see that there are two kinds of smearing present in the value of $\{x_k\}$, namely, the smearing due to the source wavelet $\{s_k\} = \{s_0, s_1\}$ and the smearing due to the multiple reflections $-a$ at lag α, a^2 at lag 2α, $-a^3$ at lag 3α, etc. The smearing due to the source wavelet acts over a time duration equal to the length of the source wavelet, for example, ϵ_k and ϵ_{k-1} are mixed together, $\epsilon_{k-\alpha}$ and $\epsilon_{k-\alpha-1}$ are mixed together, and so on. Often, this mixing of adjacent reflectors does not seriously affect the quality of the seismic record, that is, at least in the case of short-duration seismic source wavelets the source effect is not too detrimental. However, the smearing due to the multiples (multiple reflections) is much more serious, for it mixes ϵ_k, $\epsilon_{k-\alpha}$, $\epsilon_{k-2\alpha}$ ... together; the result is that the reflecting horizons which are widely separated in space are all combined in the seismogram value x_k at time k. Thus, resolution of these "desired" reflections is a serious problem.

There are two cases of practical interest, namely, the case of short-period multiples (α small) and the case of long-period multiples (α large), with of course all graduations in between. The case of short-period multiples occurs in shallow-water exploration. The period α is the two-way travel time in the water layer, so when the water depth is small the value of α is small. For example, suppose the water is shallow so $\alpha = 2$. Then the composite wavelet

$\{w_k\} = b_k * s_k$ is given by:

$$w_k = \{\overbrace{s_0, s_1}^{\text{burst}}, \overbrace{-as_0, -as_1}^{\text{burst}}, \overbrace{a^2 s_0, a^2 s_1}^{\text{burst}}, \ldots\} \qquad [4.2\text{-}14]$$

which results when equation [4.2-11] is evaluated at $\alpha = 2$. From equation [4.2-14], we see that there are no quiet intervals between the bursts of energy. As a result, such short-period multiples impart to the seismogram a burst-burst-burst-burst-burst effect which gives the appearance of sinusoidal motion. This type of short-period reverberation is called "singing" or "ringing" and makes geophysical interpretation of the raw seismogram difficult or impossible. Spike deconvolution (that is, unit-step predictive deconvolution) as described in section 4.1 provides the means to remove such reverberations from seismic records.

The case of long-period multiples occurs in deep-water exploration, as well as in any prospect in which there are one or more strong reflectors above the deep horizons of interest. It is to this case that we address ourselves, for the elimination of long-period multiples is essential to the discovery of any new deep oil reserves.

Let us now return to our simple example, and show how an α-step prediction-error operator can be determined in order to remove the long-period multiple reflections. For this example, we use a prediction operator made up of only one coefficient λ, so that:

$$\hat{x}_{k+\alpha} = \lambda x_k \qquad [4.2\text{-}15]$$

or equivalently:

$$\hat{x}_k = \lambda x_{k-\alpha} \qquad [4.2\text{-}16]$$

The constant λ is determined by minimizing the prediction-error variance Υ_0 where:

$$\begin{aligned} \Upsilon_0 &= E(x_k - \hat{x}_k)^2 = E(x_k - \lambda x_{k-\alpha})^2 \\ &= E(x_k^2) - 2\lambda E(x_k x_{k-\alpha}) + \lambda^2 E(x_{k-\alpha}^2) \qquad [4.2\text{-}17] \\ &= R_0 - 2\lambda R_\alpha + \lambda^2 R_0 \end{aligned}$$

Setting the derivative of Υ_0 with respect to λ equal to zero we obtain:

$$\partial\Upsilon_0/\partial\lambda = -2R_\alpha + 2\lambda R_0 = 0 \qquad [4.2\text{-}18]$$

which gives the normal equation:

$$\lambda R_0 = R_\alpha \qquad [4.2\text{-}19]$$

Hence, the required value of λ is:

$$\lambda = R_\alpha/R_0 \qquad [4.2\text{-}20]$$

According to our model, the seismic record is $x_k = w_k * \epsilon_k$. Because we assume that the reflection coefficient sequence $\{\epsilon_k\}$ is white noise, the power spectrum of $\{\epsilon_k\}$ is flat with height σ^2 given by:

$$E\{\epsilon_k^2\} = \sigma^2 = \text{positive constant}$$

$$= \text{power spectral density of the white noise} \qquad [4.2\text{-}21]$$

Thus, all the shape in the power spectrum $\Phi(\omega)$ of the seismic record $\{x_k\}$ is due to the composite wavelet $\{w_k\}$, that is:

$$\Phi(\omega) = |W(\omega)|^2\, \sigma^2 \qquad [4.2\text{-}22]$$

where $W(z) = \Sigma_{k=0}^{\infty} w_k z^k$ is the Laplace z-transform of the composite wavelet $\{w_k\}$. For $\{w_k\}$ a real sequence, equation [4.2-22] is equivalent to:

$$\Phi(z) = W(z)\, W(z^{-1})\, \sigma^2 \qquad [4.2\text{-}23]$$

The function $\Phi(z)$ is the Laplace z-transform of the autocorrelation coefficients R_m of the seismic record and is given by:

$$\Phi(z) = \sum_{m=-\infty}^{\infty} R_m z^m \qquad [4.2\text{-}24]$$

Similarly, the function $W(z)\, W(z^{-1})$ is the Laplace z-transform of the serial correlations ψ_m of the composite wavelet. That is:

$$W(z)\, W(z^{-1}) = \left(\sum_{k=0}^{\infty} w_k z^k\right)\left(\sum_{j=0}^{\infty} w_j z^{-j}\right)$$

$$= \sum_{m=-\infty}^{\infty} \psi_m z^m \qquad [4.2\text{-}25]$$

where the ψ_m are the serial correlation coefficients:

$$\psi_m = \sum_{k=0}^{\infty} w_k w_{k-m} = \psi_{-m}$$

Thus, equation [4.2-23] says that R_m and ψ_m are proportional, with proportionality constant σ^2. That is:

$$R_m = \sigma^2 \psi_m \qquad [4.2\text{-}26]$$

Let us now compute ψ_0 and ψ_α in the case of our simple model. The serial correlation ψ_0 is given by the dot product $\{w_k\} \cdot \{w_k\}$ such that:

$$\psi_0 = \{w_k\} \cdot \{w_k\} = \{s_0, s_1, 0, \ldots -as_0, -as_1, 0, \ldots\}$$

$$\cdot \{s_0, s_1, 0, \ldots -as_0, -as_1, 0, \ldots\}$$

$$= s_0^2 + s_1^2 + a^2 s_0^2 + a^2 s_1^2 + a^4 s_0^2 + a^4 s_1^2 + \ldots$$

$$= (s_0^2 + s_1^2)(1 + a^2 + a^4 + \ldots) = \frac{s_0^2 + s_1^2}{1 - a^2} \quad (\text{for } |a| < 1) \qquad [4.2\text{-}27]$$

Similarly, the serial correlation ψ_α is given by the dot product $\{w_k\} \cdot \{w_{k+\alpha}\}$ or $\{w_k\} \cdot \{w_{k-\alpha}\}$ such that:

$$\begin{aligned}
\psi_\alpha &= \{w_k\} \cdot \{w_{k-\alpha}\} = \{s_0, s_1, 0 \ldots -as_0, -as_1, 0, \ldots a^2 s_0, a^2 s_1, 0, \ldots\} \\
&\quad \cdot \{0, 0, 0, \ldots s_0, s_1, 0, \ldots -as_0, -as_1, 0, \ldots\} \\
&= -as_0^2 - as_1^2 - a^3 s_0^2 - a^3 s_1^2 - a^5 s_0^2 - a^5 s_1^2 - \ldots \qquad [4.2\text{-}28] \\
&= -a(s_0^2 + s_1^2)\,[1 + a^2 + a^4 + \ldots] = \frac{-a(s_0^2 + s_1^2)}{1 - a^2} \quad (\text{for } |a| < 1)
\end{aligned}$$

Therefore, the required autocorrelation values R_0 and R_α of the seismic record $\{x_k\}$ are:

$$\begin{aligned}
R_0 &= \sigma^2 \psi_0 = \sigma^2 \left(\frac{s_0^2 + s_1^2}{1 - a^2}\right) \\
R_\alpha &= \sigma^2 \psi_\alpha = -a\sigma^2 \left(\frac{s_0^2 + s_1^2}{1 - a^2}\right) \quad (\text{where } |a| < 1)
\end{aligned} \qquad [4.2\text{-}29]$$

Substitution of equation [4.2-29] in [4.2.20] gives the following result for the prediction operator λ:

$$\lambda = R_\alpha / R_0 = -a \qquad [4.2\text{-}30]$$

In order to gain physical insight, let us apply this prediction operator to the composite wavelet $\{w_k\}$, so as to predict its future value $w_{k+\alpha}$; that is, we write:

$$\hat{w}_{k+\alpha} = \lambda w_k = -a w_k \qquad [4.2\text{-}31]$$

This equation gives:

$$\begin{aligned}
\hat{w}_{k+\alpha} &= -a\{s_0, s_1, 0, \ldots -as_0, -as_1, 0, \ldots a^2 s_0, a^2 s_1, 0, \ldots\} \\
&= \{-as_0, -as_1, 0, \ldots a^2 s_0, a^2 s_1, 0, \ldots -a^3 s_0, -a^3 s_1, 0, \ldots\}
\end{aligned} \qquad [4.2\text{-}32]$$

The prediction error at time $k + \alpha$ is denoted by the symbol $\gamma_{k+\alpha}$ and defined as:

$$\gamma_{k+\alpha} \equiv w_{k+\alpha} - \hat{w}_{k+\alpha} \qquad [4.2\text{-}33]$$

In dealing with prediction-error equations of this sort, we must be careful. The prediction error $\gamma_{k+\alpha}$ and the wavelet value both occur at time $k + \alpha$, but the predicted wavelet value $\hat{w}_{k+\alpha}$ occurs at time k. At time k, of course, the future value $w_{k+\alpha}$ is unavailable, so we must wait until time $k + \alpha$ in order to compute the prediction error $\gamma_{k+\alpha}$. That is, we must delay the

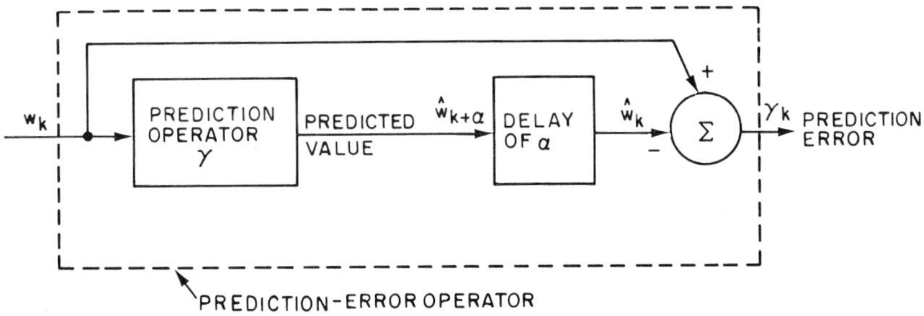

Fig. 4-6. A schematic diagram illustrating the time domain procedure used to compute the prediction error in α-step prediction.

predicted value in order to compute the prediction error. A schematic diagram of this procedure is provided in Fig. 4-6.

Now an extraordinary thing happens, namely, the delayed predicted value $\hat{w}_k$ is precisely the tail-end of the composite wavelet $\{w_k\}$. Thus, when we perform the subtraction $w_k - \hat{w}_k$, the tail-end of the composite wavelet is cancelled by $\hat{w}_k$ and we are left with the front-end of $\{w_k\}$ as the prediction error γ_k. We can see this physical result schematically as follows:

time $= k$	$=$	$0,$	$1, 2, \ldots \alpha - 1,$	$\alpha,$	$\alpha + 1,$	$\alpha + 2, \ldots$	
composite wavelet $= \{w_k\}$	$=$	$\{s_0,$	$s_1, 0, \ldots$	$0,$	$-as_0,$	$-as_1,$	$0, \ldots\}$
predicted value $= \{\hat{w}_{k+\alpha}\}$	$=$	$\{-as_0,$	$-as_1, 0, \ldots$	$0,$	$a^2s_0,$	$a^2s_1,$	$0, \ldots\}$
delayed predicted value $= \{\hat{w}_k\}$	$=$	$\{0,$	$0, 0, \ldots$	$0,$	$-as_0,$	$-as_1,$	$0, \ldots\}$
prediction error $= \{w_k - \hat{w}_k\} = \{\gamma_k\}$	$=$	$\{s_0,$	$s_1, 0, \ldots$	$0,$	$0,$	$0,$	$0, \ldots\}$

Because the front-end of the composite wavelet $\{w_k\}$ is our assumed source wavelet $\{s_k\} = \{s_0, s_1\}$, we obtain the important result that the prediction error $\{\gamma_k\}$ of the composite wavelet $\{w_k\}$ is equal to the source wavelet $\{s_k\}$. That is:

$$\{\gamma_k\} = \{s_k\} = \{s_0, s_1\} \quad [4.2\text{-}34]$$

or:

$$\{\gamma_k\} = \{\gamma_0, \gamma_1, \gamma_2, \gamma_3, \ldots\} = \{s_0, s_1, 0, 0, \ldots\} \quad [4.2\text{-}35]$$

In terms of the Laplace z-transforms, the prediction-error operator can be represented as shown in Fig. 4-7. Here $\Gamma(z)$ is the Laplace z-transform of the prediction error $\{\gamma_k\}$ of the wavelet $\{w_k\}$ where:

$$\Gamma(z) = \sum_{k=0}^{\infty} \gamma_k z^k \quad [4.2\text{-}36]$$

With reference to Fig. 4-7, we obtain:

$$\Gamma(z) = W(z)\,[1 + az^{\alpha}] \quad [4.2\text{-}37]$$

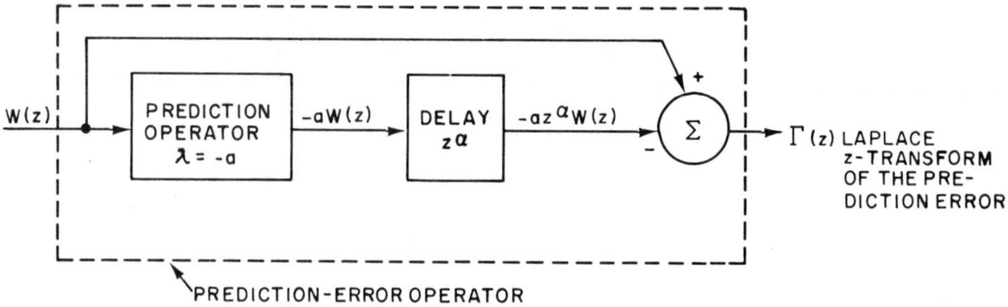

Fig. 4-7. Laplace z-transform representation of the α-step prediction-error operator.

so the Laplace z-transform $A(z)$ of the α-step prediction-error operator $\{a_k\}$ is:

$$A(z) = 1 + az^{\alpha} \quad [4.2\text{-}38]$$

In the time domain, the α-step prediction-error operator is:

$$\begin{aligned} \{a_k\} &= \{a_0, a_1, a_2, \ldots a_{\alpha}, a_{\alpha+1}, \ldots\} \\ &= \{1, 0, 0, \ldots a, \quad 0, \ldots\} \end{aligned} \quad [4.2\text{-}39]$$

Now since $\{w_k\} = s_k * b_k$, we have:

$$W(z) = S(z)\,B(z) \quad [4.2\text{-}40]$$

so that:

$$\begin{aligned} \Gamma(z) &= W(z)\,A(z) \\ &= [S(z)\,B(z)] \cdot A(z) = S(z) \cdot [B(z)\,A(z)] \end{aligned} \quad [4.2\text{-}41]$$

But since $B(z)\,A(z) = 1$, then:

$$\Gamma(z) = S(z) \quad [4.2\text{-}42]$$

Thus, again we have shown that the prediction error occurring in predicting the composite wavelet $\{w_k\}$ is equal to the source wavelet $\{s_k\}$. This represents an important result for the understanding of α-step deconvolution. That is, when the α-step prediction-error operator $\{a_k\}$ is applied to the composite wavelet $\{w_k\}$, we obtain the source wavelet $\{s_k\}$. In symbols:

$$a_k * w_k = \{\gamma_k\} = \{s_k\} \quad [4.2\text{-}43]$$

Returning to the physics of this situation, we see that the Laplace z-transform $A(z) = 1 + az^{\alpha}$ of the α-step prediction-error operator $\{a_k\}$ is equal to the characteristic polynomial of the assumed reverberatory system.

Let us now summarize our results on the simple example of the reverberatory system characterized by the characteristic polynomial $A(z) = 1 + az^{\alpha}$. Given the seismic record $\{x_k\}$, we estimate its autocorrelation values R_0 and R_{α}. We then compute the prediction operator coefficient

$\lambda = R_0/R_\alpha$, which, as we have seen, is equal to $\lambda = -a$. Thus, the prediction-error operator $\{a_k\} = \{a_0, a_1, a_2, \ldots a_\alpha\} = \{1, 0, 0, \ldots a\}$. Let us now apply the prediction-error operator to the seismic record to obtain the deconvolved record $\{y_k\}$, given by:

$$\begin{aligned}\{y_k\} &= a_k * x_k = a_k * w_k * \epsilon_k \\ &= \gamma_k * \epsilon_k = s_k * \epsilon_k\end{aligned} \quad [4.2\text{-}44]$$

That is, the deconvolved seismic record $\{y_k\}$ is equal to the convolution of the source wavelet $\{s_k\}$ with the reflection coefficient sequence $\{\epsilon_k\}$. Thus:

$$\begin{aligned}\{y_k\} &= s_k * \epsilon_k = \{s_0, s_1\} * \epsilon_k \\ &= s_0\epsilon_k + s_1\epsilon_{k-1}\end{aligned} \quad [4.2\text{-}45]$$

Hence, the deconvolved record is made up of the reflection coefficients smeared by the source wavelet only; all the harmful smearing done by the multiple reflections has been eliminated. Since the prediction-error operator converts (or in geophysical terms "compresses") the composite wavelet into the source pulse, it follows that the prediction-error operator replaces each composite wavelet in the seismic record with the source wavelet. Since the source pulse is shorter than the composite wavelet, seismic resolution is increased, i.e. our ability to correctly identify the desired reflections $\{\epsilon_1, \epsilon_2, \ldots \epsilon_N\}$ is improved.

Up to this point, we have assumed that the source wavelet $\{s_k\} = \{s_0, s_1\}$ is minimum-delay. But nowhere in our treatment have we made use of the minimum-delay property of the source wavelet, and in fact there is no need to assume the source wavelet is minimum-delay. Thus, the above result that the α-step prediction-error operator $\{a_k\}$ "compresses" the composite wavelet into the source pulse is true even though the source pulse is non-minimum-delay. That is, *the method of predictive deconvolution removes the long-period multiple reflections and thereby increases seismic resolution even under the hypothesis of a non-minimum-delay source pulse*. We have now come to the end of our simple example of α-step deconvolution of the reverberatory system with the characteristic polynomial $A(z) = 1 + az^\alpha$.

Let us now develop the general mathematical apparatus required for α-step deconvolution.

Part 1: Linear prediction with arbitrary prediction distance α based on the infinite past; autocorrelation known

The problem is to find the best (in the mean-square sense) estimate of the future value $x_{k+\alpha}$ of the zero mean real stationary time series $\{x_k\}$, where α is the prediction distance, when we know the infinite past $\{\ldots x_{k-2}, x_{k-1}, x_k\}$. In other words, the prediction problem is one of finding a causal prediction filter $\{\lambda_0, \lambda_1, \lambda_2, \ldots\}$ which acts on the input $\{x_k\}$ to yield the output $\hat{x}_{k+\alpha}$ such that the prediction-error variance:

$$\Upsilon_0 = E\{(x_{k+\alpha} - \hat{x}_{k+\alpha})^2\} \qquad [4.2\text{-}46]$$

is a minimum. Using the fact that the output $\hat{x}_{k+\alpha}$ of the linear prediction filter is the convolution of the filter (operator) coefficients $\{\lambda_0, \lambda_1, \lambda_2, \ldots\}$ with the input $\{x_k\}$, we obtain:

$$\Upsilon_0 = E\{(x_{k+\alpha} - \sum_{n=0}^{\infty} \lambda_n x_{k-n})^2\} \qquad [4.2\text{-}47]$$

We carry out the minimization of equation [4.2-47] by differentiating Υ_0 with respect to each of the operator coefficients and setting the result equal to zero. Doing so we get:

$$\frac{\partial \Upsilon_0}{\partial \lambda_l} = E\{2(x_{k+\alpha} - \sum_{n=0}^{\infty} \lambda_n x_{k-n})\, x_{k-l}\} = 0 \qquad [4.2\text{-}48]$$

or:

$$\sum_{n=0}^{\infty} \lambda_n E\{x_{k-n} x_{k-l}\} = E\{x_{k+\alpha} x_{k-l}\}, \quad \text{for } l = 0, 1, 2, \ldots \qquad [4.2\text{-}49]$$

If $R_m = E\{x_k x_{k-m}\}$ defines the autocorrelation at lag m of the stationary, real, zero mean time series $\{x_k\}$, then equation [4.2-49] is equivalently:

$$\sum_{n=0}^{\infty} \lambda_n R_{l-n} = R_{l+\alpha}, \quad \text{for } l = 0, 1, 2, \ldots \qquad [4.2\text{-}50]$$

We thus obtain the infinite set (equation [4.2-50]) of simultaneous equations, called the *normal equations*. An important conclusion can be drawn from an inspection of the normal equations. The only way in which the time series $\{x_k\}$ enters these equations is through its autocorrelation function R_m defined above. Because the autocorrelation function and power spectrum of a stationary time series are a Fourier transform pair, it follows that the only information needed to solve the prediction problem for a stationary time series $\{x_k\}$ is its power spectrum $\Phi(\omega)$. However, two different stationary time series with the same power spectrum will lead to the same optimal prediction operator and to the same minimum mean-squared error (or variance for the case of zero mean stationary time series). An entirely non-deterministic stationary time series $\{x_k\}$ with a given power spectrum $\Phi(\omega)$ has a infinite number of causal representations of the form:

$$\{x_k\} = \sum_{n=0}^{\infty} w_n^{(j)} \epsilon_{k-n}^{(j)} \qquad [4.2\text{-}51]$$

where $w_k^{(j)}$ is a stable causal sequence (i.e wavelet) and $\epsilon_k^{(j)}$ is the corresponding white noise sequence. Let $\sigma^2 = E\{(\epsilon_k^{(j)})^2\}$ and $W^{(j)}(z) = \Sigma_{k=0}^{\infty} w_k^{(j)} z^k$. Then the only requirement on the wavelet $w_k^{(j)}$ is that $\Phi(\omega) = |W^{(j)}(\omega)|^2\sigma^2$ in order for the representation in equation [4.2-51] to hold. Thus, in the linear least-squares prediction problem, we can replace any representation of the time series $\{x_k\}$ by any other representation as long as the wavelets in each representation have the same amplitude spectrum, namely:

$$|W^{(j)}(\omega)| = \frac{1}{\sigma}[\Phi(\omega)]^{\frac{1}{2}}$$

Such a change will not alter the optimum prediction filter or the minimum mean-square error in any way. This is because the solution of the prediction problem depends only upon the autocorrelation function (or equivalently, the power spectrum) of the observed stationary time series. For the representation in equation [4.2-51], the autocorrelation R_m of the time series $\{x_k\}$ is given by:

$$R_m = \sigma^2 \psi_m \qquad [4.2\text{-}52]$$

where $\sigma^2 = E\{(\epsilon_k^{(j)})^2\}$ and the serial correlations $\psi_m = \Sigma_{k=0}^{\infty} w_k^{(j)} w_{k-m}^{(j)}$.

Of all the possible representations of a stationary time series, there is one which is distinguished in that it possesses various extremal properties and hence it can serve as a common reference point for all other representations. This particular representation is the minimum-delay one; that is, it is the representation in which the wavelet $\{w_k^{(j)}\}$ is minimum-delay. Let us designate this minimum-delay wavelet by $\{w_k^{(0)}\}$ and the corresponding white noise sequence by $\{\epsilon_k^{(0)}\}$, so that the minimum-delay representation is designated by:

$$\{x_k\} = \sum_{n=0}^{\infty} w_n^{(0)} \epsilon_{k-n}^{(0)} \qquad [4.2\text{-}53]$$

The interpretation of the representation in equation [4.2-53] in terms of linear filter theory is that the stationary time series $\{x_k\}$ is the output of a filter with memory function $\{w_k^{(0)}\}$ (the minimum-delay wavelet) and input $\{\epsilon_k^{(0)}\}$ (the white noise process with variance σ^2) (see Fig. 4-8). A minimum-delay wavelet has a causal inverse. Hence, the inverse of $\{w_k^{(0)}\}$ is realizable in the sense that only the present and past values (and no future values) of the input are required to yield the present value of the output. What is actually available in the prediction problem is the time series $\{x_k\}$ up to the present time k. However, this knowledge is equivalent to the knowledge of the white noise sequence $\{\epsilon_k^{(0)}\}$ up to the present time k, since the filter $\{w_k^{(0)}\}$ has a causal inverse $\{q_k\} \equiv \{w_k^{(0)}\}^{-1} = \{q_0, q_1, q_2, \ldots\}$ and we can pass the available time series $\{x_k\}$ through the inverse filter $\{q_k\}$ to obtain the white noise $\{\epsilon_k^{(0)}\}$ as depicted in Fig. 4-9. In other words, we deconvolve the minimum-delay representation of the stationary time series $\{x_k\}$ by the

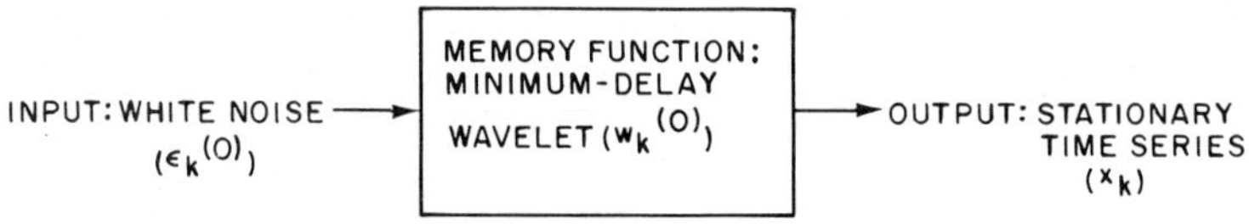

Fig. 4-8. Minimum-delay representation of a stationary time series.

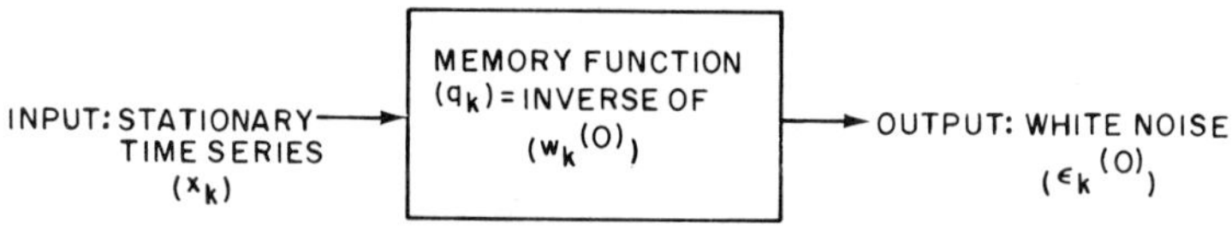

Fig. 4-9. Deconvolution of the minimum-delay representation of a stationary time series.

causal minimum-delay inverse operator $\{q_0, q_1, q_2, \ldots\}$ in order to obtain the white noise sequence $\{\epsilon_k^{(0)}\}$.

The prediction problem therefore is equivalent to asking what operator should be applied to the white noise $\{\epsilon_k^{(0)}\}$ in order to approximate the future value $x_{k+\alpha}$ of the time series $\{x_k\}$ in the least-squares sense. This question may be answered as follows. White noise $\{\epsilon_k^{(0)}\}$ (for $-\infty \leqslant k \leqslant \infty$) represents a series of spikes (i.e. short impulses) of random amplitudes. The random amplitudes of these spikes are uncorrelated with each other. The minimum-delay representation:

$$\{x_k\} = \sum_{n=0}^{\infty} \epsilon_n^{(0)} w_{k-n}^{(0)} \qquad [4.2\text{-}54]$$

says that each of the impulses $\epsilon_n^{(0)}$ entering the filter produces an output corresponding to the impulse response function $w_{k-n}^{(0)}$ of the filter, and that the time series $\{x_k\}$ is the sum of these elementary responses. By deconvolution of the time series up to the present time k, we are able to obtain the white noise impulses $\epsilon_n^{(0)}$ up to the present time k (i.e. $\epsilon_n^{(0)}$ for $-\infty \leqslant n \leqslant k$). But we know nothing about the impulses after the present time k, for these have not yet occurred.

Let us now write the minimum-delay representation of the time series $\{x_k\}$ at the future time instant $k + \alpha$. It is:

$$x_{k+\alpha} = \sum_{n=-\infty}^{k+\alpha} \epsilon_n^{(0)} w_{k+\alpha-n}^{(0)} \qquad [4.2\text{-}55]$$

This expression may be separated into two parts, namely:

$$x_{k+\alpha} = \sum_{n=-\infty}^{k} \epsilon_n^{(0)} w_{k+\alpha-n}^{(0)} + \sum_{n=k+1}^{k+\alpha} \epsilon_n^{(0)} w_{k+\alpha-n}^{(0)} \qquad [4.2\text{-}56]$$

The first part is made up of the tails of the elementary responses due to impulses that have already occurred (i.e. $\epsilon_n^{(0)}$ for $-\infty \leqslant n \leqslant k$) while the second part is due to impulses which will occur in the time interval from the present time k to the future time $k + \alpha$ (i.e. $\epsilon_n^{(0)}$ for $k \leqslant n \leqslant k + \alpha$). The first part is entirely predictable; however, the second part is completely unpredictable, for it is uncorrelated with our available information at the present time k.

Now the first part of equation [4.2-56] can be obtained by passing $\{\epsilon_k^{(0)}\}$ into a filter whose impulse response is the tail of the wavelet $\{w_k^{(0)}\}$ moved ahead α time units; that is, the memory function of this new filter should be the wavelet $\{w_\alpha^{(0)}, w_{\alpha+1}^{(0)}, w_{\alpha+2}^{(0)}, \ldots\}$, where $w_\alpha^{(0)}$ is the amplitude value at $k = 0$, $w_{\alpha+1}^{(0)}$ is the amplitude value at $k = 1$, and so on. That is, the memory function of the new filter is the minimum-delay wavelet shifted left by α time units and with the first α coefficients chopped off. This new filter responds to an impulse as the filter $\{w_0^{(0)}, w_1^{(0)}, w_2^{(0)}, \ldots\}$ will respond in α time units. Hence, if the white noise sequence $\epsilon_k^{(0)}$ is used as an input to the new filter, the output at the present time k will be:

$$\{x_k\} = \sum_{n=-\infty}^{k} \epsilon_n^{(0)} w_{k+\alpha-n}^{(0)} \qquad [4.2\text{-}57]$$

But this is the predictable part of $x_{k+\alpha}$. In short, if $\epsilon_k^{(0)}$ is used as an input to the new filter $\{w_\alpha^{(0)}, w_{\alpha+1}^{(0)}, w_{\alpha+2}^{(0)}, \ldots\}$, the output now will be the predictable part of the future response α time units from now of the filter $\{w_0^{(0)}, w_1^{(0)}, w_2^{(0)}, \ldots\}$ to the same input $\{\epsilon_k^{(0)}\}$.

The second, or unpredictable, part of equation [4.2-56], corresponding to impulses yet to occur, cannot be constructed at the present time instant k. However, the mean value of this part is known to be zero, for future impulses are as likely to be of one sign as the other. Hence, the mean value of the future response is just the first part of the future response, namely, the predictable part given by the output of the filter $\{w_\alpha^{(0)}, w_{\alpha+1}^{(0)}, w_{\alpha+2}^{(0)}, \ldots\}$. But it is well known that the mean value of a random variable is the point about which the mean-squared error is the least. Thus, the output of the filter $\{w_\alpha^{(0)}, w_{\alpha+1}^{(0)}, w_{\alpha+2}^{(0)}, \ldots\}$ with the white noise sequence $\{\epsilon_k^{(0)}\}$ as the input is the required prediction of $x_{k+\alpha}$.

The input to the filter $\{w_\alpha^{(0)}, w_{\alpha+1}^{(0)}, w_{\alpha+2}^{(0)}, \ldots\}$ is the white noise sequence $\{\epsilon_k^{(0)}\}$. Actually, our given data is the time series $\{x_k\}$. Thus, the best operation on the given data is:

$$\{w_\alpha^{(0)}, w_{\alpha+1}^{(0)}, w_{\alpha+2}^{(0)}, \ldots\} * \{q_0, q_1, q_2, \ldots\} * x_k \qquad [4.2\text{-}58]$$

where the minimum-delay inverse operator $\{q_0, q_1, q_2, \ldots\}$ reduces the time series $\{x_k\}$ to white noise $\{\epsilon_k^{(0)}\}$, and the operation $\{w_\alpha^{(0)}, w_{\alpha+1}^{(0)}, w_{\alpha+2}^{(0)}, \ldots\}$ performs the optimum prediction based upon the white noise sequence $\{\epsilon_k^{(0)}\}$. In other words, the optimum prediction operator $\{\lambda_0, \lambda_1, \lambda_2, \ldots\}$ for the prediction of a stationary time series $\{x_k\}$ α time units into the future is given by the convolution:

$$\{\lambda_k\} = \sum_{n=0}^{\infty} q_n w_{k+\alpha-n}^{(0)} \qquad [4.2\text{-}59]$$

where $\{w_k^{(0)}\}$ is the minimum-delay operator in the minimum-delay representation of the time series $\{x_k\}$ and $\{q_k\}$ is the inverse of $\{w_k^{(0)}\}$. The minimum-delay operator $\{w_k^{(0)}\}$ can be found by factoring the power spectrum $\Phi(\omega)$ of the time series $\{x_k\}$ (see Chapter 3).

In terms of Laplace z-transforms, equation [4.2-59] has the representation:

$$\Lambda(z) = Q(z) \cdot \sum_{k=\alpha}^{\infty} w_k^{(0)} z^{k-\alpha} \qquad [4.2\text{-}60]$$

where $W^{(0)}(z) = \sum_{k=0}^{\infty} w_k^{(0)} z^k = \dfrac{1}{Q(z)}$

$$Q(z) = \sum_{k=0}^{\infty} q_k z^k = \frac{1}{W^{(0)}(z)}$$

$$\Lambda(z) = \sum_{k=0}^{\infty} \lambda_k z^k$$

The α-step prediction-error operator is shown in Fig. 4-10. The α-step prediction-error operator (see Fig. 4-10) therefore has the Laplace z-transform $A(z)$ given by:

$$A(z) = 1 - z^{\alpha}\Lambda(z) \qquad [4.2\text{-}61]$$

$$= \frac{W^{(0)}(z) - \left(\sum_{k=\alpha}^{\infty} w_k^{(0)} z^{k-\alpha}\right) z^{\alpha}}{W^{(0)}(z)}$$

or:

$$A(z) = \frac{\sum_{k=0}^{\alpha-1} w_k^{(0)} z^k}{W^{(0)}(z)} = \sum_{k=0}^{\infty} a_k z^k \qquad [4.2\text{-}62]$$

In the time domain, the α-step prediction-error operator is:

$$\{a_k\} = w_k^{(0)}(\alpha) * q_k \qquad [4.2\text{-}63]$$

where $w_k^{(0)}(\alpha) = \{w_0^{(0)}, w_1^{(0)}, \ldots w_{\alpha-1}^{(0)}, 0, 0, \ldots\}$ = α-length front-end of $\{w_k^{(0)}\}$.

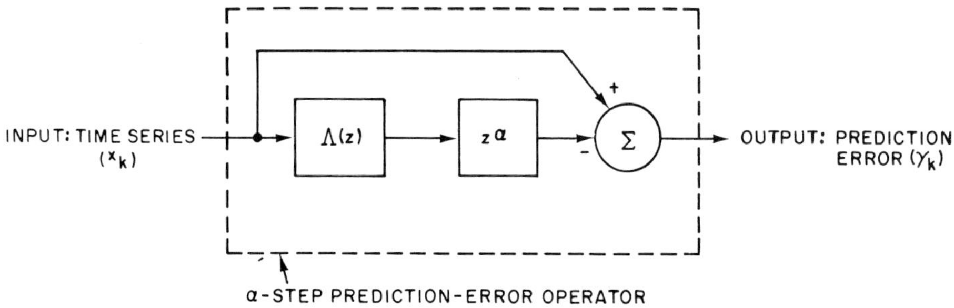

Fig. 4-10. Laplace z-transform representation of the α-step prediction-error operator.

The theory of prediction for a stationary time series $\{x_k\}$ is actually isomorphic to the theory of prediction for a wavelet $\{w_0, w_1, w_2, \ldots\}$, whose autocorrelation ψ_m is proportional to the autocorrelation R_m of the time series $\{x_k\}$, that is, $R_m = \sigma^2 \psi_m$. Here we wish briefly to discuss the salient features involved in predicting the future values of a wavelet from its past values by means of a causal digital filter.

Suppose that the wavelet $\{w_0, w_1, w_2, \ldots\}$ is the input to a causal digital filter $\{\lambda_0, \lambda_1, \lambda_2, \ldots\}$. At the output of the filter we desire the predicted values of the input wavelet. If we denote the prediction distance by the integer α, then the desired output wavelet is simply a replica of the input wavelet advanced in time by α time units. The time correspondence between the input $\{w_k\} = \{w_0, w_1, w_2, \ldots\}$, desired output $\{w_{k+\alpha}\}$, actual output $\{c_k\}$, and error $\{w_{k+\alpha}\} - \{c_k\}$ are shown in Table 4-1. From this table we see that the energy of the error signal for prediction distance α is given by:

$$I_0 = w_0^2 + w_1^2 + \ldots + w_{\alpha-1}^2 + (w_\alpha - c_0)^2 + (w_{\alpha+1} - c_1)^2 + (w_{\alpha+2} - c_2)^2 + \ldots \qquad [4.2\text{-}64]$$

That is, there is a contribution $w_0^2 + w_1^2 + \ldots + w_{\alpha-1}^2$ due to the fact that the filter is causal (and hence it can produce no output before the input, which starts at time zero) plus a contribution $(w_\alpha - c_0)^2 + (w_{\alpha+1} - c_1)^2 + \ldots$ which we will now discuss.

Essentially, a causal time-invariant linear filter can only do one thing, namely, it can form the sum of delayed replicas of the input weighted by the filter coefficients. That is, the actual output:

$$\{c_k\} = \sum_{n=0}^{\infty} \lambda_n w_{k-n} \quad (\text{where } k = 0, 1, 2, 3, \ldots) \qquad [4.2\text{-}65]$$

is the sum of the delayed replicas w_{k-n} (where $n = 0, 1, 2, \ldots$ represents the delays) weighted by the coefficients λ_n. In other words, the actual output $\{c_k\}$ is a weighted sum of the delayed replicas w_{k-n} for $n = 0, 1, 2, \ldots$

TABLE 4-1

Time correspondence between the input $\{w_k\} = \{w_0, w_1, w_2, \ldots\}$, desired output $\{w_{k+\alpha}\}$, actual output $\{c_k\}$, and error $\{w_{k+\alpha}\} - \{c_k\}$

Input	0	0	0	0	...	0	w_0	w_1	w_2	...
Desired output	0	0	w_0	w_1	...	$w_{\alpha-1}$	w_α	$w_{\alpha+1}$	$w_{\alpha+2}$	...
Actual output	0	0	0	0	...	0	c_0	c_1	c_2	...
Error	0	0	w_0	w_1	...	$w_{\alpha-1}$	$w_\alpha - c_0$	$w_{\alpha+1} - c_1$	$w_{\alpha+2} - c_2$	...
Time instant	$-\alpha - 2$	$-\alpha - 1$	$-\alpha$	$-\alpha + 1$	...	-1	0	1	2	...

Now let us appeal to numerical approximation theory. Let us consider the class of all wavelets (i.e. the class of all causal time sequences with finite energy). A set of wavelets is called *complete* if any wavelet whatsoever may be expressed as a linear combination of members of this set.

It can be proved that the set of delayed replicas w_{k-n} for $n = 0, 1, 2, \ldots$ is a complete set if the input wavelet is minimum-delay, whereas this set is not complete if the input wavelet is not minimum-delay.

Let us first consider the case when the input wavelet is minimum-delay, so we will represent it as $\{w_1^{(0)}, w_2^{(0)}, w_3^{(0)}, \ldots\}$ according to our convention of letting the superscript 0 designate "minimum-delay", whereas the subscript designates the time index. The desired output before time zero is unpredictable; the desired output from time zero on is the wavelet $\{w_\alpha^{(0)}, w_{\alpha+1}^{(0)}, w_{\alpha+2}^{(0)}, \ldots\}$. This wavelet can be expressed as a linear combination of the delayed replicas of the minimum-delay input wavelet. In other words, if the input wavelet is minimum-delay, then we can find filter coefficients $\{\lambda_0, \lambda_1, \lambda_2, \ldots\}$ such that the actual output wavelet $\{c_0, c_1, c_2, \ldots\}$ is exactly equal to the wavelet $\{w_\alpha^{(0)}, w_{\alpha+1}^{(0)}, w_{\alpha+2}^{(0)}, \ldots\}$. Therefore, if the input wavelet is minimum-delay, then we can find an operator $\{\lambda_0, \lambda_1, \lambda_2, \ldots\}$ such that the contribution $(w_\alpha^{(0)} - c_0)^2 + (w_{\alpha+1}^{(0)} - c_1)^2 + (w_{\alpha+2}^{(0)} - c_2)^2 + \ldots$ is zero, and hence the prediction-error variance becomes:

$$I_\alpha = [w_0^{(0)}]^2 + [w_1^{(0)}]^2 + \ldots + [w_{\alpha-1}^{(0)}]^2 \quad \text{(minimum-delay case)} \qquad [4.2\text{-}66]$$

This represents the least possible prediction-error variance and hence the operator $\{\lambda_0, \lambda_1, \lambda_2, \ldots\}$ which achieves this minimum is the optimum prediction operator. We recognize this least prediction-error variance as the energy build-up of the minimum-delay input wavelet up to time $\alpha - 1$.

In case the input wavelet is not minimum-delay, then there will be a discrepancy between the actual output wavelet $\{c_0, c_1, c_2, \ldots\}$ and the wavelet $\{w_\alpha, w_{\alpha+1}, w_{\alpha+2}, \ldots\}$. However, the same optimum operator coefficients as in the minimum-delay case yield the minimum value of the prediction-error variance. This minimum value is numerically the same as the minimum value of the prediction-error variance in the minimum-delay case, except that now both contributions to the prediction-error variance are present. The contribution due to the energy build-up is less than in the minimum-delay case (as the energy build-up of a non-minimum-delay wavelet falls below the energy build-up curve of the minimum-delay wavelet) whereas the contribution due to the discrepancy between the wavelets $\{w_\alpha, w_{\alpha+1}, w_{\alpha+2}, \ldots\}$ and $\{c_0, c_1, c_2, \ldots\}$ exactly makes up for the decrease in the first contribution.

Specifically, the optimum operator $\{\lambda_0, \lambda_1, \lambda_2, \ldots\}$ may be found by minimizing the error energy I where:

$$I = \sum_{k=-\alpha}^{\infty} (w_{k+\alpha} - c_k)^2 \qquad [4.2\text{-}67]$$

If we carry out the minimization, we obtain the *normal equations:*

$$\sum_{n=0}^{\infty} \lambda_n \psi_{l-n} = \psi_{l+\alpha}, \quad \text{for } l = 0, 1, 2, \ldots \qquad [4.2\text{-}68]$$

where ψ_l is the serial correlation of the input wavelet. We see that these normal equations are the same as equation [4.2-50] since $R_m = \sigma^2 \psi_m$, as we would expect from the isomorphism.

Part 2: Linear prediction with arbitrary prediction distance α based on the finite past; autocorrelation function known

We suppose $\{x_k\}$ is a zero mean real stationary process, where we know the autocorrelations $R_0, R_1, \ldots R_{n+\alpha}$. Minimization of the prediction-error variance:

$$E\{(x_k - \hat{x}_k)^2\} = E\{(x_k - \sum_{j=0}^{n} \lambda_j x_{k-\alpha-j})^2\} \qquad [4.2\text{-}69]$$

with respect to $\lambda_0, \lambda_1, \ldots \lambda_n$ gives the normal equations:

$$E\{x_k x_{k-\alpha-q}\} - \sum_{j=0}^{n} \lambda_j E\{x_{k-\alpha-j} x_{k-\alpha-q}\} = 0, \quad \text{for } q = 0, 1, 2, \ldots n \qquad [4.2\text{-}70]$$

or:

$$R_{\alpha+q} - \sum_{j=0}^{n} \lambda_j R_{q-j} = 0, \quad \text{for } q = 0, 1, 2, \ldots n \qquad [4.2\text{-}71]$$

In matrix notation, the normal equations are:

$$\begin{bmatrix} R_0 & R_1 & \ldots & R_n \\ R_1 & R_0 & \ldots & R_{n-1} \\ \vdots & & & \vdots \\ R_n & R_{n-1} & \ldots & R_0 \end{bmatrix} \begin{bmatrix} \lambda_0 \\ \lambda_1 \\ \vdots \\ \lambda_n \end{bmatrix} = \begin{bmatrix} R_\alpha \\ R_{\alpha+1} \\ \vdots \\ R_{\alpha+n} \end{bmatrix} \qquad [4.2\text{-}72]$$

where we have used the fact that $R_{-q} = R_q$ for real processes. The solution to these normal equations yields the prediction operator $\{\lambda_0, \lambda_1, \ldots \lambda_n\}$, which in turn gives the prediction-error operator $\{a_k\} = \{1, 0, \ldots 0, -\lambda_0, -\lambda_1, \ldots -\lambda_n\}$. (When the prediction distance $\alpha = 1$, then the prediction-error operator $\{a_k\}$ reduces to the unit-step prediction-error operator (or spiking operator) $\{a_0, a_1, \ldots a_{n+1}\}$ of order $n + 1$, as described in section 4.1).

In computing the prediction-error operator *for the case of finite data and unknown autocorrelation*, we must obtain estimates of the autocorrelation values $R_0, R_1, \ldots R_{n+\alpha}$ as discussed in section 4.1.

In closing this section, let us now look at the deconvolution of long-period multiple reflections from a theoretical point of view. We no longer restrict ourselves to the two-term source pulse and the elementary reverberation wavelet given in the simple example considered at the beginning of this section. Let us consider the model:

$\{s_0, s_1, \ldots\}$ = source pulse (causal and stable but not necessarily minimum-delay)

$\{b_0, b_1, \ldots\}$ = reverberation wavelet (causal and stable and necessarily minimum-delay)

$\{w_k\} = s_k * b_k$ = composite wavelet (causal and stable)

$\{\epsilon_k\}$ = uncorrelated random reflection series (white noise)

$x_k = w_k * \epsilon_k$ = observed seismic record (stationary time series)

The reverberation wavelet is minimum-delay by physical reasoning, as we have discussed in Chapter 2. The source pulse may or may not be minimum-delay.

In our analysis, we want to make use of the canonical representation of an arbitrary wavelet. The canonical representation is given in Robinson (1962) and states that any wavelet (i.e. any causal stable sequence) can be represented as the convolution of a minimum-delay wavelet with a causal all-pass system. Let the canonical representation of the source pulse be:

$$\{s_k\} = p_k * s_k^{(0)} \qquad [4.2\text{-}73]$$

where $\{p_k\}$ is a causal all-pass operator and $\{s_k^{(0)}\}$ is a minimum-delay source wavelet. The composite wavelet $\{w_k\}$ is then:

$$\{w_k\} = s_k * b_k = p_k * s_k^{(0)} * b_k \qquad [4.2\text{-}74]$$

If we define the minimum-delay wavelet $\{w_k^{(0)}\}$ as:

$$\{w_k^{(0)}\} = s_k^{(0)} * b_k \qquad [4.2\text{-}75]$$

then it follows (due to the uniqueness property of the canonical representation) that the canonical representation of the composite wavelet is:

$$\{w_k\} = p_k * w_k^{(0)} \qquad [4.2\text{-}76]$$

Thus, the same all-pass operator occurs both in the canonical representation of the source pulse and in the canonical representation of the composite wavelet, as we would expect. Now we shall define $w_k^{(0)}(\alpha)$ as:

$$\begin{aligned} w_k^{(0)}(\alpha) &= \{w_0^{(0)}, w_1^{(0)}, \ldots w_{\alpha-1}^{(0)}\} \\ &= \alpha\text{-length initial section of } \{w_k^{(0)}\} \end{aligned} \qquad [4.2\text{-}77]$$

That is, $\{w_k^{(0)}(\alpha)\}$ is the front-end section of the minimum-delay wavelet appearing in the canonical representation of the composite wavelet. Also, let $\{q_k\} = [w_k^{(0)}]^{-1}$ denote the causal minimum-delay inverse of the minimum-delay wavelet $\{w_k^{(0)}\}$; that is:

$$q_k * w_k^{(0)} = \delta_k \qquad [4.2\text{-}78]$$

where both $\{q_k\}$ and $\{w_k^{(0)}\}$ are minimum-delay (i.e. causal stable operators with minimum-negative-phase or minimum-phase-lag). Then, as we have derived earlier in this section, the prediction-error operator for prediction distance α is given by:

$$\{a_k\} = w_k^{(0)}(\alpha) * q_k \qquad [4.2\text{-}79]$$

where $\{q_k\}$ is defined in equation [4.2-78]. This prediction-error operator represents the optimum infinitely long causal operator: it reduces to a finite operator in those cases when $\{q_k\}$ is finite. If we apply this prediction-error operator to the seismic trace $\{x_k\}$, we obtain the deconvolved trace $\{y_k\}$ given by:

$$\{y_k\} = a_k * x_k = w_k^{(0)}(\alpha) * q_k * x_k \qquad [4.2\text{-}80]$$

But since $\{x_k\} = w_k * \epsilon_k$ and $\{w_k\} = w_k^{(0)} * p_k$, then:

$$\begin{aligned} \{y_k\} &= w_k^{(0)}(\alpha) * q_k * w_k * \epsilon_k \\ &= w_k^{(0)}(\alpha) * q_k * w_k^{(0)} * p_k * \epsilon_k \end{aligned} \qquad [4.2\text{-}81]$$

With the aid of equation [4.2-78], equation [4.2-81] simplifies to:

$$\{y_k\} = w_k^{(0)}(\alpha) * p_k * \epsilon_k \qquad [4.2\text{-}82]$$

Thus, the deconvolved seismic record $\{y_k\}$ is equal to the convolution of the random reflection series $\{\epsilon_k\}$ with the wavelet $p_k * w_k^{(0)}(\alpha)$. Except in special cases, the wavelet $p_k * w_k^{(0)}(\alpha)$ will be infinitely long even though $w_k^{(0)}(\alpha)$ is of finite length α. Thus, the deconvolved trace is equal to the wanted reflection coefficient sequence $\{\epsilon_k\}$ filtered by the wavelet $p_k * w_k^{(0)}(\alpha)$.

Let us now consider from this new point of view the case of long-period multiple reflections discussed earlier in this section. The basic requirement in order to handle long-period multiples is that the length of the source pulse $\{s_k\}$ is less than or equal to the reverberation period. Let the reverberation period be given by α; more specifically, let us assume that the initial portion of the reverberation wavelet $\{b_k\}$ is made up of an initial unit spike at time $k = 0$ followed by $\alpha - 1$ zero values. That is, after the initial unit spike, the next non-zero value of $\{b_k\}$ cannot occur before time α. We also assume that the source wavelet $\{s_k\}$ is at most of length α; that is, $\{s_k\}$ can have non-zero values only from time $k = 0$ to time $\alpha - 1$. Thus, the case of long-period multiple reflections represents a very nice dovetailing. The composite wavelet $\{w_k\}$ is the convolution of $\{s_k\}$ and $\{b_k\}$. Hence, the initial α-length part of the composite wavelet is equal to the source pulse; that is, $\{w_k(\alpha)\} = \{s_k\}$, where $w_k(\alpha)$ represents the first α values of the composite wavelet $\{w_k\}$. Moreover, since the source wavelet $\{s_k\}$ is of finite length, its minimum-delay counterpart $\{s_k^{(0)}\}$ is also of the same finite length. Since $\{w_k^{(0)}\} = s_k^{(0)} * b_k$, the initial α-length part of the minimum-delay wavelet $\{w_k^{(0)}\}$ is equal to the minimum-delay wavelet $\{s_k^{(0)}\}$; that is, $w_k^{(0)}(\alpha) = s_k^{(0)}$. Thus, in the case of long-period multiple reflections (reverberations), the prediction-error operator is:

$$\{a_k\} = s_k^{(0)} * q_k \qquad [4.2\text{-}83]$$

where $q_k * w_k^{(0)} = \delta_k$. Now $\{a_k\}$ convolved with the seismic trace $\{x_k\}$ gives the deconvolved trace:

$$\begin{aligned}\{y_k\} &= a_k * x_k = (s_k^{(0)} * q_k) * (p_k * w_k^{(0)} * \epsilon_k) \\ &= p_k * s_k^{(0)} * \epsilon_k = s_k * \epsilon_k\end{aligned} \qquad [4.2\text{-}84]$$

Equation [4.2-84] represents the same result as that obtained earlier with the simple example in the first part of this section, namely, when applied to the seismic trace $\{x_k\} = w_k * \epsilon_k$, the prediction-error operator $\{a_k\}$ replaces each seismic wavelet $\{w_k\}$ with the source pulse $\{s_k\}$. In other words, in the case of long-period multiple reflections, the deconvolved trace $\{y_k\}$ represents the wanted reflection series $\{\epsilon_k\}$ filtered by the finite-length source pulse $\{s_k\}$. Thus, the method of predictive deconvolution removes the multiple reflections despite the fact that the source pulse $\{s_k\}$ is not minimum-delay.

If, in fact, the source pulse is minimum-delay, then the all-pass wavelet $\{p_k\}$ is a unit spike δ_k and $\{s_k\} = s_k^{(0)} * p_k = s_k^{(0)} * \delta_k = s_k^{(0)}$. For this case, the deconvolved trace $\{y_k\}$ represents the desired reflection coefficient sequence $\{\epsilon_k\}$ filtered by the minimum-delay finite-length source pulse $\{s_k\} = \{s_k^{(0)}\}$. Since a minimum-delay source pulse has a sharper leading edge than any of its non-minimum-delay counterparts, our ability to resolve primary reflections from two closely spaced interfaces will be enhanced. Thus, greater seismic resolution can be expected in the minimum-delay case. In fact, when the source pulse $\{s_k\}$ is an ideal unit spike, we obtain perfect resolution, i.e. there is a one-to-one relationship between the deconvolved sequence $\{y_1, y_2, \ldots y_k\}$ and the corresponding interfaces 1, 2, 3, . . . k, such that we are assured that the value y_k represents the reflection coefficient from interface k.

We have shown that the method of predictive deconvolution is a valid method to remove long-period multiple reflections, even in the case when the source pulse is not minimum-delay. Of course, greater resolution can be expected in the case of a minimum-delay source pulse. In either case, if the actual shape of the source pulse can be ascertained, then an optimum shaping filter can be obtained to transform the source pulse (whether or not it is minimum-delay) into some desired shape that would increase resolution. Here we can use subroutine SHAPER (Robinson, 1967a, p. 84) which computes the waveshaping filter for the optimum positioning in time of the desired output waveform with respect to the source pulse.

Now let us consider the method of predictive deconvolution in the case of short-period reverberations or multiple reflections. In this case, the source wavelet $\{s_k\}$, again assumed to be of finite length, has length greater than the period of reverberation α. There will only be partial dovetailing between source wavelet $\{s_k\}$ and reverberation wavelet $\{b_k\}$. If we let $\{s_k(\alpha)\}$ represent the α-length initial portion of the source pulse $\{s_k\}$, then the α-length initial portion of the composite wavelet $\{w_k\} = s_k * b_k$ is equal to $\{s_k(\alpha)\}$; that is, $\{w_k(\alpha)\} = \{s_k(\alpha)\}$. Since the minimum-delay counterpart $\{s_k^{(0)}\}$ is also of finite length, and since $\{w_k^{(0)}\} = s_k^{(0)} * b_k$, it follows that $\{w_k^{(0)}(\alpha)\} = \{s_k^{(0)}(\alpha)\}$. Thus, in the case of short-period reverberations, the prediction-error operator $\{a_k\}$ is:

$$\{a_k\} = s_k^{(0)}(\alpha) * q_k \qquad [4.2\text{-}85]$$

where $q_k * w_k^{(0)} = \delta_k$. The deconvolved trace is then:

$$\begin{aligned}\{y_k\} &= a_k * x_k = s_k^{(0)} * q_k * w_k * \epsilon_k \\ &= p_k * s_k^{(0)}(\alpha) * \epsilon_k\end{aligned} \qquad [4.2\text{-}86]$$

where $\{w_k\} = p_k * w_k^{(0)}$ and $\{x_k\} = w_k * \epsilon_k$. That is, the deconvolved trace is equal to the random reflection series $\{\epsilon_k\}$ filtered by the wavelet $p_k * s_k^{(0)}(\alpha)$. Now the wavelet $p_k * s_k^{(0)}(\alpha)$ is equal to the source pulse $\{s_k\}$ up to time $k = \alpha$, but following time α the wavelet $p_k * s_k^{(0)}(\alpha)$ has a distorted tail; that is:

$$p_k * s_k^{(0)}(\alpha) = \{s_k(\alpha)\} + \text{distorted tail}$$

This distorted tail can extend to infinity and thereby adversely affect seismic resolution. In the case of a minimum-delay source wavelet, the distorted tail is zero, and thus the resolution is better. However, in any case, resolution can be increased by waveform shaping, such as by subroutine SHAPER (Robinson, 1967a), provided the shape of the source pulse is known.

In conclusion, the method of predictive deconvolution with a proper choice of α is a valid method for removing either long-period or short-period reverberations in the case of a minimum-delay or non-minimum-delay source pulse. The best seismic resolution is attained for the case of long-period reverberations and a minimum-delay source pulse; the worst seismic resolution for the case of short-period reverberations and a non-minimum-delay source pulse (see Fig. 4-11). In any case, if the shape of the source pulse is known, seismic resolution can be improved by a waveshaping filter.

The ultimate resolution can theoretically be obtained in the case of a minimum-delay source pulse with an infinitely long prediction-error operator with unit prediction distance ($\alpha = 1$). This operator is:

$$\{a_k\} = w_k^{(0)}(1) * q_k \qquad [4.2\text{-}88]$$

where $q_k * w_k^{(0)} = \delta_k$. The quantity $\{w_k^{(0)}(1)\}$ is simply the initial value of the composite wavelet $w_k^{(0)}$. Thus, except for a constant, the prediction-error operator for unit prediction distance is the spike operator $\{q_k\}$. For simplicity, let us assume that this constant is unity. Then the deconvolved trace $\{y_k\}$ is:

$$\{y_k\} = a_k * x_k = w_k^{(0)}(1) * q_k * w_k^{(0)} * \epsilon_k = \{\epsilon_k\} \qquad [4.2\text{-}89]$$

where we have assumed that $w_0^{(0)} = w_k^{(0)}(1) = 1$.

	LONG-PERIOD REVERBERATION	SHORT-PERIOD REVERBERATION
MINIMUM-DELAY SOURCE PULSE	REVERBERATION ELIMINATED. SOURCE PULSE PRESERVED.	REVERBERATION ELIMINATED. INITIAL SECTION OF SOURCE PULSE PRESERVED WITH NO TAIL FOLLOWING.
NONMINIMUM-DELAY SOURCE PULSE	REVERBERATION ELIMINATED. SOURCE PULSE PRESERVED.	REVERBERATION ELIMINATED. INITIAL SECTION OF SOURCE PULSE PRESERVED BUT WITH DISTORTED TAIL FOLLOWING.

Fig. 4-11. Predictive deconvolution with infinite-length operators and prediction distance equal to the reverberation period.

Thus, in this ideal case, the deconvolved trace $\{y_k\}$ is the wanted random reflection series $\{\epsilon_k\}$. However, in practice, spiking filters are often associated with a build-up of high-frequency noise in the deconvolved trace. This condition can be improved by various techniques such as filtering the deconvolved trace by a low-pass filter. As we have seen, the output of prediction-error operators for a prediction distance greater than one ($\alpha > 1$) represents a filtering of the random reflection series $\{\epsilon_k\}$, and thereby has certain favorable signal-to-noise ratio characteristics.

In this section, we have treated α-step predictive deconvolution. Except for a scale factor, spike deconvolution is the same as unit-step predictive deconvolution. The method of α-step predictive deconvolution is shown to be a valid method for eliminating either long-period or short-period multiple reflections (reverberations), whether or not the source pulse is minimum-delay. Computational methods for obtaining the α-step prediction-error operator to deconvolve field records, including how to properly select a value for α, are given in Chapter 5.

4.3. KEPSTRAL DECONVOLUTION

In section 3.3, we introduced new properties of the kepstrum, which bring out its essential features and symmetries. We showed how the basic idea of the kepstrum appeared in the classical work of Poisson (1823) and Schwarz (1872) and discussed the relationship between the kepstrum and the spectral factorization method of Szegö (1915) and Kolmogorov (1939). In this section we shall relate the kepstrum to the deconvolution process, and in particular, we will discuss the physical significance of kepstral deconvolution, i.e. its relationship to our layered earth model.

The spectrum, power spectrum, and other techniques based on Fourier analysis have evolved as valuable signal-processing operations on time series arising in many different fields of research. Robinson (1954) introduced the idea of the kepstrum to geophysical analysis in his discussion of the spectral factorization problem. Bogert et al. (1963) were the first to use the kepstrum as a signal-processing operation. Bogert et al. investigated the problem of trying to determine the depth of a deep seismic source. Knowing the depth of a seismic disturbance is useful for differentiating between man-made and seismic events. In particular, an earthquake or underground nuclear explosion results in the propagation of seismic waves or phases. The arrival of these waves at a seismic receiver can be regarded more or less as distorted "echoes". Moreover, the time interval between the arrival of certain waves (phases) provides depth information useful to the geophysicist. Hence, let us begin our discussion of the kepstrum as a signal-processing technique by reviewing the work of Bogert et al. (1963) on the problem of echo determination.

Let us consider the problem of determining the time delay and strength of a single echo. For this case, the values of a continuous time series $s(t)$ are multiplied by a constant b (which may be negative), delayed by a time difference τ, and added to the original time series $s(t)$ resulting in a new time series:

$$x(t) = s(t) + bs(t-\tau) \qquad [4.3\text{-}1]$$

where b and τ are the echo parameters and represent the strength and time delay of the echo respectively. We shall now convert the continuous representation in equation [4.3-1] to discrete form by letting $t = k\Delta t$, where Δt is properly chosen in accordance with the sampling theorem. Further, we assume that $\tau = m\Delta t$ or that the time delay is an integer multiple of Δt. Under these conditions, equation [4.3-1] becomes:

$$x_k = s_k + bs_{k-m} \qquad [4.3\text{-}2]$$

where $x(k\Delta t) = x_k$, $s(k\Delta t) = s_k$, and without loss of generality, $\Delta t = 1$. The Laplace z-transform of equation [4.3-2] yields:

$$X(z) = S(z) + bz^m S(z) = S(z)[1 + bz^m] \qquad [4.3\text{-}3]$$

Evaluating $X(z)$ on the unit circle, i.e. $z = \mathrm{e}^{-i\omega}$, we obtain the *spectrum* $X(\omega)$ of the sequence $\{x_k\}$ where:

$$X(\omega) = X(\mathrm{e}^{-i\omega}) = S(\omega)\,[1 + b\,\mathrm{e}^{-im\omega}] \qquad [4.3\text{-}4]$$

which may be written as:

$$X(\omega) = |X(\omega)|\,\mathrm{e}^{i\theta_x(\omega)} \qquad [4.3\text{-}5]$$

with $|X(\omega)| = |S(\omega)| \{1 + b^2 + 2b \cos m\omega\}^{\frac{1}{2}}$

$$\theta_x(\omega) = \theta_s(\omega) + \tan^{-1}\left\{\frac{-\sin m\omega}{1 + b\cos m\omega}\right\}$$

$\theta_s(\omega)$ = phase spectrum of the sequence $\{s_k\}$.

Bogert et al. considered the *power spectrum* (energy spectral density) of $\{x_k\}$, denoted by $\Phi_x(\omega) = |X(\omega)|^2$. Under this condition:

$$\Phi_x(\omega) = \Phi_s(\omega)\,[1 + b^2 + 2b \cos m\omega] \qquad [4.3\text{-}6]$$

with $\Phi_s(\omega) = |S(\omega)|^2$ = energy spectral density of the sequence $\{s_k\}$.

Taking the natural logarithm of the power spectrum $\Phi_x(\omega)$, we get:

$$\log \Phi_x(\omega) = \log \Phi_s(\omega) + \log\,[1 + b^2 + 2b \cos m\omega] \qquad [4.3\text{-}7]$$

where it is assumed that both $\Phi_x(\omega)$ and $\Phi_s(\omega)$ do not vanish on the unit circle.

If we assume that $b \ll 1$, then equation [4.3-7] may be approximated by:

$$\log \Phi_x(\omega) \approx \log \Phi_s(\omega) + \log\,[1 + 2b \cos m\omega] \qquad [4.3\text{-}8]$$

With the aid of the series expansion:

$$\log\,(1 + a) = \sum_{n=1}^{\infty} \frac{(-1)^{n+1} a^n}{n}, \quad \text{for } |a| < 1 \qquad [4.3\text{-}9]$$

we can rewrite equation [4.3-8] as:

$$\log \Phi_x(\omega) \approx \log \Phi_s(\omega) + 2b \cos m\omega - \frac{4b^2}{2}\cos^2 m\omega + \frac{8b^3}{3}\cos^3 m\omega - \ldots \qquad [4.3\text{-}10]$$

Neglecting higher-order terms in equation [4.3-10], we obtain:

$$\log \Phi_x(\omega) \approx \log \Phi_s(\omega) + 2b \cos m\omega \qquad [4.3\text{-}11]$$

Thus, in terms of the *logarithm of the power spectrum*, Bogert et al. recognized that the effect of adding a single echo to the original sequence $\{s_k\}$ contributed a cosinusoidal variation with frequency in the log power spectrum $\log \Phi_x(\omega)$ in addition to the variation due to $\log \Phi_s(\omega)$. As a result, they considered the log power spectrum $\log \Phi_x(\omega)$ as a "frequency" series in which they *treated* the independent frequency variable ω as an independent time variable. Now we know that the cosinusoidal quantity $A \cos(\omega t + \theta)$ is described by its magnitude A, frequency ω, and phase θ when we consider time as the independent variable. Bogert et al. considered the frequency ω as the independent variable and for the cosinusoidal variation $B \cos(m\omega + \phi)$ defined the gamnitude B, quefrency m, and saphe (relative to some frequency origin) ϕ. Thus, we see that these quantities are just the analogs of what in a "time" series would be described as magnitude, frequency, and phase. In particularly, the parameters of the frequency series $2b \cos m\omega$ in equation [4.3-11] are $B = 2b$ and the "quefrency" m, which describe the desired echo parameters of echo strength (b) and time delay (m). In order to obtain an estimate of the time delay m (quefrency), they computed the power spectrum of the log power spectrum $\log \Phi_x(\omega)$ and called this quantity the cepstrum, which is the analog of the power spectrum for a time series. A peak at a certain quefrency in the cepstrum suggests the existence of an echo with the corresponding time delay of the original time series. Let us now consider a simple example to illustrate the basic ideas behind the cepstrum, as proposed by Bogert et al. (1963).

Example 4.3-1. We shall consider a unit spike $s_k = \delta_k$ with an echo consisting of strength $b = 0.1$ and time delay $m = 1$. Thus, the signal δ_k and its echo $0.1\delta_{k-1}$ is additively combined to form the time series:

$$x_k = \delta_k + 0.1\delta_{k-1} \qquad [1]$$

The Laplace z-transform of [1] yields:

$$X(z) = 1 + 0.1z \qquad [2]$$

which is also minimum-delay, i.e. $X(z)$ contains no zeroes or poles inside the unit circle. Evaluating $X(z)$ on the unit circle ($z = e^{-i\omega}$), we get the *spectrum:*

$$X(\omega) = 1 + 0.1\, e^{-i\omega} \qquad [3]$$

If we let $\Phi_x(\omega) = |X(\omega)|^2$ denote the *power spectrum* of the sequence $\{x_k\}$, then:

$$\begin{aligned} \Phi_x(\omega) &= (1 + 0.1 \cos \omega)^2 + (0.1 \sin \omega)^2 \\ &= 1 + 0.01 + 0.2 \cos \omega \end{aligned} \qquad [4]$$

which can be approximated by:

$$\Phi_x(\omega) \approx 1 + 0.2 \cos \omega \qquad [5]$$

where ω is the continuous frequency variable with dimensions of radians/sample interval.

Now the logarithm of the power spectrum $\Phi_x(\omega)$ or *log power spectrum* is approximately:

$$\log \Phi_x(\omega) \approx \log [1 + 0.2 \cos \omega] \approx 0.2 \cos \omega \qquad [6]$$

We note that the log power spectrum $\log \Phi_x(\omega)$ is a periodic function in the continuous frequency variable ω, although it is usually only considered over one period, i.e. $-\pi \leqslant \omega \leqslant \pi$. If we treat the frequency series in equation [6] as a time series, then the autocorrelation of equation [6] is given by $C(\tau)$, where:

$$C(\tau) = \tfrac{1}{2}\,(0.2)^2 \cos\tau \qquad [7]$$

where τ is actually a frequency lag although we are treating τ as a time lag. We shall define the Fourier transform of the autocorrelation $C(\tau)$ by:

$$C(w) = \int_{-\infty}^{\infty} C(\tau)\, \mathrm{e}^{-iw\tau}\, \mathrm{d}\tau \qquad [8]$$

and note that w is the quefrency continuous variable with dimensions of radians/radians per sample interval or sample interval (time). Substitution of equation [7] in equation [8] yields:

$$C(w) = \int_{-\infty}^{\infty} \tfrac{1}{2}\,(0.2)^2 \cos\tau\, \mathrm{e}^{-iw\tau} \mathrm{d}\tau = \int_{-\infty}^{\infty} 0.01\, \mathrm{e}^{-i(w-1)\tau} \mathrm{d}\tau + \int_{-\infty}^{\infty} 0.01\, \mathrm{e}^{-i(w+1)\tau} \mathrm{d}\tau$$
$$= 0.01\delta(w-1) + 0.01\delta(w+1) \qquad [9]$$

where we see that $C(w)$ is actually the *power spectrum of the log power spectrum* $\log \Phi_x(\omega)$, and was referred to as the cepstrum by Bogert et al. (1963). Fig. 4-12 depicts the original time series $\{x_k\}$, its spectrum $X(\omega)$, power spectrum $\Phi_x(\omega)$, log power spectrum $\log \Phi_x(\omega)$, and power spectrum of the log power spectrum (cepstrum) $C(w)$. We note that the peak (spike) at $w = 1$ corresponds to the time delay of the echo, and thus illustrates the usefulness of the cepstrum in echo determination.

However, the cosinusoidal variation in the log power spectrum is not so clearly defined in practical situations as it was in example 4.3-1. In fact, a "ripple" of this kind in the log power spectrum will in practice be obscured to a greater or lesser degree by irregularities present in the power spectrum $\Phi_s(\omega)$, the effects of finite data records in computing power spectra, and noise present in the data. In order to improve the cepstral computation, Bogert et al. considered the idea of filtering the log power spectrum, namely applying low-pass, bandpass, or high-pass filters to the log power spectrum to possibly remove the effects of noise or the shape of $\log \Phi_s(\omega)$. In addition to paraphrasing the terms magnitude (gamnitude), frequency (quefrency), and phase (saphe), they paraphrased the filtering operation and referred to a low-pass filter as a low-pass "lifter". Filtering of the log power spectrum $\log \Phi_x(\omega)$ generally involved a trial and error procedure and no set filtering pattern was generally used. In addition, a priori knowledge of the frequency content of the noise and/or the log power spectrum $\log \Phi_s(\omega)$ was required in order to effectively "smooth" the log power spectrum $\log \Phi_x(\omega)$ before computing the cepstrum.

Let us briefly summarize the work of Bogert et al. (1963) and the use of their cepstrum. We see that the cepstrum involved no phase spectrum information, i.e. it utilized magnitude spectrum and power spectrum information only. It incorporated the logarithm as a homomorphic transformation, useful

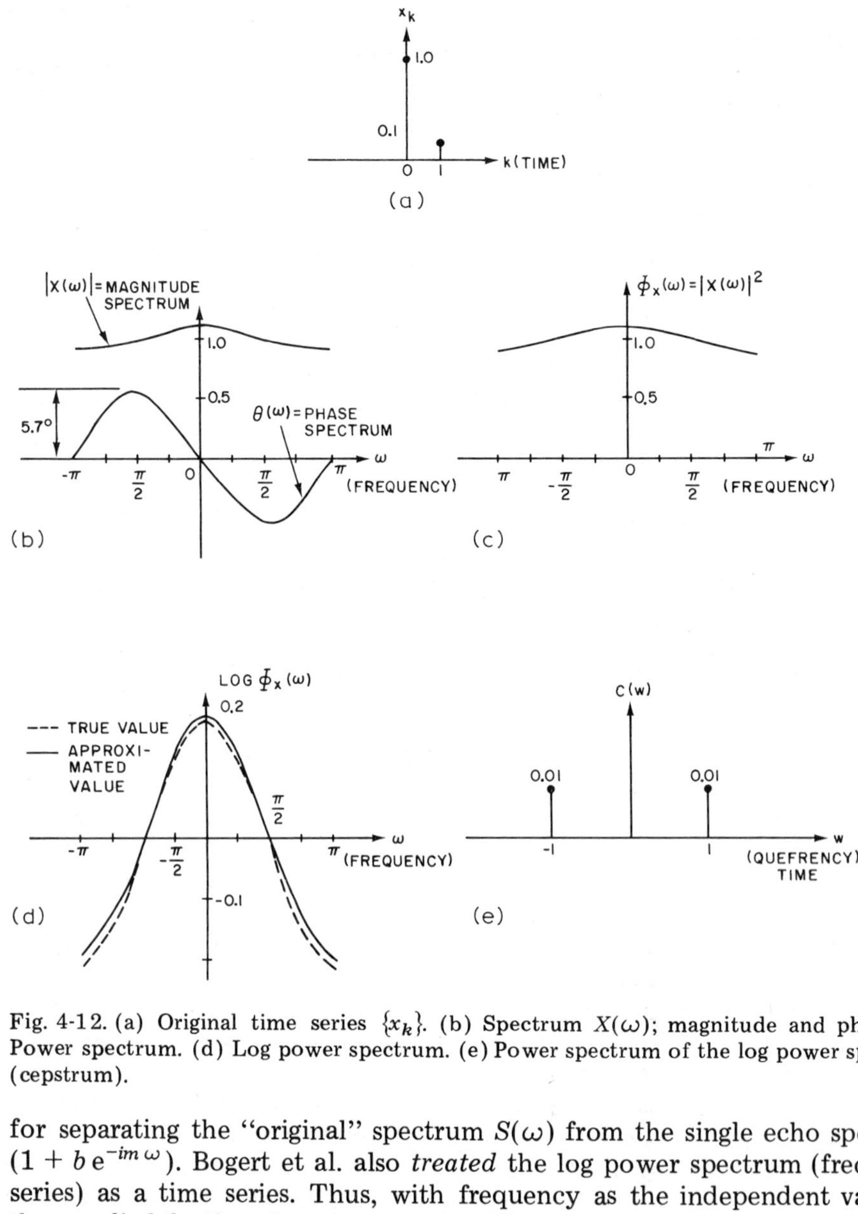

Fig. 4-12. (a) Original time series $\{x_k\}$. (b) Spectrum $X(\omega)$; magnitude and phase. (c) Power spectrum. (d) Log power spectrum. (e) Power spectrum of the log power spectrum (cepstrum).

for separating the "original" spectrum $S(\omega)$ from the single echo spectrum $(1 + b\,\mathrm{e}^{-im\omega})$. Bogert et al. also *treated* the log power spectrum (frequency series) as a time series. Thus, with frequency as the independent variable, they applied the time domain concepts of linear filtering (liftering) and spectral analysis to frequency series as if they were time series. Hence, used in conjunction with judicious filtering (liftering) of the log power spectrum $\log \Phi_x(\omega)$, their cepstrum was effective in detecting *single* echoes, even when other methods (such as the autocovariance) were ineffective. However, it is important to note that Bogert et al. recognized the limitations of their

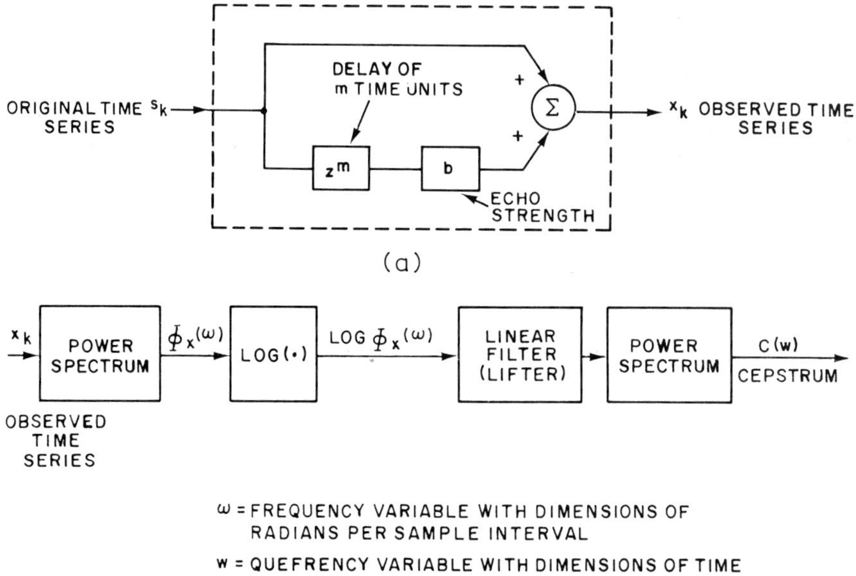

Fig. 4-13. (a) Single-echo (noise-free) time-series model used by Bogert et al. (1963). (b) Signal-processing operations for the cepstrum as proposed by Bogert et al. (1963).

technique and the difficulties encountered in the multiple-echo determination problem. In fact, no numerical analysis of the multiple-echo detection problem is given in their original work. We summarize the cepstral operations of Bogert et al. (1963) in Fig. 4-13.

Thus, we see that in cepstrum analysis one looks for periodicities in the estimated log power spectrum log $\Phi_x(\omega)$ which, if found, are assumed to have been caused by the addition of an echo to a signal $\{s_k\}$ with a "different" log power spectrum log $\Phi_s(\omega)$, i.e. one not having that periodicity. Cepstral analysis does not require precise a priori knowledge of the power spectra $\Phi_s(\omega)$ or log $\Phi_s(\omega)$, but needs only assume that these spectra do not produce cosinusoidal variations which might be attributed to an echo. Although first used in geophysical applications, the cepstrum has made its way into other fields of research. In acoustics, for example, the technique has been applied to the determination of the fundamental pitch frequency in voiced speech sounds.

Schafer (1969) was concerned with the recovery of signals generated by a convolutional process as opposed to the determination of echoes and called his method homomorphic deconvolution. Specifically, let us consider the convolution model discussed in section 1.4. If $\{s_k\}$ denotes the source wavelet and $\{f_k\}$ the impulse response of the earth, then the noise-free response $\{x_k\}$ due to the excitation $\{s_k\}$ is:

$$x_k = s_k * f_k = \sum_{n=0}^{\infty} f_n s_{k-n} \qquad [4.3\text{-}12]$$

Thus, given the observed time series $\{x_k\}$, we now consider the problem of recovering the signal $\{s_k\}$, i.e. estimating the source wavelet $\{s_k\}$. Recall from our discussion on the kepstrum (section 3.3), we defined the composition of homomorphisms K by:

$$K \equiv \zeta_L^{-1} \circ \log \circ \zeta_L \qquad [4.3\text{-}13]$$

as the kepstral operator where:

$$\zeta_L(\cdot) \equiv \sum_{k=0}^{\infty} (\cdot)\, z^k \qquad [4.3\text{-}14]$$

defines the one-sided Laplace z-transform and:

$$\zeta_L^{-1}(\cdot) = \frac{1}{2\pi i} \oint_C (\cdot)\, z^{-k-1}\, dz \qquad [4.3\text{-}15]$$

defines the inverse Laplace z-transform with C being a circular clockwise contour with radius determined by the region of convergence of $(\cdot)$ in the complex z-plane. Recalling the properties of the kepstral operator K, the kepstrum of the response $\{x_k\}$ in equation 4.3-12 is $K(x_k)$ where:

$$K(x_k) = K(s_k * f_k) = K(s_k) + K(f_k) \qquad [4.3\text{-}16]$$

or:

$$\{\chi_k\} = \{\sigma_k\} + \{\phi_k\} \qquad [4.3\text{-}17]$$

where we have assumed that $S(z)$, $F(z)$, and $\log[F(z)]$ and $\log[S(z)]$ are well-behaved Laplace z-transforms.

Hence, if the kepstrum $\{\sigma_k\}$ of the source wavelet $\{s_k\}$ and the kepstrum $\{\phi_k\}$ of the impulse response $\{f_k\}$ are disjoint in kepstral time, i.e. the kepstra $\{\sigma_k\}$ and $\{\phi_k\}$ are non-overlapping, then we have a means of separating or deconvolving the signals $\{s_k\}$ and $\{f_k\}$. The details of this procedure, which describes kepstral deconvolution, are as follows:

(1) We calculate the kepstrum $\{\chi_k\}$ of the observed noise-free response $\{x_k\}$.

(2) Assuming that the convolution model properly characterizes the earth as a linear time-invariant system, then the kepstra of the convolutional components $\{s_k\}$ and $\{f_k\}$ are additive, i.e. $\chi_k = \sigma_k + \phi_k$.

(3) Since we do not have precise knowledge of $\{f_k\}$ or $\{s_k\}$ and their corresponding kepstra $\{\phi_k\}$ and $\{\sigma_k\}$, we need some additional information in order to solve the problem. For example, if $\{s_k\}$ has an anti-causal kepstrum and $\{f_k\}$ a causal kepstrum, or if $\{\sigma_k\}$ and $\{\phi_k\}$ occupy disjoint kepstral time

intervals, then we can use this information to eliminate the kepstrum $\{\phi_k\}$, leaving the kepstrum $\{\sigma_k\}$. This operation is referred to as either high-pass or low-pass filtering or sometimes called "liftering", a term borrowed from the work of Bogert et al. (1963).

(4) Assuming we have properly eliminated the kepstrum $\{\phi_k\}$, we determine the inverse kepstrum $K^{-1}\{\sigma_k\}$ yielding the desired source wavelet $\{s_k\}$. The entire procedure is described in Fig. 4-14.

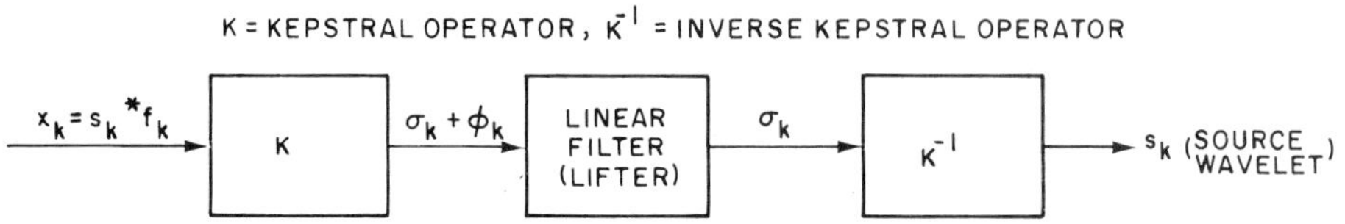

Fig. 4-14. Schematic diagram of kepstral deconvolution and the process of wavelet estimation.

Computational realizations of kepstral deconvolution utilize the fast Fourier transform algorithm, and unlike the kepstral analysis of Bogert et al., both the magnitude and phase spectra are essential parts of kepstral deconvolution. The kepstrum $\{\chi_k\}$ is also called the complex cepstrum (Schafer, 1969; Oppenheim and Schafer, 1975), where the term cepstrum is retained from the work of Bogert et al. (1963) and the term "complex" was added by Oppenheim and Schafer to stress the fact that the complex spectrum, i.e. both magnitude and phase, of the signal $\{x_k\}$ are considered. As shown in section 3.2, when the signal $\{x_k\}$ is minimum-delay and real, we can determine the kepstrum $\{\chi_k\}$ by the relations:

$$\chi_0 = \frac{1}{2\pi}\int_{-\pi}^{\pi} \log|X(\omega)|\, d\omega$$

$$\chi_k = \frac{1}{\pi}\int_{-\pi}^{\pi} \cos\omega k \log|X(\omega)|\, d\omega, \quad k = 1, 2, 3, \ldots \qquad [4.3\text{-}18]$$

where χ_0 and χ_k are real. Further, we have the result for determining the kepstra of minimum-delay signals which are also complex. Hence, for a minimum-delay complex sequence $\{x_k\}$, the corresponding kepstrum $\{\chi_k\}$ is determined by:

$$\chi_k' = \frac{1}{2\pi}\int_{-\pi}^{\pi} \log|X(\omega)| e^{i\omega k}\, d\omega, \quad k = 0, \pm 1, \pm 2, \ldots \qquad [4.3\text{-}19]$$

and:

$$\chi_0 = \chi_0' \quad \text{for } k = 0$$

$$\chi_k = 2\chi_k' \quad \text{for } k = 1, 2, 3, \ldots$$

$$\chi_k = 0 \quad \text{for } k = -1, -2, -3, \ldots$$

as discussed in section 3.2. For a detailed account of the computational procedures for calculating the kepstra of non-minimum-delay systems, the reader is referred to Schafer (1969).

With respect to the convolution model in equation [4.3-12], we see that the kepstral deconvolution method is generally suited for decomposing the source wavelet $\{s_k\}$ and impulse response $\{f_k\}$, assuming that the *kepstra of these signals do not overlap*, i.e. that $\{\chi_k\}$ can be properly filtered (liftered) to yield $\{\sigma_k\}$. However, to apply kepstral deconvolution per se to a more detailed physical model, such as our layered earth model, does not appear to offer any advantages simply because the technique of kepstral deconvolution does not exploit any "fine grain" structure of the physical model, i.e. the reflection coefficient sequence, system reverberation, random reflection coefficient hypothesis or small reflection coefficient conditions. Example 3.3-4 of section 3.3 demonstrates this idea and illustrates the complicated kepstral expressions obtained by considering the kepstrum of a general rational function. Hence, interpretation of the kepstrum for "fine grain" model characteristics becomes quite difficult. Moreover, in the filtering (liftering) procedure associated with kepstral deconvolution, one is faced with the difficult task of trying to sort out desired physical information (reflection coefficients) from the effects of the system reverberations. Nevertheless, we have seen the usefulness of the kepstrum for estimating the time delay of a single echo, as proposed by Bogert et al. (1963). In this case, we concern ourselves only with the amplitude spectrum of a given signal. In the problem of deconvolution, the kepstrum is a means of estimating the source wavelet for the case of possibly a priori knowledge of the source kepstrum $\{\sigma_k\}$ or non-overlapping kepstra. In this case, we concern ourselves with both the magnitude and phase spectrum of a given signal. Current research in the area of deconvolution is investigating the idea of combining the "fine grain" model features of predictive deconvolution with the source wavelet estimation capabilities of kepstral deconvolution.

4.4. STATE SPACE FILTERING

In section 4.1 we defined the concept of a system and classified systems as either linear or non-linear. In addition to inputs, outputs, and rules of operation, systems generally require the *concept of state* for a complete description.

It became evident in the early 1950's that the classical approaches for modeling and analyzing systems, i.e. control systems and systems with multiple inputs and outputs, were nearing a peak in their development and consequently offered little promise for future growth. A different viewpoint was needed. The motivations for this different viewpoint were:

(1) The recognition of the classical approach limitations.

(2) The need for a general framework to handle multi-input and multi-output systems.

(3) The general availability of powerful vector space concepts, matrix analysis, and a wealth of related theorems concerning the analytic structure of system response.

Thus, the researchers in various scientific fields turned to the time domain and the use of vector space concepts in hopes of overcoming the limitations of classical systems analysis. This different approach is often called the state space approach and it utilizes extensively the algebra of matrices. The application of the state space approach was further stimulated by the need for techniques to analyze the highly complex systems being spawned by modern technology and the rise to prominence of the general-purpose digital computer. With the above ideas in mind, let us discuss the concept of state.

Dynamic systems may be described by difference equations or ordinary or partial differential equations that can be of a determinstic or stochastic nature. The dynamic features appear in these equations in terms of the derivatives or their respective difference expressions. Now any nth-order difference equation (or ordinary differential equation) may be transformed into a set of n first-order difference equations (differential equations). For example, consider the nth-order difference equation:

$$x_k + a_1 x_{k-1} + a_2 x_{k-2} + \ldots + a_n x_{k-n} = \epsilon_k \qquad [4.4\text{-}1]$$

where $\{1, a_1, a_2, \ldots a_n\}$ are the stable difference equation operator coefficients discussed in section 1.5. We shall define the n state variables $x_k^{(1)}$, $x_k^{(2)}, \ldots x_k^{(n)}$ by:

$$\begin{aligned}
x_k^{(1)} &\equiv x_k \\
x_k^{(2)} &\equiv x_{k-1} = x_{k-1}^{(1)} \\
x_k^{(3)} &\equiv x_{k-2} = x_{k-1}^{(2)} \\
&\;\;\cdot \\
&\;\;\cdot \\
x_k^{(n)} &\equiv x_{k-(n-1)} = x_{k-1}^{(n-1)}
\end{aligned} \qquad [4.4\text{-}2]$$

The state variables that we have selected are also referred to as phase variables. That is, corresponding to an nth-order difference equation, the dependent variable and its first $n-1$ differences form the set of n phase variables. Solving equation [4.4-1] for x_k we obtain:

$$x_k = -a_1 x_{k-1} - a_2 x_{k-2} - \ldots - a_n x_{k-n} + \epsilon_k \qquad [4.4\text{-}3]$$

Writing equation [4.4-3] in state variable notation we get:

$$x_k^{(1)} = -a_1 x_{k-1}^{(1)} - a_2 x_{k-1}^{(2)} - \ldots - a_n x_{k-1}^{(n)} + \epsilon_k \qquad [4.4\text{-}4]$$

Combining equations [4.4-2] and [4.4-4], we obtain the set of equations:

$$\begin{aligned} x_k^{(1)} &= -a_1 x_{k-1}^{(1)} - a_2 x_{k-1}^{(2)} - \ldots - a_n x_{k-1}^{(n)} + \epsilon_k \\ x_k^{(2)} &= x_{k-1}^{(1)} \\ x_k^{(3)} &= x_{k-1}^{(2)} \\ &\;\;\vdots \\ x_k^{(n)} &= x_{k-1}^{(n-1)} \end{aligned} \qquad [4.4\text{-}5]$$

which can be expressed in matrix form as:

$$\begin{bmatrix} x_k^{(1)} \\ x_k^{(2)} \\ \vdots \\ \cdot \\ \vdots \\ x_k^{(n)} \end{bmatrix} = \begin{bmatrix} -a_1 & -a_2 & \ldots & \ldots & -a_n \\ 1 & 0 & \ldots & \ldots & 0 \\ \vdots & & & & \\ 0 & 1 & \ldots 0 & \ldots & 0 \\ \vdots & & & & \\ 0 & 0 & \ldots & \ldots 1 & 0 \end{bmatrix} \begin{bmatrix} x_{k-1}^{(1)} \\ x_{k-1}^{(2)} \\ \vdots \\ \cdot \\ \vdots \\ x_{k-1}^{(n)} \end{bmatrix} + \begin{bmatrix} 1 \\ 0 \\ \vdots \\ 0 \\ \vdots \\ 0 \end{bmatrix} \epsilon_k \qquad [4.4\text{-}6]$$

The $n \times 1$ column vector $\boldsymbol{x}_k$ defined by:

$$\boldsymbol{x}_k \equiv \begin{bmatrix} x_k^{(1)} \\ x_k^{(2)} \\ \vdots \\ x_k^{(n)} \end{bmatrix} \quad \text{(components of } \boldsymbol{x}_k \text{ are the state variables)} \qquad [4.4\text{-}7]$$

is called the *state vector* of the system and equation [4.4-6] is called the *state equation* of the system. We can write the state equation in matrix form as:

$$\boldsymbol{x}_k = \boldsymbol{\Phi} \boldsymbol{x}_{k-1} + \boldsymbol{\gamma}\, \epsilon_k \qquad [4.4\text{-}8]$$

where:

$$\boldsymbol{\Phi} = \underbrace{\begin{bmatrix} -a_1 & -a_2 & \ldots & \ldots & -a_n \\ 1 & 0 & \ldots & \ldots & 0 \\ 0 & 1 & 0 & \ldots & 0 \\ \vdots & & & & \\ 0 & 0 & \ldots & \ldots 1 & 0 \end{bmatrix}}_{n \times n}, \quad \boldsymbol{\gamma} = \underbrace{\begin{bmatrix} 1 \\ 0 \\ 0 \\ \vdots \\ 0 \end{bmatrix}}_{n \times 1}$$

We note that the difference equation [4.4-1] describes the so-called "all-pole" model. Thus, equations [4.4-6] and [4.4-7] form the state space description of an all-pole model. Further, if we assume that the sequence $\{\epsilon_k\}$ is a white noise sequence, i.e. a sequence of uncorrelated random variables, with zero mean ($E(\epsilon_k) = 0$) and constant variance or average power ($E(\epsilon_k^2) = \sigma^2$), then equation [4.4-1] also describes an nth-order *autoregressive process*. Hence, if we consider $\{\epsilon_k\}$ as a zero mean white noise sequence with average power σ^2, then equations [4.4-6] and [4.4-7] also provide the state space description of an nth-order autoregressive process.

For the system described by equation [4.4-1], it was convenient to choose the state variables as a set of phase variables. But in more general terms, the state variables are defined as a set of variables which, along with knowledge of the inputs to the system, are necessary and sufficient to specify the response of the system. Thus, the above definition of the state variables illustrates the fact that the set of state variables is not unique. The number of state variables necessary is fixed by the order of the system, but many different sets of state variables may be chosen. Nevertheless, the set of phase variables constitutes a convenient choice of state variables for all pole models or autoregressive processes.

The state space representation of a system is a fundamental concept in modern control theory and forms the basis of Kalman filtering introduced by Kalman (1960). Although state space representations of systems have been discussed in some statistical literature (for example, Whittle, 1969; Akaike, 1971), it has not yet been fully exploited by statisticians. This may be partly due to the somewhat abstract definition of the concept of state, as described by Kalman and others. The state is sometimes vaguely understood as a condensed representation of information from the present and past, such that the future behavior of the system can be completely described by the knowledge of the present state and future input. This idea finds a precise mathematical formulation in the statistical context and when the system is stochastic, i.e. when the input $\{\epsilon_k\}$ and output $\{x_k\}$ are stochastic processes, equation [4.4-8] is generally referred to by statisticians as a Markovian representation of a stochastic system. Thus, control theory engineers speak of state space representations of stochastic or deterministic systems, whereas statisticians call the state equation a Markov process. Nevertheless, the autoregressive model of equation [4.4-1] and the state space model of equation [4.4-8] represent the same system; the only thing that differs is our point of view. Let us now proceed to develop a state space model of our layered earth system.

When the magnitudes of the reflection coefficients are small (which is generally true in practice), we showed (section 2.4) that the transfer function of an N-layered earth $F^{(N)}(z)$ was:

$$F^{(N)}(z) = \frac{E^{(N)}(z)}{A^{(N)}(z)} \approx \frac{\epsilon_1 z + \epsilon_2 z^2 + \ldots + \epsilon_N z^N}{1 + \phi_1^{(N)} z + \phi_2^{(N)} z^2 + \ldots + \phi_N^{(N)} z^N} \qquad [4.4\text{-}9]$$

where $\{\epsilon_k\} = \{r_1, r_2, \ldots r_N\}$ are precisely the reflection coefficients and $\phi_1^{(N)}, \phi_2^{(N)}, \ldots \phi_N^{(N)}$ are the serial correlations given by $\phi_m^{(N)} = \Sigma_{n=1}^{N} r_n r_{n+m}$; $m, N \geqslant 1$. If a source wavelet $\{s_k\}$ excites this system at $k = 0$, then the Laplace z-transform of the noise-free output $\{x_k\}$ is given by:

$$X(z) = S(z)F^{(N)}(z) = \frac{S(z)\,E^{(N)}(z)}{A^{(N)}(z)} \qquad [4.4\text{-}10]$$

The time domain version of equation [4.4-10] in difference equation form is:

$$x_k + \phi_1^{(N)} x_{k-1} + \phi_2^{(N)} x_{k-2} + \ldots + \phi_N^{(N)} x_{k-N} = s_0\epsilon_k + s_1\epsilon_{k-1} + \ldots + s_N\epsilon_{k-N} \qquad [4.4\text{-}11]$$

where $\{s_k\} = \{s_0, s_1, s_2, \ldots s_N\}$ comprise the elements of the source sequence, which usually consist of less than N values.

We would like a state space description of our layered earth system in the form $\boldsymbol{x}_k = \boldsymbol{\Phi x}_{k-1} + \boldsymbol{\gamma}\epsilon_k$ (see equation [4.4-8]), but due to the higher-order differences $\epsilon_{k-1}, \epsilon_{k-2}, \ldots \epsilon_{k-N}$ we cannot choose our state variables as the phase variables, since this choice of state variables would not yield the desired state space form. Thus, instead of the phase variables, we define the following set of state variables:

$$\begin{aligned}
x_k^{(1)} &\equiv -(\phi_1^{(N)} x_{k-1}^{(1)} + \phi_2^{(N)} x_{k-1}^{(2)} + \ldots + \phi_N^{(N)} x_{k-1}^{(N)}) + \gamma_1\epsilon_k \\
x_k^{(2)} &\equiv x_{k-1}^{(1)} + \gamma_2\epsilon_k \\
x_k^{(3)} &\equiv x_{k-1}^{(2)} + \gamma_3\epsilon_k \qquad [4.4\text{-}12] \\
&\;\,\cdot \\
&\;\,\cdot \\
x_k^{(N)} &\equiv x_{k-1}^{(N-1)} + \gamma_N\epsilon_k
\end{aligned}$$

and:

$$x_k \equiv x_k^{(1)} + \gamma_{N+1}\epsilon_k$$

where $x_k^{(1)}, x_k^{(2)}, \ldots x_k^{(N)}$ are the state variables, x_k is the system output at time k, and $\gamma_1, \gamma_2, \gamma_3, \ldots \gamma_{N+1}$ are arbitrary coefficients.

With this choice of state variables, equation [4.4-12] can be expressed in matrix form as:

$$\begin{bmatrix} x_k^{(1)} \\ x_k^{(2)} \\ \cdot \\ \cdot \\ x_k^{(N)} \end{bmatrix} = \begin{bmatrix} -\phi_1^{(N)} & -\phi_2^{(N)} & \cdots & -\phi_N^{(N)} \\ & & & 0 \\ & \mathrm{I} & & 0 \\ & & & \vdots \\ & & & 0 \end{bmatrix} \begin{bmatrix} x_{k-1}^{(1)} \\ x_{k-1}^{(2)} \\ \cdot \\ \cdot \\ x_{k-1}^{(N)} \end{bmatrix} + \begin{bmatrix} \gamma_1 \\ \gamma_2 \\ \cdot \\ \cdot \\ \gamma_N \end{bmatrix} \epsilon_k \qquad [4.4\text{-}13]$$

where $\mathrm{I} =$ the $(N-1) \times (N-1)$ identity matrix and $x_k = \boldsymbol{h}^{\mathrm{T}}\boldsymbol{x}_k + \gamma_{N+1}\epsilon_k$

with the $(N \times 1)$ vector $\boldsymbol{h}$ defined as:

$$\boldsymbol{h} \equiv \begin{bmatrix} 1 \\ 0 \\ 0 \\ \vdots \\ 0 \end{bmatrix}, \quad \boldsymbol{h}^{\mathrm{T}} = [1 \quad 0 \quad 0 \quad \ldots \quad 0]$$

(Unless stated otherwise, we shall refer to the state vector at time k by $\boldsymbol{x}_k$ and the system (scalar) output at time k by x_k. All column or row vectors will be represented by boldfaced, italicized, lower-case letters, and matrices by boldfaced capitalized letters. Thus, $\mathbf{X}$ denotes a matrix, $\boldsymbol{x}$ a column vector, and $\boldsymbol{x}^{\mathrm{T}}$ a row vector.)

Let us now proceed to relate the coefficients $\{\gamma_1, \gamma_2, \ldots \gamma_{N+1}\}$ to the physical parameters of our model. Now the system (scalar) output x_k is:

$$x_k \equiv x_k^{(1)} + \gamma_{N+1}\epsilon_k \qquad [4.4\text{-}14]$$

or:

$$x_{k-1} = x_{k-1}^{(1)} + \gamma_{N+1}\epsilon_{k-1} \qquad [4.4\text{-}15]$$

But from equation [4.4-12], we obtain the relation:

$$x_{k-1}^{(1)} = x_k^{(2)} - \gamma_2\epsilon_k \qquad [4.4\text{-}16]$$

Now substitution of equation [4.4-16] in [4.4-15] yields:

$$x_{k-1} = x_k^{(2)} - \gamma_2\epsilon_k + \gamma_{N+1}\epsilon_{k-1} \qquad [4.4\text{-}17]$$

or:

$$x_{k-2} = x_{k-1}^{(2)} - \gamma_2\epsilon_{k-1} + \gamma_{N+1}\epsilon_{k-2} \qquad [4.4\text{-}18]$$

Similarly, from equation [4.4-12] we get:

$$x_{k-1}^{(2)} = x_k^{(3)} - \gamma_3\epsilon_k \qquad [4.4\text{-}19]$$

Substitution of equation [4.4-19] in [4.4-18] gives:

$$x_{k-2} = x_k^{(3)} - \gamma_3\epsilon_k - \gamma_2\epsilon_{k-1} + \gamma_{N+1}\epsilon_{k-2} \qquad [4.4\text{-}20]$$

or:

$$x_{k-3} = x_{k-1}^{(3)} - \gamma_3\epsilon_{k-1} - \gamma_2\epsilon_{k-2} + \gamma_{N+1}\epsilon_{k-3} \qquad [4.4\text{-}21]$$

If we continue this procedure to find x_{k-N}, we obtain the result:

$$x_{k-N} = x_{k-1}^{(N)} - \gamma_N\epsilon_{k-1} - \gamma_{N-1}\epsilon_{k-2} - \ldots - \gamma_2\epsilon_{k-(N-1)} + \gamma_{N+1}\epsilon_{k-N} \qquad [4.4\text{-}22]$$

Substitution of equations [4.4-14], [4.4-15], [4.4-18], [4.4-21], and similar equations up to and including equation [4.4-22] into equation [4.4-11] yields:

$$(x_k^{(1)} + \gamma_{N+1}\epsilon_k) + \phi_1^{(N)}(x_{k-1}^{(1)} + \gamma_{N+1}\epsilon_{k-1}) + \phi_2^{(N)}(x_{k-1}^{(2)} - \gamma_2\epsilon_{k-1} + \gamma_{N+1}\epsilon_{k-2})$$
$$+ \ldots + \phi_N^{(N)}(x_{k-1}^{(N)} - \gamma_N\epsilon_{k-1} - \gamma_{N-1}\epsilon_{k-2} - \ldots - \gamma_2\epsilon_{k-(N-1)} + \gamma_{N+1}\epsilon_{k-N})$$
$$= s_0\epsilon_k + s_1\epsilon_{k-1} + \ldots + s_N\epsilon_{k-N} \qquad [4.4\text{-}23]$$

Substitution of $x_k^{(1)}$ from equation [4.4-12] in equation [4.4-23] gives:

$$(\gamma_1 + \gamma_{N+1})\epsilon_k + (\phi_1^{(N)}\gamma_{N+1} - \phi_2^{(N)}\gamma_2 - \ldots - \phi_N^{(N)}\gamma_N)\epsilon_{k-1}$$
$$+ (\phi_2^{(N)}\gamma_{N+1} - \phi_3^{(N)}\gamma_2 - \ldots - \phi_N^{(N)}\gamma_{N-1})\epsilon_{k-2} + \ldots + (\phi_N^{(N)}\gamma_{N+1})\epsilon_{k-N}$$
$$= s_0 + s_1\epsilon_{k-1} + s_2\epsilon_{k-2} + \ldots + s_N\epsilon_{k-N} \qquad [4.4\text{-}24]$$

Thus, equating coefficients of $\epsilon_k, \epsilon_{k-1}, \ldots \epsilon_{k-N}$ in equation [4.4-24], we obtain the relations:

$$\begin{aligned}
s_0 &= \gamma_1 + \gamma_{N+1} \\
s_1 &= -\phi_2^{(N)}\gamma_2 - \phi_3^{(N)}\gamma_3 - \ldots - \phi_N^{(N)}\gamma_N + \phi_1^{(N)}\gamma_{N+1} \\
s_2 &= -\phi_3^{(N)}\gamma_2 - \phi_4^{(N)}\gamma_3 - \ldots - \phi_N^{(N)}\gamma_{N-1} + \phi_2^{(N)}\gamma_{N+1} \\
s_3 &= -\phi_4^{(N)}\gamma_2 - \phi_5^{(N)}\gamma_3 - \ldots - \phi_N^{(N)}\gamma_{N-2} + \phi_3^{(N)}\gamma_{N+1} \\
&\;\;\vdots \\
s_N &= \phi_N^{(N)}\gamma_{N+1}
\end{aligned} \qquad [4.4\text{-}25]$$

or in matrix form:

$$\begin{bmatrix} s_0 \\ s_1 \\ s_2 \\ s_3 \\ \vdots \\ \cdot \\ s_N \end{bmatrix} = \begin{bmatrix} 1 & 0 & 0 & \ldots & 0 & 1 \\ 0 & -\phi_2^{(N)} & -\phi_3^{(N)} & \ldots & -\phi_N^{(N)} & \phi_1^{(N)} \\ 0 & -\phi_3^{(N)} & -\phi_4^{(N)} & \ldots & 0 & \phi_2^{(N)} \\ 0 & -\phi_4^{(N)} & -\phi_5^{(N)} & \ldots & 0 & \phi_3^{(N)} \\ \vdots & \vdots & & & & \vdots \\ \cdot & -\phi_N^{(N)} & 0 & \ldots & 0 & \phi_{N-1}^{(N)} \\ 0 & 0 & \ldots & \ldots & 0 & \phi_N^{(N)} \end{bmatrix} \begin{bmatrix} \gamma_1 \\ \gamma_2 \\ \gamma_3 \\ \gamma_4 \\ \vdots \\ \cdot \\ \gamma_{N+1} \end{bmatrix} \qquad [4.4\text{-}26]$$

Thus, we have the γ_n coefficients in terms of the physical parameters $\{s_0, s_1, s_2, \ldots s_N\}$ = the amplitude values of the source sequence and $\{\phi_1^{(N)}, \phi_2^{(N)}, \ldots \phi_N^{(N)}\}$ = the serial correlations, which are a function of the reflection coefficients. As a practical consideration, the last coefficient of the source sequence s_N may be taken as zero, since the source wavelet is a finite-duration signal. Under this condition, $\gamma_{N+1} = 0$, and we obtain the following state space representation of our layered system:

$$\begin{aligned}
\boldsymbol{x}_k &= \boldsymbol{\Phi}\boldsymbol{x}_{k-1} + \boldsymbol{\gamma}\epsilon_k \\
x_k &= \boldsymbol{h}^{\mathrm{T}}\boldsymbol{x}_k
\end{aligned} \qquad [4.4.27]$$

with Φ and h^{T} defined in equation [4.4-13], γ defined by the relations in equations [4.4-25] or [4.4-26], the state vector x_k at time k defined by equation [4.4-12], and x_k the system output at time k.

Let us *treat* the sequence $\{\epsilon_k\} = \{0, r_1, r_2, \ldots r_N\}$ as an uncorrelated sequence, and thus consider the random reflection coefficient model discussed in section 2.4. Further, we shall now consider the situation where our noise-free observation $\{x_k\}$ is degraded by additive noise $\{n_k\}$, possibly introduced by the measurement process. Thus, we shall define our noisy measurement $\{y_k\} \equiv \{x_k\} + \{n_k\}$, where $\{n_k\}$ is assumed to be an uncorrelated zero mean Gaussian process with variance R. Since we are *treating* the reflection coefficient sequence as a random sequence, let us assume that $\{\epsilon_k\}$ represents an uncorrelated zero mean Gaussian process with variance Q. With these assumptions, our state space model becomes:

$$
\begin{aligned}
x_k &= \Phi x_{k-1} + \gamma \epsilon_k \\
y_k &= h^{\mathrm{T}} x_k + n_k
\end{aligned}
\qquad [4.4\text{-}30]
$$

where $E\{\epsilon_k\} = E\{n_k\} = E\{\epsilon_k n_k\} = 0$; $E\{\epsilon_k^2\} = Q$; and $E\{n_k^2\} = R$.

It is important to note that the state space representation in equation [4.4-30] does not represent a new physical model, but rather a different viewpoint of the same physical model. In fact, the noisy time series $\{y_k\}$ can also be represented by:

$$
\{y_k\} = (s_k * b_k) * \epsilon_k + n_k = w_k * \epsilon_k + \{n_k\} \qquad [4.4\text{-}31]
$$

where $\{b_k\}$ represents the system reverberations and $\{w_k\}$ the composite wavelet discussed earlier. Thus, equations [4.4-30] and [4.4-31] are different mathematical realizations of the same physical model. For the convolution model in equation [4.4-31], we used the theory of linear prediction and Gauss least squares to generate estimates of the reflection coefficient sequence $\{\epsilon_k\}$. In using the theory of linear prediction, we made the following basic assumptions:

(1) The layered earth represents a linear time-invariant system.

(2) The source wavelet is minimum-delay.

(3) The reflection coefficients $\{\epsilon_k\}$ are *treated* as zero mean uncorrelated random variables and the magnitude of any reflection coefficient is much less than one.

Let us now consider using the state space representation of our layered earth model as a basis for generating estimates of the reflection coefficient sequence $\{\epsilon_k\}$.

The concept of state space filtering

From a historical viewpoint, the linear filtering methods developed over the last century have been essentially applications and extensions of the work of Gauss (1809) and other 19th century mathematicians. A notable contribution in the present century was made by Kalman (1960) who introduced the concept of state into the engineering filtering problems.

The rapid expansion which took place in the present century in communications and electronics, motivated largely by the development programs of World War II, led to a new class of problems in which research workers were faced with the need to design *electrical filters* which could estimate signals in the presence of noise. Neither the classical methods of estimation theory nor existing filter-design techniques based on the Fourier integral were adequate, and so an entirely new approach was developed, independently by Kolmogorov (1941a, b) and Wiener (1942), in which both the signal and noise were treated as random variables. The criterion on which their work was based is known as the least mean-squared error criterion which requires that the ensemble expectation of the squared difference between the desired value and its estimate be a minimum (see section 4.1). This approach was a departure from the classical method of Gauss least squares.

The Wiener-Kolmogorov theory has served both as an end in itself and as the motivation for related theories which are designed to avoid the problems encountered in solving the Wiener-Hopf integral equation as well as the practical problem of synthesizing the theoretically optimum filter from its impulse response. An alternative approach to signal filtering and prediction has been suggested, which essentially avoids the Wiener-Hopf integral equation by substituting an equivalent differential equation. Of more practical interest is the fact that the differential equation technique has the property that the optimum filter can be synthesized in a sequential fashion and, thus, often readily implemented on the digital computer. In 1960 and 1961 an extension was made to the Kolmogorov-Wiener theory by Kalman (1960) and Kalman and Bucy (1961) who addressed themselves to the estimation of random variables which satisfy a linear difference equation (Kalman's formulation) and a linear differential equation (Bucy's formulation). Both Kalman and Bucy independently recognized that, rather than attack the Wiener-Hopf integral equation directly with the attendant problems of factorization, it was desirable to convert this integral equation into a non-linear differential equation whose solution yields the covariance matrix of the minimum filtering error. In turn, this matrix contains all the necessary information for the design of the optimum filter, or Kalman filter. Let us now look at the basic components of Kalman filtering and its relationship to the state space representation of our layered earth model.

Let us consider the following:

(1) Let $\boldsymbol{y}_j^{\mathrm{T}} \equiv [y_1 \quad y_2 \quad \ldots \quad y_j]$ be a vector of random variables and let $\boldsymbol{x}_k$ also be a vector of random variables, all of which share the joint probability density function $p(\boldsymbol{x}_k, \boldsymbol{y}_j)$.

(2) Let $\hat{\boldsymbol{x}}_{k/j}$ denote the estimate of $\boldsymbol{x}_k$ at time k given the j samples of $\boldsymbol{y}_j$. Then the error in the estimate is:

$\tilde{x}_{k/j} \equiv x_k - \hat{x}_{k/j}$ = error in estimate of x_k given y_j

(3) The function $L(\tilde{x}_{k/j})$ is termed a *loss function* if it satisfies the following conditions:

$L(0) = 0$, $L(\tilde{x}_{k/j})$ is monotonic, $L(\tilde{x}_{k/j})$ is symmetric

(4) Then the *optimal estimator*, relative to a chosen loss function, is the one (if it exists) that minimizes $E\{L(\tilde{x}_{k/j})\}$, i.e. the expected loss.

Kalman (1960) presented the very powerful result from estimation theory:

Theorem. Let the conditional probability density function for x_k given the sample values y_j be $p(x_k/y_j)$ and assume that it is (a) unimodal, (b) symmetric about its conditional expectation $E\{x_k/y_j\}$, then the optimal estimate of x_k based on the vector of observations y_j with respect to the chosen loss function $L(\tilde{x}_{k/j})$ is precisely:

$\hat{x}_{k/j} = E\{x_k/y_j\}$ = optimal estimator

Thus, the conditional expectation of x_k given the observations y_j is the optimal estimate. Since this conditional expectation depends only on the form of the conditional probability density function $p(x_k/y_j)$ and does not depend on the choice for $L(\tilde{x}_{k/j})$, it follows that subject to (a) and (b) in the above theorem, *every loss function $L(\tilde{x}_{k/j})$ is minimized by the conditional expectation of x_k.* For example, in the case of Gaussian statistics, the Gaussian conditional density function is again Gaussian and so it satisfies (a) and (b) of the theorem. We note that in general, $E\{x_k/y_j\}$ is a non-linear function of the observations y_j, which may be very difficult to compute. In the event that we are interested only in the *quadratic* loss function, i.e. the squared-error criterion discussed in section 4.1, then conditions (a) and (b) can be relaxed. It is now only necessary that $p(x_k/y_j)$ have a finite second moment in order that the optimal estimate with respect to the quadratic loss function be given by $E\{x_k/y_j\}$. We also add the well-known result in estimation theory that for quadratic loss functions of Gaussian processes, the optimal estimator of x_k is a projection on a Hilbert space generated from the observable y_j. With these basic concepts in mind, let us develop the state space filtering approach.

In control theory applications, the state space representation:

$$x_k = \Phi x_{k-1} + \gamma \epsilon_k \qquad [4.4\text{-}32]$$

is referred to as the *plant* model (refer to equation [4.4-30]), and:

$$y_k = h^{\mathrm{T}} x_k + n_k \qquad [4.4\text{-}33]$$

is referred to as the *measurement* model. The classic state space problem is to determine the best estimate of x_k given the measurement y_k. But for the case of Gaussian statistics and a quadratic loss function, we showed that the optimal estimator $\hat{x}_{k/j}$ is:

$$\hat{x}_{k/j} = E\{x_k/y_j\} \qquad [4.4\text{-}34]$$

with the error in our estimate given by:

$$\tilde{x}_{k/j} = x_k - \hat{x}_{k/j} \qquad [4.4\text{-}35]$$

For Gaussian statistics, our error has zero mean, i.e.:

$$E\{\tilde{x}_{k/j}\} = E\{x_k - \hat{x}_{k/j}\} = 0 \qquad [4.4\text{-}36]$$

which further implies that:

$$E\{\hat{x}_{k/j}\} = E\{x_k\} \qquad [4.4\text{-}37]$$

Thus, equation [4.4-37] indicates that our estimate is also *unbiased*. Further, the estimate in equation [4.4-34] is also a *minimum-variance* estimate. Moreover, if we restrict ourselves to the class of *linear* estimators, then equation [4.4-34] also represents a linear, unbiased, minimum-variance estimate of x_k. With these assumptions, let us attempt a one-step *prediction* of x_k given $k-1$ samples y_{k-1}. We shall denote the one-step predicted value of x_k by $\hat{x}_{k/k-1}$ and $y_{k-1}^{\mathrm{T}} = [y_1 \quad y_2 \quad \ldots y_{k-1}]$.

First, we know that the optimum predictor is given by:

$$\hat{x}_{k/k-1} = E\{x_k/y_{k-1}\} = E\{x_k/(y_{k-1}, y_{k-2})\} \qquad [4.4\text{-}38]$$

where $y_{k-2}^{\mathrm{T}} = [y_1 \quad y_2 \quad \ldots \; y_{k-2}]$. But by employing the orthogonal projection principle (see Deutsch, 1965), we can rewrite equation [4.4-38] as:

$$\hat{x}_{k/k-1} = E\{x_k/\tilde{y}_{k-1}\} + E\{x_k/y_{k-2}\} \qquad [4.4\text{-}39]$$

where $\tilde{y}_{k-1} = y_{k-1} - E\{y_{k-1}/y_{k-2}\} = y_{k-1} - \hat{y}_{k-1/k-2}$ = innovation process. But our measurement model in equation [4.4-33] is also expressed as:

$$y_{k-1} = h^{\mathrm{T}} x_{k-1} + n_{k-1} \qquad [4.4\text{-}40]$$

Using equation [4.4-40], the innovation process $\tilde{y}_{k-1}$ is given by:

$$\begin{aligned} \tilde{y}_{k-1} &= y_{k-1} - E\{h^{\mathrm{T}} x_{k-1} + n_{k-1}/y_{k-2}\} \\ &= y_{k-1} - E\{h^{\mathrm{T}} x_{k-1}/y_{k-2}\} - E\{n_{k-1}/y_{k-2}\} \end{aligned} \qquad [4.4\text{-}41]$$

But since n_{k-1} is uncorrelated with all the past measurements y_{k-2}, then $E\{n_{k-1}/y_{k-2}\}$ is really not conditioned on y_{k-2}. Thus:

$$\tilde{y}_{k-1} = y_{k-1} - E\{h^{\mathrm{T}} x_{k-1}/y_{k-2}\} - E\{n_{k-1}\} \qquad [4.4\text{-}42]$$

Now since n_k was assumed to be a zero mean uncorrelated Gaussian sequence, then:

$$\begin{aligned} \tilde{y}_{k-1} &= y_{k-1} - h^{\mathrm{T}} E\{x_{k-1}/y_{k-2}\} \\ &= y_{k-1} - h^{\mathrm{T}} \hat{x}_{k-1/k-2} \end{aligned} \qquad [4.4\text{-}43]$$

Now using our plant model (equation [4.4-32]), equation [4.4-39] becomes:

$$\hat{x}_{k/k-1} = E\{x_k/\tilde{y}_{k-1}\} + E\{\Phi x_{k-1} + \gamma \epsilon_k / y_{k-2}\} \quad [4.4\text{-}44]$$

or:

$$\hat{x}_{k/k-1} = E\{x_k/\tilde{y}_{k-1}\} + E\{\Phi x_{k-1}/y_{k-2}\} + E\{\gamma \epsilon_k / y_{k-2}\} \quad [4.4\text{-}45]$$

But ϵ_k is uncorrelated with all past measurements and is not conditioned on either $\tilde{y}_{k-1}$ or y_{k-2}. Thus:

$$E\{\gamma \epsilon_k / y_{k-2}\} = \gamma E\{\epsilon_k\} = 0 \quad [4.4\text{-}46]$$

where we have assumed that the *plant noise* ≡ reflection coefficient sequence is uncorrelated and a zero mean Gaussian process. With these assumptions, equation [4.4-45] becomes:

$$\hat{x}_{k/k-1} = \Phi \hat{x}_{k-1/k-2} + \Delta \tilde{y}_{k-1} \quad [4.4\text{-}47]$$

where $\hat{x}_{k-1/k-2} = E\{x_{k-1}/y_{k-2}\}$; $\Delta \tilde{y}_{k-1} = E\{x_k/\tilde{y}_{k-1}\}$; and Δ is an $N \times 1$ proportionality vector. Substitution of equation [4.4-43] in [4.4-47] gives:

$$\hat{x}_{k/k-1} = \Phi \hat{x}_{k-1/k-2} + \Delta [y_{k-1} - h^T \hat{x}_{k-1/k-2}] \quad [4.4\text{-}48]$$

But for Gaussian processes:

$$E\{x_k/\tilde{y}_{k-1}\} = \Delta \tilde{y}_{k-1} = E\{x_k \tilde{y}_{k-1}\}[E\{\tilde{y}_{k-1}^2\}]^{-1} \tilde{y}_{k-1} \quad [4.4\text{-}49]$$

or:

$$\Delta = E\{x_k \tilde{y}_{k-1}\}[E\{\tilde{y}_{k-1}^2\}]^{-1} \quad [4.4\text{-}50]$$

But if we assume that $E\{\epsilon_k n_k\} = 0$, i.e. that the measurement noise and plant noise are uncorrelated, then it is easy to show that:

$$E\{x_k \tilde{y}_{k-1}\} = \Phi \mathbf{P}_{k-1/k-2} h \quad [4.4\text{-}51]$$

where $\mathbf{P}_{k-1/k-2} \equiv E\{\tilde{x}_{k-1/k-2} \tilde{x}_{k-1/k-2}^T\}$ = error covariance matrix for prediction of x_{k-1} given y_{k-2}. Similarly, one can show that:

$$E\{\tilde{y}_{k-1} \tilde{y}_{k-1}^T\} = h^T \mathbf{P}_{k-1/k-2} h + R = \text{scalar} \quad [4.4\text{-}52]$$

where $R = E\{n_k^2\}$ = scalar. Thus, the proportionality vector Δ becomes:

$$\Delta = \Phi \mathbf{P}_{k-1/k-2} h [h^T \mathbf{P}_{k-1/k-2} h + R]^{-1} \quad [4.4\text{-}53]$$

Finally, the *state space filter* (Kalman filter) for the one-step prediction of x_k given y_{k-1} is:

$$\boxed{\hat{x}_{k/k-1} = \Phi \hat{x}_{k-1/k-2} + \mathbf{K}_{k-1/k-2} [y_{k-1} - h^T \hat{x}_{k-1/k-2}] \quad [4.4\text{-}54]}$$

$$\mathbf{K}_{k-1/k-2} = [\mathbf{\Phi} \mathbf{P}_{k-1/k-2} h] \, [h^{\mathrm{T}} \mathbf{P}_{k-1/k-2} h + R]^{-1}$$
$$= \text{Kalman gain}$$

It is important to note that the computation of $\hat{x}_{k/k-1}$ requires knowledge of $\mathbf{\Phi}$, which is dependent on the serial correlations $\{\phi_1^{(N)}, \phi_2^{(N)}, \ldots \phi_N^{(N)}\}$ of our physical model and is unknown. (Note that in the case of linear prediction by Gauss least squares, it was possible to predict x_k given past values of the observation.) Mendel (1976) has derived an optimal smoothed estimate of the white noise sequence $\{\epsilon_k\}$ via the state space filtering approach. However, his estimate is also dependent on $\mathbf{\Phi}$ and γ which are both unknown, i.e. $\mathbf{\Phi}$ depends on the serial correlations and γ on the source wavelet and serial correlations (refer to equation [4.4-26]). Thus, although we have introduced a new point of view to the seismic problem, we have also introduced more unknowns into the problem. Further, implicit in our state space (Kalman) filtering formulation is the assumption of Gaussian statistics. This assumption implies that the amplitudes of the reflection coefficients are Gaussian random variables, which might not be physically correct. Nevertheless, the problem of estimating the reflection coefficients of a layered earth via state space filtering is a challenging one.

Chapter 5

COMPUTER PROGRAMS FOR FILTERING AND SPECTRAL ANALYSIS

5.1. INTRODUCTION

A signal refers to a quantity that varies with time. Because time represents a unidirectional flow from past to future, the field of signal analysis has developed as a particular mathematical discipline. In turn, this distinguishing characteristic of time must be carried and incorporated in digital computer programs for signal analysis.

In this chapter, we use the FORTRAN IV computer language, and we presuppose that the reader has a knowledge of that language equivalent to that in a standard textbook.

When we speak of Fortran programs, we mean more specifically Fortran subroutines. In other words, everything that we present is in subroutine form; this means that our programming efforts are, as nearly as possible, independent of any particular computer, and can be used in conjunction with a main routine on practically any digital computer. The purpose of the main routine is to bring data into the machine, to call the subroutines that perform the necessary calculations, and to get the results out of the machine. As we have said, we are only interested in the subroutines; the main routine, which is to a large extent dependent on the input-output hardware available on a particular machine, is left for the user to devise.

5.2. THE "STANDARD" PACKAGE OF SUBROUTINES

The subroutines in this package are so basic that we have chosen to call them "standard". The names of the standard subroutines are:

ZERO	(zero)
MOVE	(move)
IMPULS	(impulse)
SCALE	(scale)
DOT	(dot or inner product)
DOTR	(dot product reverse)
SYMDOT	(symmetrical dot product)
FOLD	(polynomial multiply or convolve)
PAC	(partial autocorrelation from autocorrelation)
CROSS	(cross-correlate)
CROSST	(cross-correlate)

TUKEY	(Tukey autocorrelation)
MINSN	(minimum, with sign)
MAXSN	(maximum, with sign)
NORME	(normalize with respect to RMS energy)
NORM1	(normalize with respect to first element)
REVERS	(reverse vector)
POLYDV	(polynomial divide or deconvolve)
MAINE	(symmetric matrix inverse)
CROUT	(Crout method for solving simultaneous equations)
POLYEV	(polynomial evaluation, for a complex value of its argument)
GENSYM	(generate a symmetric vector given one side)

We frequently deal with a group of variables that form a single collection. If these variables can be related to each other by subscript notation, we then call the collection an array. The purpose of subroutine ZERO is to put the floating point number zero (0.0) into each storage location of an array. In other words, this subroutine clears out any old numbers that might be in these storage locations, and replaces them by clean, fresh zeroes.

Let LX be a Fortran integer variable, and X be an array of Fortran floating point variables, which more specifically may be written as:

X(1), X(2), X(3), . . . X(LX)

That is, LX (called the length of X) is the number of elements in the array X. The call statement of subroutine ZERO is:

CALL ZERO (LX, X)

The result is that each element of the array X becomes 0.0; that is:

X(1) = 0.0, X(2) = 0.0, . . . X(LX) = 0.0

The program for subroutine ZERO is:

```
      SUBROUTINE ZERO(LX,X)
C     ZERO IS TO STORE ZERO IN AN ARRAY
      DIMENSION X(2)
      IF(LX)30,30,10
   10 DO 20 I=1,LX
   20 X(I)=0.0
   30 RETURN
      END
```

The purpose of subroutine MOVE is to move an array to another storage location. The call statement is:

CALL MOVE (LX, X, Y)

The result is that the numbers that were in an array X are now in both array X and array Y. The program is:

```
      SUBROUTINE MOVE(LX,X,Y)
C     MOVE IS MOVE ARRAY TO ANOTHER STORAGE LOCATION
      DIMENSION X(2),Y(2)
      DO 10 I=1,LX
   10 Y(I)=X(I)
      RETURN
      END
```

The purpose of subroutine IMPULS is to put the Kronecker impulse function into an array. The call statement is:

CALL IMPULS (LD, D, K)

The result is that each element of array D, of length LD, is zero except element D(K) which is unity. The program is:

```
      SUBROUTINE IMPULS(LD,D,K)
C     IMPULS IS KRONECKER IMPULSE FUNCTION
      DIMENSION D(2)
      DO 10 I=1,LD
   10 D(I)=0.0
      D(K)=1.0
      RETURN
      END
```

The purpose of subroutine SCALE is to multiply each element of an array by a scale factor. The call statement is:

CALL SCALE (S, LX, X)

The result is that each element of array X, of length LX, has been multiplied by the scale factor S. The program is:

```
      SUBROUTINE SCALE(S,LX,X)
C     SCALE IS SCALE AN ARRAY BY MULTIPLYING EACH
C       ELEMENT BY A SCALE FACTOR
      DIMENSION X(2)
      DO 10 I=1,LX
   10 X(I)=S*X(I)
      RETURN
      END
```

The *dot product* of the vector $x = (x_1, x_2, \ldots x_n)$ with the vector $y = (y_1, y_2, \ldots y_n)$ is defined as:

$$x_1 y_1 + x_2 y_2 + \ldots + x_n y_n$$

The purpose of subroutine DOT is to form the dot product of two vectors. The call statement is:

CALL DOT (L, X, Y, ANS)

The x vector is in array X of length L = n; the y vector is in array Y also of length L = n. Here L, X, and Y represent subroutine inputs; the subroutine output is ANS (standing for "answer"), which is the required dot product. The program is:

```
      SUBROUTINE DOT(L,X,Y,ANS)
C     DOT IS DOT PRODUCT
      DIMENSION X(2),Y(2)
      ANS=0.0
      IF(L)30,30,10
   10 DO 20 I=1,L
   20 ANS=ANS+X(I)*Y(I)
   30 RETURN
      END
```

The dot product reverse of the vector $x = (x_1, x_2, \ldots x_n)$ with the vector $y = (y_1, y_2, \ldots y_n)$ is defined as:

$$x_1 y_n + x_2 y_{n-1} + \ldots + x_{n-1} y_2 + x_n y_1$$

The purpose of subroutine DOTR is to form the dot product reverse of two vectors. The call statement is:

CALL DOTR (L, X, Y, ANS)

Here L, X, and Y are the subroutine inputs and ANS, the required dot product reverse, is the subroutine output. The program is:

```
      SUBROUTINE DOTR(L,X,Y,ANS)
C     DOTR IS DOT PRODUCT REVERSE
      DIMENSION X(2),Y(2)
      ANS=0.0
      IF(L)30,30,10
   10 DO 20 I=1,L
      J=L-I+1
   20 ANS=ANS+X(I)*Y(J)
   30 RETURN
      END
```

Subroutine SYMDOT computes a symmetrical dot product. The program, with the explanation given by comment cards, is:

```
      SUBROUTINE SYMDOT(Y,F,N,D)
C     SYMDOT IS SYMMETRICAL DOT PRODUCT
C     COMPUTES THE FOLLOWING SUMMATION(SUM)
C                  N-1
C     D=F(N)*Y(N)+SUM(F(N-I)*(Y(N-I)+Y(N+I)))
C                  I=1
C     AND DOES SO RATHER EFFICIENTLY.
C     FOR NORMAL USE,THE FOLLOWING MEANINGS APPLY
C     Y=DATA(DATA OF LENGTH 2*N-1 IS OPERATED ON)
C     F=ONE LOBE+CENTER POINT OF A SYMMETRICAL FILTER
C     N=LENGTH OF F(ONE LOBE+CENTER)
C     F(N)=CENTER , F(1)=END
C     D=DOT PRODUCT(A SINGLE NUMBER)
      DIMENSION Y(100),F(100)
      D=0.0
      M=N-1
      K=N+M
C     COMPUTES FROM THE OUTSIDE INTO THE CENTER
      DO 10 J=1,M
      D=F(J)*(Y(J)+Y(K))+D
      K=K-1
   10 CONTINUE
      D=D+F(N)*Y(N)
      RETURN
      END
```

The *convolution* (or folding) of the vector $a = (a_0, a_1, \ldots a_m)$ with the vector $b = (b_0, b_1, \ldots b_n)$ is defined as the vector $c = (c_0, c_1, \ldots c_{m+n})$ whose elements are given by the formula:

$$c_k = \sum_{n=0}^{k} a_n b_{k-n} = a_0 b_k + a_1 b_{k-1} + a_2 b_{k-2} + \ldots + a_k b_0$$

where the elements of any vector outside the given range are taken to be zero.

Alternatively, we may think of the elements c_k as the coefficients of the polynomial formed by the product of the polynomials whose coefficients are the elements of the two given vectors; that is:

$$c_0 + c_1 z + c_2 z^2 + \ldots + c_{m+n} z^{m+n}$$
$$= (a_0 + a_1 z + a_2 z^2 + \ldots + a_m z^m)(b_0 + b_1 z + b_2 z^2 + \ldots + b_n z^n)$$

The purpose of subroutine FOLD is to calculate the convolution of two vectors. The call statement is:

CALL FOLD (LA, A, LB, B, LC, C)

The subroutine inputs are LA, A, LB, B; the outputs are LC, C. The a vector is in array A of length LA $= m + 1$; the b vector is in array B of length LB $= n + 1$. The required c vector is found in array C of length LC $= m + n + 1$. A serious and inexcusable defect of the Fortran language is that the integers that are used as subscripts are not allowed to have numerical values that are zero or negative. Consequently, whenever a mathematical vector starts with the subscript 0, we have no choice but to make the Fortran vector start with the subscript 1.

Thus, in the present case we have:

$a_0 =$ A(1),	$a_1 =$ A(2),	$\ldots a_m$	= A(LA),	where LA $= m + 1$
$b_0 =$ B(1),	$b_1 =$ B(2),	$\ldots b_n$	= B(LB),	where LB $= n + 1$
$c_0 =$ C(1),	$c_1 =$ C(2),	$\ldots c_{n+m}$	= C(LC),	where LC $= m + n + 1$

The program for subroutine FOLD is:

```
      SUBROUTINE FOLD(LA,A,LB,B,LC,C)
C     FOLD IS FOLDING(CONVOLUTION)
      DIMENSION A(2),B(2),C(2)
      LC=LA+LB-1
      CALL ZERO(LC,C)
      DO 10 I=1,LA
      DO 10 J=1,LB
      K=I+J-1
   10 C(K)=A(I)*B(J)+C(K)
      RETURN
      END
```

It is seen that this program is based on the analogy of multiplication of polynomials.

Subroutine PAC computes the partial autocorrelation coefficient from the normalized autocorrelation coefficient (i.e. normalized so that the zero-lag autocorrelation coefficient is unity):

```
      SUBROUTINE PAC(LP,P,A,C)
C     P = NORMALIZED AUTOCORRELATION FOR LAGS 1 TO LP
C     A = WORKING SPACE
C     C = PARTIAL AUTOCORRELATION FOR INDICES 1 TO LP
      DIMENSION P(LP),A(LP),C(LP)
      DO 6 I=1,LP
      IF(I.GE.2) GO TO 1
      Q=P(1)
      V=1.0-Q*Q
      GO TO 4
    1 Q=U/V
      V=V*(1.0-Q*Q)
      L=(I-1)/2
      IF(L.EQ.0) GO TO 3
      DO 2 J=1,L
      HOLD=A(J)
      K=I-J
      A(J)=A(J)-Q*A(K)
    2 A(K)=A(K)-Q*HOLD
    3 IF(2*L.LT.I-1) A(L+1)=A(L+1)*(1.0-Q)
    4 A(I)=-Q
      U=P(I+1)
      DO 5 J=1,I
    5 U=U+A(J)*P(I+1-J)
    6 C(I)=Q
      RETURN
      END
```

The *cross-product* (or cross-correlation) of the vector $(x_0, x_1, \ldots x_m)$ with the vector $(y_0, y_1, \ldots y_n)$ is defined as the vector $(c_0, c_1, \ldots c_l)$ whose elements are given by the formula:

$$c_k = \sum_{j=0}^{m} x_j y_{j-k}$$

where values of x_j outside the range $j = 0, 1, 2, \ldots m$ and values of y_i outside the range $i = 0, 1, 2, \ldots n$ are taken to be 0.0.

The number of elements in the c vector is $l + 1$; the integer l is arbitrary and is specified in advance. The purpose of subroutine CROSS is to compute the cross-product of two vectors. The call statement is:

CALL CROSS (LX, X, LY, Y, LC, C)

The subroutine inputs are LX, X, LY, Y, LC; the output is C. The x vector is in array X, of length LX $= m + 1$; the y vector is in array Y, of length LY $= n + 1$. The required c vector is found in array C, of length LC $= l + 1$. The program for subroutine CROSS is:

```
      SUBROUTINE CROSS(LX,X,LY,Y,LC,C)
C     CROSS IS CROSS PRODUCT
      DIMENSION X(2),Y(2),C(2)
      DO 10 I=1,LC
   10 CALL DOT(MIN0(LY+I-1,LX)-I+1,X(I),Y,C(I))
      RETURN
      END
```

If X and Y are the same array, CROSS gives us the *autocorrelation*.

A cross-correlation subroutine based on FOLD is CROSST. The calling statement is:

CALL CROSST (N, X, M, Y, C)

where the subroutine inputs are:

N = length of X
X = one vector to be cross-correlated
M = length of Y
Y = other vector to be cross-correlated

and the subroutine output is:

C = cross-correlation

The program is (note: subroutine REVERS is given later in this section):

```
      SUBROUTINE CROSST(N,X,M,Y,C)
C     CROSST IS CROSS CORRELATION BASED ON SUBROUTINE FOLD
      DIMENSION X(100),Y(100),C(100)
      CALL REVERS(M,Y)
      CALL FOLD(N,X,M,Y,LC,C)
      CALL REVERS(M,Y)
      RETURN
      END
```

Subroutine TUKEY computes the Tukey autocorrelation. The program with its explanation is:

```
      SUBROUTINE TUKEY(LX,X,LACOR,ACOR,S)
C     TUKEY IS TUKEY AUTOCORRELATION
C     TUKEY COMPUTES THE AUTOCORRELATION OF X USING THE
C       TUKEY APPROXIMATION
C     INPUTS ARE LX=LENGTH OF X , X=THE TIME SERIES ,
C       LACOR=LARGEST LAG
C     OUTPUT IS ACOR WHICH IS TUKEY AUTOCORRELATION(NORMALIZED)
      DIMENSION X(2),ACOR(2)
      DO 15 IT=1,LACOR
      ACOR(IT)=0.0
      MM=LX-IT+1
      DO 10 J=1,MM
      I=J+IT-1
   10 ACOR(IT)=X(I)*X(J)+ACOR(IT)
      SCALE=1.0/FLOAT(LX-IT+1)
   15 ACOR(IT)=SCALE*ACOR(IT)
      S=ACOR(1)
      SCALE=1.0/ACOR(1)
      DO 20 I=1,LACOR
   20 ACOR(I)=ACOR(I)*SCALE
      RETURN
      END
```

The purpose of subroutine MINSN is to find the minimum element of an array, taking into account the algebraic signs of the elements. The calling statement is:

CALL MINSN (LX, X, XMIN, INDEX)

The inputs are LX, X; the outputs are XMIN, INDEX. The given array X is of length LX. The minimum element of this array is XMIN = X(INDEX), that is, INDEX is the subscript of the minimal element XMIN. The program is:

```
      SUBROUTINE MINSN(LX,X,XMIN,INDEX)
      DIMENSION X(2)
      XMIN=X(1)
      DO 10 I=1,LX
   10 XMIN=AMIN1(XMIN,X(I))
      DO 20 J=1,LX
      INDEX=J
      IF(X(J)-XMIN)20,30,20
   20 CONTINUE
   30 RETURN
      END
```

The subroutine MAXSN is just like MINSN, except that it finds the maximum element. The program is:

```
      SUBROUTINE MAXSN(LX,X,XMAX,INDEX)
      DIMENSION X(2)
      XMAX=X(1)
      DO 10 I=1,LX
   10 XMAX=AMAX1(XMAX,X(I))
      DO 20 J=1,LX
      INDEX=J
      IF(X(J)-XMAX)20,30,20
   20 CONTINUE
   30 RETURN
      END
```

The purpose of subroutine NORME is to normalize an array by dividing each element by the RMS energy of the array. The calling statement is:

CALL NORME (LX, X)

The inputs are LX, X, where LX is the number of elements in array X; the output is X, now normalized. The program is:

```
      SUBROUTINE NORME(LX,X)
C     NORME IS NORMALIZATION OF ARRAY BY ITS RMS ENERGY
      DIMENSION X(2)
      IF(LX)30,30,10
   10 CALL DOT(LX,X,X,XX)
      Z=SQRT(XX)
      DO 20 I=1,LX
   20 X(I)=X(I)/Z
   30 RETURN
      END
```

A similar program is NORM1, which normalizes an array by its first element. The program is:

```
      SUBROUTINE NORM1(LX,X)
      DIMENSION X(2)
      IF(LX)30,30,10
   10 XN=X(1)
      DO 20 I=1,LX
   20 X(I)=X(I)/XN
   30 RETURN
      END
```

The purpose of subroutine REVERS is to reverse, that is, interchange, the order of the elements of an array. The calling statement is:

CALL REVERS (N, X)

The inputs are N, X, where N is the number of elements in array X; the output is X, which now has the same elements but in the reverse order. The program is:

```
      SUBROUTINE REVERS(N,X)
      DIMENSION X(100)
      NN=N/2
      DO 10 I=1,NN
      J=N-I
      TEMP=X(I)
      X(I)=X(J+1)
      X(J+1)=TEMP
   10 CONTINUE
      RETURN
      END
```

The purpose of subroutine POLYDV is to divide one polynomial by another, that is, to *deconvolve* one vector by another. The program, with the explanation given by comment cards, is:

```
      SUBROUTINE POLYDV(N,DVS,M,DVD,L,Q)
C     PERFORM DIVISION OF POWER SERIES OF THE FORM
C     (DVD(1)+X*DVD(2)+X*X*DVD(3)+...+)/
C     (DVS(1)+X*DVS(2)+X*X*DVS(3)+...+).
C     THE RESULT IS = Q(1)+X*Q(2)+X*X*Q(3)+...
C     N IS THE NUMBER OF COEFFICIENTS IN THE DIVISOR POLYNOMIAL
C     M IS THE NUMBER OF COEFFICIENTS IN THE DIVIDEND POLYNOMIAL
C     L IS THE NUMBER OF COEFFICIENTS IN THE QUOTIENT POLYNOMIAL
C     DVS IS THE INPUT DIVISOR POLYNOMIAL COEFFICIENTS
C     DVD IS THE INPUT DIVIDEND POLYNOMIAL COEFFICIENTS
C     Q IS THE OUTPUT QUOTIENT POLYNOMIAL COEFFICIENTS
C     DVS(1) MUST BE NONZERO
C     THIS COULD BE REPROGRAMMED TO ALLOW EQUIVALENCE(DVD,Q)
      DIMENSION DVS(10),DVD(10),Q(10)
      NM=N-1
      DO 8 I=1,L
    8 Q(I)=0.0
C     MOVE THE USED PORTION OF DVD TO Q
      MML=MINO(M,L)
      DO 10 I=1,MML
   10 Q(I)=DVD(I)
      DO 50 I=1,L
```

```
      Q(I)=Q(I)/DVS(1)
      IF(I-L)30,20,30
   20 RETURN
   30 K=I
      ISUB=MINO(NM,L-I)
      DO 40 J=1,ISUB
      K=K+1
      Q(K)=Q(K)-Q(I)*DVS(J+1)
   40 CONTINUE
   50 CONTINUE
C     PROGRAM NEVER GETS HERE
      RETURN
      END
```

The purpose of subroutine MAINE is to invert a symmetric matrix. The call statement is:

CALL MAINE (N, A, B)

The inputs are N, A; the output is B. The symmetric matrix has order N, and is stored in array A. The inverse of this matrix is found stored in array B. The matrix A and its inverse B are each stored closely packed columnwise as square matrices; that is, in storage, the first column is followed by the second column, and so on, until the last column of the matrix. The program is:

```
      SUBROUTINE MAINE(N,A,B)
C     DIMENSION A(N*N),B(N*N)
      DIMENSION A(2),B(2)
      B(1)=1.0/A(1)
      IF(N-1)60,60,2
    2 CALL ZERO(N*N-1,B(2))
      DO 50 M=2,N
      K=M-1
      MM=M+(M-1)*N
      EK=A(MM)
      DO 10 I=1,K
      DO 10 J=1,K
      MI=M+(I-1)*N
      IJ=I+(J-1)*N
      JM=J+(M-1)*N
   10 EK=EK-A(MI)*B(IJ)*A(JM)
      B(MM)=1.0/EK
      DO 30 I=1,K
      IM=I+(M-1)*N
      DO 20 J=1,K
      IJ=I+(J-1)*N
      JM=J+(M-1)*N
   20 B(IM)=B(IM)-B(IJ)*A(JM)/EK
      MI=M+(I-1)*N
   30 B(MI)=B(IM)
      DO 40 I=1,K
      IM=I+(M-1)*N
      DO 40 J=1,K
      MJ=M+(J-1)*N
      IJ=I+(J-1)*N
   40 B(IJ)=B(IJ)+B(IM)*B(MJ)*EK
   50 CONTINUE
   60 RETURN
      END
```

Simultaneous equations may be solved by the Crout method. Subroutine CROUT is:

```
      SUBROUTINE CROUT(M,N,A,ALPHA,NSYS,X)
      DIMENSION A(M,N),ALPHA(M,N),X(M,NSYS)
      DO 1 I=1,M
    1 ALPHA(I,1)=A(I,1)
      DO 2 J=2,N
    2 ALPHA(1,J)=A(1,J)/ALPHA(1,1)
      IF(M.EQ.1)GO TO 9
      DO 6 K=2,M
      KP=K+1
      KM=K-1
      DO 4 I=K,M
      SUM=0.0
      DO 3 L=1,KM
    3 SUM=SUM+ALPHA(I,L)*ALPHA(L,K)
    4 ALPHA(I,K)=A(I,K)-SUM
      DO 6 J=KP,N
      SUM=0.0
      DO 5 L=1,KM
    5 SUM=SUM+ALPHA(K,L)*ALPHA(L,J)
    6 ALPHA(K,J)=(A(K,J)-SUM)/ALPHA(K,K)
      DO 8 IT=1,NSYS
      J=M+IT
      X(M,IT)=ALPHA(M,J)
      DO 8 I=2,M
      K=M+1-I
      KP=K+1
      SUM=0.0
      DO 7 L=KP,M
    7 SUM=SUM+ALPHA(K,L)*X(L,IT)
    8 X(K,IT)=ALPHA(K,J)-SUM
      RETURN
    9 DO 10 IT=1,NSYS
   10 X(1,IT)=ALPHA(1,IT+1)
      RETURN
      END
```

Suppose that we are given q systems of simultaneous equations, each with the same matrix of coefficients on the left-hand side but with a different right-hand side. Each system has m equations and m unknowns. Let system t be given by:

$$a_{11}x_{1t} + a_{12}x_{2t} + \ldots + a_{1m}x_{mt} = a_{1,m+t}$$

$$a_{21}x_{1t} + a_{22}x_{2t} + \ldots + a_{2m}x_{mt} = a_{2,m+t}$$

$$\vdots$$

$$a_{m1}x_{1t} + a_{m2}x_{2t} + \ldots + a_{mm}x_{mt} = a_{m,m+t}$$

where $t = 1, 2, \ldots q$. Subroutine CROUT computes the solutions of these q systems. The call statement is:

CALL CROUT (M, N, A, ALPHA, NSYS, X)

where the subroutine inputs are:

M $= m$
N $= n$, where n is defined as $m + q$

$$A = \text{initial array} = \begin{bmatrix} a_{11} & a_{12} & \ldots a_{1n} \\ a_{21} & a_{22} & \ldots a_{2n} \\ \vdots & \vdots & \\ a_{m1} & a_{m2} & \ldots a_{mn} \end{bmatrix}$$

NSYS = q = number of systems

and where the subroutine outputs are:

$$\text{ALPHA} = \text{Crout array} = \begin{bmatrix} \alpha_{11} & \alpha_{12} & \ldots \alpha_{1n} \\ \alpha_{21} & \alpha_{22} & \ldots \alpha_{2n} \\ \vdots & \vdots & \\ \alpha_{m1} & \alpha_{m2} & \ldots \alpha_{mn} \end{bmatrix}$$

$$X = \text{solution array} = \begin{bmatrix} x_{11} & x_{12} & \ldots x_{1q} \\ x_{21} & x_{22} & \ldots x_{2q} \\ \vdots & \vdots & \\ x_{m1} & x_{m2} & \ldots x_{mq} \end{bmatrix}$$

where column t of the solution array is the solution to system t.

Numerical example (1). We are given the system of simultaneous equations:

$$\begin{aligned} 2x_{11} + 3x_{21} + 8x_{31} &= 4 \\ 4x_{11} + 9x_{21} + 10x_{31} &= 20 \\ 3x_{11} + 5x_{21} + 13x_{31} &= 6 \end{aligned}$$

The subroutine inputs are:

$M = m = 3$
$N = n = m + q = 4$

$$A = \begin{bmatrix} 2 & 3 & 8 & 4 \\ 4 & 9 & 10 & 20 \\ 3 & 5 & 13 & 6 \end{bmatrix}$$

NSYS $= q = 1$

Then the call statement:

CALL CROUT (M, N, A, ALPHA, NSYS, X)

computes the subroutine outputs:

$$\text{ALPHA} = \begin{bmatrix} 2 & 1.5 & 4.0 & 2.0 \\ 4 & 3.0 & -2.0 & 4.0 \\ 3 & 0.5 & 2.0 & -1.0 \end{bmatrix}$$

$$X = \begin{bmatrix} x_{11} \\ x_{21} \\ x_{31} \end{bmatrix} = \begin{bmatrix} 3 \\ 2 \\ -1 \end{bmatrix}$$

Numerical example (*2*). We are now given two systems of simultaneous equations, where system 1 is given by:

$$\begin{aligned} 2x_{11} + 3x_{21} + 8x_{31} &= 4 \\ 4x_{11} + 9x_{21} + 10x_{31} &= 20 \\ 3x_{11} + 5x_{21} + 13x_{31} &= 6 \end{aligned}$$

and where system 2 is given by:

$$\begin{aligned} 2x_{12} + 3x_{22} + 8x_{32} &= 5 \\ 4x_{12} + 9x_{22} + 10x_{32} &= 25 \\ 3x_{12} + 5x_{22} + 13x_{32} &= 8 \end{aligned}$$

The subroutine inputs are:

$$\text{M} = m = 3$$
$$\text{N} = n = m + q = 5$$

$$\text{A} = \begin{bmatrix} 2 & 3 & 8 & 4 & 5 \\ 4 & 9 & 10 & 20 & 25 \\ 3 & 5 & 13 & 6 & 8 \end{bmatrix}$$

$$\text{NSYS} = q = 2$$

Then the call statement:

CALL CROUT (3, 5, A, ALPHA, 2, X)

computes the subroutine outputs:

$$\text{ALPHA} = \begin{bmatrix} 2 & 1.5 & 4.0 & 2.0 & 2.5 \\ 4 & 3.0 & -2.0 & 4.0 & 5.0 \\ 3 & 0.5 & 2.0 & -1.0 & -1.0 \end{bmatrix}$$

$$X = \begin{bmatrix} x_{11} & x_{12} \\ x_{21} & x_{22} \\ x_{31} & x_{32} \end{bmatrix} = \begin{bmatrix} 3 & 2 \\ 2 & 3 \\ -1 & -1 \end{bmatrix}$$

Numerical example (*3*). In certain cases we might not wish to store the solutions in a separate array X, but instead want to store the solutions in the right-hand portion of array ALPHA. Subroutine CROUT has the property that we can make the array X equivalent to the right-hand portion of array ALPHA. Hence, for example, if:

$$A = \begin{bmatrix} 2 & 3 & 8 & 4 & 5 \\ 4 & 9 & 10 & 20 & 25 \\ 3 & 5 & 13 & 6 & 8 \end{bmatrix}$$

then the call statement:

CALL CROUT (3, 5, A, ALPHA, 2, ALPHA (1, 4))

produces:

$$\text{ALPHA} = \begin{bmatrix} 2 & 1.5 & 4.0 & 3 & 2 \\ 4 & 3.0 & -2.0 & 2 & 3 \\ 3 & 0.5 & 2.0 & -1 & -1 \end{bmatrix}$$

Numerical example (4). In other instances, we may not be interested in saving the values in array A, but instead want to store the left-hand portion (i.e. the left-most m columns) of the CROUT array in the left-hand portion of array A and the solution array X in the right-hand portion of array A. Subroutine CROUT has the property that we can make array ALPHA equivalent to array A. Hence, given the numerical values of array A in the previous example, the call statement:

CALL CROUT (3, 5, A, A, 2, A (1, 4))

produces:

$$A = \begin{bmatrix} 2 & 1.5 & 4.0 & 3 & 2 \\ 4 & 3.0 & -2.0 & 2 & 3 \\ 3 & 0.5 & 2.0 & -1 & -1 \end{bmatrix}$$

The purpose of subroutine POLYEV is to evaluate a polynomial for a complex value of its argument. The program, with the explanation given by comment cards, is:

```
      SUBROUTINE POLYEV(N,C,X,A)
C     POLYEV IS EVALUATE POLYNOMIAL
C     COMPUTES THE FORMULA
C
C     A=C(1)+C(2)*X+C(3)*(X**2)+...+C(N)*(X**(N-1))
C
C     EXAMPLE , IF N=5,THE FORMULA IS EVALUATED IN THE FOLLOWING WAY
C
C     A=C(1)+X*(C(2)+X*(C(3)+X*(C(4)+X*C(5))))
C
C     THUS,ONLY N-1 ADDS AND N-1 MULTIPLIES ARE REQUIRED.
      DIMENSION C(100)
      COMPLEX X,A
      A=0.0
      DO 10 I=1,N
      J=N-I
   10 A=X*A+C(J+1)
      RETURN
      END
```

Subroutine GENSYM generates a symmetric vector given one side. Its program is:

```
      SUBROUTINE GENSYM(N,A,B)
C     GENSYM GENERATES A SYMMETRIC VECTOR GIVEN ONE SIDE
C     INPUTS   ANY SERIES A OF LENGTH N
C     OUTPUTS   THE SERIES B DEFINED AS FOLLOWS
C       B(I)=A(N-I+1) , FOR I=1 TO I=N
C       B(I)=A(I-N+1) , FOR I=N TO I=N+N-1
C     THUS,FOR EXAMPLE,IF A IS THE CENTER PLUS ONE LOBE OF AN
C       AUTOCORRELATION,THEN B IS THE WHOLE AUTOCORRELATION WITH
C       CENTER AT B(N).
C     EQUIVALENCE(A,B) IS ALLOWED
      DIMENSION A(100),B(100)
      DO 10 I=1,N
      J=N+I
   10 B(J-1)=A(I)
      IN=N+N
      DO 20 I=1,N
      J=IN-I
   20 B(I)=B(J)
      RETURN
      END
```

5.3. THE FILTER PACKAGE

The subroutines in this package are:

EUREKA	(general Toeplitz recursion)
PEO	(auxiliary Toeplitz recursion)
INVTOP	(inverse Toeplitz matrix)
SHAPE	(shaping filter)
SPIKE	(spiking filter)
SIDE	(Simpson sideways iteration)
SPIKER	(spiking filter, more efficient than spike)
SHAPER	(shaping filter for optimum positioning)
WVPRED	(wavelet prediction)
BNDPAS	(band-pass filter)
BNDPS2	(band-pass filter)
TREN	(traveling energy)
TRAP	(traveling power)
TRAF	(traveling correlation filter)
REMAV	(remove arithmetic average)
RMSDEV	(root-mean-square deviation)
BURG	(Burg algorithm)

The solution of the least-squares optimum filtering problem involves solving a set of simultaneous equations called the *normal equations*. In general, there will be one equation for each coefficient in the filter. The requirements for computer time and computer storage space for solving these equations by use of a standard simultaneous equations routine is prohibitive, except in the case of a small number of filter coefficients. The Toeplitz recursion gives a more efficient scheme for arriving at the desired filter coefficients.

This scheme makes use of the special form of the autocorrelation matrix **R**, called the Toeplitz form. This form can be written as:

$$\mathbf{R} = \begin{bmatrix} R_0 & R_1 & R_2 & \ldots R_m \\ R_1 & R_0 & R_1 & \ldots R_{m-1} \\ R_2 & R_1 & R_0 & \ldots R_{m-2} \\ \vdots & & & \\ R_m & R_{m-1} & R_{m-2} & \ldots R_0 \end{bmatrix}$$

All terms along each diagonal are the same. Thus, given the entries in the top row of **R**, the matrix **R** is fully specified.

The Toeplitz recursion involves initially finding a filter of length one, using this filter to find a filter of length two, and so on, until the desired length filter is reached. The principal advantages of using the recursive technique are time and space savings. The standard solution of simultaneous equations requires time proportional to m^3 and space proportional to m^2. The recursive technique reduces these requirements to m^2 and m for time and space respectively. An important side benefit for using this scheme is that we can compute the mean-square error at each step of the process. This allows us to formulate a criterion for determining the length of the filter. As the filter becomes longer, the prediction-error variance v will decrease and then level off at some value.

The Toeplitz recursion is a classic recursion occurring in the theory of polynomials orthogonal on the unit circle. The scheme given here is the one given by Wiggins and Robinson (1965), which includes a recursive formula for the prediction-error variance as well as for the filter coefficients.

The normal equations for a scalar process are:

$$\sum_{i=0}^{n} f_i R_{j-i} = g_j \quad (j = 0, 1, 2, \ldots n)$$

where the f_i are the filter coefficients, the R_{j-i} are the autocorrelation coefficients, and the g_j are the right-hand side coefficients. Associated with the normal equations are the normal equations for the unit-step prediction-error operator a_i:

$$\sum_{i=0}^{n} a_i R_{j-i} = \begin{cases} v, & \text{when } j = 0 \\ 0, & \text{when } j = 1, 2, \ldots n \end{cases}$$

where $a_0 \equiv 1$ and v is the prediction-error variance.

The hindsight operator is one which "predicts" past values of a time series from future values. For the scalar case, the unit-step hindsight-error operator b_j is just the reverse of the unit-step prediction-error operator since **R** is symmetric; that is:

$$\begin{bmatrix} b_0 \\ b_1 \\ \vdots \\ b_n \end{bmatrix} = \begin{bmatrix} a_n \\ a_{n-1} \\ \vdots \\ a_0 \end{bmatrix}$$

The scheme for extending the unit-step prediction-error operator $a = (a_0, a_1, \ldots a_n)$ to the new unit-step prediction-error operator $a_0', a_1', \ldots a_{n+1}'$ involving one more coefficient is first to extend a by adding a zero to the right end of a:

$$\begin{bmatrix} R_0 & \ldots & R_{n+1} \\ & \ldots & \\ R_{n+1} & \ldots & R_0 \end{bmatrix} \begin{bmatrix} a_0 \\ \ldots \\ a_n \\ 0 \end{bmatrix} = \begin{bmatrix} v \\ 0 \\ \ldots \\ 0 \\ d \end{bmatrix}$$

The quantity d is called the discrepancy, defined as the dot product

$$d = R_{n+1}a_0 + R_n a_1 + \ldots + R_1 a_n$$

If the discrepancy is zero, then the extended prediction-error operator is the correct one. Generally, the discrepancy will not be zero, so the next step is to modify the coefficients of the extended operator so as to cancel out the discrepancy. We do this by adding a weighted version of the extended hindsight-error operator to the extended prediction-error operator. Thus, we obtain:

$$\begin{bmatrix} R_0 & \ldots & R_{n+1} \\ & \ldots & \\ R_{n+1} & \ldots & R_0 \end{bmatrix} \begin{bmatrix} a_0 \\ a_1 - ca_n \\ \ldots \\ a_n - ca_0 \\ -ca_0 \end{bmatrix} = \begin{bmatrix} v - cd \\ 0 \\ \ldots \\ 0 \\ d - cv \end{bmatrix}$$

The quantity $d - cv$ is set equal to zero in order to determine the value of c:

$$c = d/v$$

That is, c is equal to the ratio of the discrepancy d to the prediction-error variance v. The new operator is thus:

$$\begin{bmatrix} a_0' \\ a_1' \\ \vdots \\ a_n' \\ a_{n+1}' \end{bmatrix} = \begin{bmatrix} a_0 \\ a_1 - ca_n \\ \vdots \\ a_n - ca_1 \\ -ca_0 \end{bmatrix} = \begin{bmatrix} 1 \\ a_1 - ca_n \\ \vdots \\ a_n - ca_1 \\ -c \end{bmatrix}$$

and the new prediction-error variance is:

$$v' = v - cd = v - c(cv) = v(1 - c^2)$$

We now use the new prediction-error operator to extend the length of the filter $f = (f_0, f_1, \ldots f_n)$. As before, we make a first approximation to f' by adding a zero to the right end of f:

$$\begin{bmatrix} R_0 & \ldots & R_{n+1} \\ & \ldots & \\ R_{n+1} & \ldots & R_0 \end{bmatrix} \begin{bmatrix} f_0 \\ f_1 \\ \ldots \\ f_n \\ 0 \end{bmatrix} = \begin{bmatrix} g_0 \\ g_1 \\ \cdot \\ g_n \\ q \end{bmatrix}$$

where q is defined as the dot product:

$$q = R_{n+1}f_0 + R_nf_1 + \ldots + R_1f_n$$

If we weight and add the new hindsight-error operator to the above extended filter we get:

$$\begin{bmatrix} R_0 & \ldots & R_{n+1} \\ & \ldots & \\ R_{n+1} & \ldots & R_0 \end{bmatrix} \begin{bmatrix} f_0 - sa_{n+1}' \\ f_1 - sa_n' \\ \ldots \\ f_n - sa_1' \\ -sa_0' \end{bmatrix} = \begin{bmatrix} g_0 \\ g_1 \\ \ldots \\ g_n \\ q - sv' \end{bmatrix}$$

Now we choose the constant s such that:

$$q - sv' = g_{n+1}$$

That is, if we define s as:

$$s = \frac{q - g_{n+1}}{v'}$$

then the new filter is:

$$\begin{bmatrix} f'_0 \\ f'_1 \\ \cdot \\ \cdot \\ f'_n \\ f'_{n+1} \end{bmatrix} = \begin{bmatrix} f_0 - sa'_{n+1} \\ f_1 - sa'_n \\ \cdot \\ \cdot \\ f_n - sa'_1 \\ - sa'_0 \end{bmatrix} = \begin{bmatrix} f_0 - sa'_{n+1} \\ f_1 - sa'_n \\ \cdot \\ \cdot \\ f_n - sa'_1 \\ - s \end{bmatrix}$$

Subroutine EUREKA solves the least-squares normal equations for the coefficients of the filter f as well as for the coefficients of the unit-step prediction-error operator a. The program for this general Toeplitz recursion is:

```
      SUBROUTINE EUREKA(LR,R,G,F,A)
C     EUREKA SOLVES THE LEAST SQUARES NORMAL EQUATIONS
C       FOR F(I) AS WELL AS THE PREDICTION-ERROR OPERATOR A(I).
      DIMENSION R(LR),G(LR),F(LR),A(LR)
      V=R(1)
      A(1)=1.0
      F(1)=G(1)/V
      IF(LR.EQ.1) RETURN
      DO 6 L=2,LR
      D=0.0
      Q=0.0
      L3=L-1
      DO 1 J=1,L3
      K=L-J+1
      D=D+A(J)*R(K)
    1 Q=Q+F(J)*R(K)
      C=D/V
      IF(L.EQ.2) GO TO 4
      L1=(L-2)/2
      L2=L1+1
      IF(L2.LT.2) GO TO 3
      DO 2 J=2,L2
      HOLD=A(J)
      K=L-J+1
      A(J)=A(J)-C*A(K)
    2 A(K)=A(K)-C*HOLD
      IF(2*L1.EQ.L-2)GO TO 4
    3 LT3=L2+1
      A(LT3)=A(LT3)-C*A(LT3)
    4 A(L)=-C
      V=V-C*D
      S=(Q-G(L))/V
      DO 5 J=1,L3
      K=L-J+1
    5 F(J)=F(J)-S*A(K)
    6 F(L)=-S
      RETURN
      END
```

Each step of this general Toeplitz scheme requires the computation of two dot products, namely d and q, instead of three dot products, namely d, q, and also one for v as required by the Levinson method. Criticisms as to the accuracy of the Levinson method are due to the errors resulting from computing the variance v as a dot product, which is unnecessary in the scheme given here.

In the special case when one desires the prediction-error operator (or, equivalently, the prediction operator), then with the recognition that a is the desired result, one can eliminate the computation of the dot product q as well as s and f. This recognition essentially reduces the computation by one-half for prediction operators as compared to general filter operators. Subroutine PEO is the required modification of EUREKA to compute only the prediction-error operator a. Subroutine PEO may be described as the auxiliary Toeplitz recursion.

```
      SUBROUTINE PEO(LR,R,A)
C     PEO IS THE AUXILLARY TOEPLITZ RECURSION.
C     IT GIVES THE PREDICTION-ERROR OPERATOR A(I).
      DIMENSION R(LR),A(LR)
      V=R(1)
      A(1)=1.0
      IF(LR.EQ.1)RETURN
      DO 6 L=2,LR
      D=0.0
      L3=L-1
      DO 1 J=1,L3
      K=L-J+1
    1 D=D+A(J)*R(K)
      C=D/V
      IF(L.EQ.2)GO TO 4
      L1=(L-2)/2
      L2=L1+1
      IF(L2.LT.2)GO TO 3
      DO 2 J=2,L2
      HOLD=A(J)
      K=L-J+1
      A(J)=A(J)-C*A(K)
    2 A(K)=A(K)-C*HOLD
      IF(2*L1.EQ.L-2)GO TO 4
    3 LT3=L2+1
      A(LT3)=A(LT3)-C*A(LT3)
    4 A(L)=-C
    6 V=V-C*D
      RETURN
      END
```

The calling statement for EUREKA is:

CALL EUREKA (LR, R, G, F, A)

where the subroutine inputs are:

LR = length of R = length of G = length of F = length of A
R = autocorrelation from lag zero to lag LR-1
G = cross-correlation

and the subroutine outputs are:

F = filter coefficients
A = prediction-error operator coefficients

The calling statement for PEO is:

CALL PEO (LR, R, A)

where the subroutine inputs are:

LR = length of R = length of A
R = autocorrelation from lag zero to lag LR-1

and the subroutine outputs are:

A = prediction-error operator coefficients

One example of the use of subroutine EUREKA is for the calculation of the inverse of a Toeplitz matrix. This is accomplished by subroutine INVTOP, which calls EUREKA. The calling statement is:

CALL INVTOP (LR, R, RI, SPACE)

The inputs are LR, R; the output is RI; working space is SPACE. LR is the order of the Toeplitz matrix and R is its first row. The inverse matrix is found in RI. The program is:

```
      SUBROUTINE INVTOP(LR,R,RI,SPACE)
C     INVTOP IS INVERSE TOEPLITZ MATRIX
C     INVERSE MATRIX IS STORED BY ROWS OR COLUMNS IN R(I)
C     I=1,LR*LR
C     WORKING SPACE(I) , I=1,2*LR
      DIMENSION R(5),RI(25),SPACE(10)
      DO 10 K=1,LR
      CALL IMPULS(LR,SPACE,K)
      J=LR*(K-1)+1
      LT=LR+1
      CALL EUREKA(LR,R,SPACE,RI(J),SPACE(LT))
   10 CONTINUE
      RETURN
      END
```

The numerical examples computed by subroutine INVTOP are:

Example (1)

LR = 5

R =	10.00000	4.00000	−1.00000	−4.00000	−4.00000
RI =	0.14899	−0.05657	0.02727	0.03232	0.02677
	−0.05657	0.16566	−0.07273	0.01010	0.03232
	0.02727	−0.07273	0.16364	−0.07273	0.02727
	0.03232	0.01010	−0.07273	0.16566	−0.05657
	0.02677	0.03232	0.02727	−0.05657	0.14899

Example (2)

LR = 5

R =	10.00000	4.00000	1.00000	4.00000	4.00000
RI =	0.15323	−0.06452	0.03763	−0.06452	−0.01344
	−0.06452	0.17921	−0.08602	0.06810	−0.06452
	0.03763	−0.08602	0.16129	−0.08602	0.03763
	−0.06452	0.06810	−0.08602	0.17921	−0.06452
	−0.01344	−0.06452	0.03763	−0.06452	0.15323

Example (*3*)

LR =	5				
R =	16.72605	0.	10.75437	0.	2.44141
RI =	0.12867	0.	−0.12045	0.	0.05867
	0.	0.10192	0.	−0.06553	0.
	−0.12045	0.	0.21468	0.	−0.12045
	0.	−0.06553	0.	0.10192	0.
	0.05867	0.	−0.12045	0.	0.12867

Another example of the use of subroutine EUREKA is for the calculation of shaping filters. This is accomplished by subroutine SHAPE. The calling statement is:

CALL SHAPE (LB, B, LD, D, LA, A, LC, C, ASE, SPACE)

where the subroutine inputs are:

LB = length of B
B = wavelet that is input to shaping filter
LD = length of D
D = wavelet that is desired output of filter
LA = length of A

and the subroutine outputs are:

A = coefficients of shaping filter
LC = length of C
C = wavelet that is actual output of filter
ASE = average-squared error between desired and actual outputs of filter
SPACE = working space

The program is:

```
      SUBROUTINE SHAPE(LB,B,LD,D,LA,A,LC,C,ASE,SPACE)
C     SHAPE IS SHAPING FILTER
C     INPUTS ARE LB,B,LD,D,LA
C     OUTPUTS ARE A,LC,C,ASE
C     SPACE(I),I=1,3*LA (WORKING SPACE)
C     MAY BE EQUIVALENCE(C,D)
      DIMENSION B(2),D(1),A(4),C(5),SPACE(16)
      LT1=LA+1
      LT2=2*LA+1
      CALL CROSS(LB,B,LB,B,LA,SPACE)
      CALL CROSS(LD,D,LB,B,LA,SPACE(LT1))
      CALL EUREKA(LA,SPACE,SPACE(LT1),A,SPACE(LT2))
      CALL DOT(LD,D,D,DD)
      CALL DOT(LA,A,SPACE(LT1),AG)
      ASE=(DD-AG)/DD
      CALL FOLD(LA,A,LB,B,LC,C)
      RETURN
      END
```

Some numerical examples computed by subroutine SHAPE are:

Example (1).

LB	=	2				
B	=	−1.25000	1.00000	(minimum-delay)		
LD	=	5				
D	=	2.00	1.00	0.00	1.00	2.00
LA	=	2				
A	=	−1.08040	−1.01483			
LC	=	3				
C	=	1.35050	0.18813	−1.01483		
ASE	=	0.71109				
SPACE	=	2.56250	−1.25000	−1.50000	−1.25000	

Example (2).

LB	=	2				
B	=	1.00000	−1.25000	(maximum-delay)		
LD	=	5				
D	=	2.00	1.00	0.00	1.00	2.00
LA	=	2				
A	=	0.63388	0.69945			
LC	=	3				
C	=	0.63388	−0.09290	−0.87432		
ASE	=	0.88251				
SPACE	=	2.56250	−1.25000	0.75000	1.00000	

The above two examples show that it is more difficult to shape the maximum-delay input into this desired output than the minimum-delay input.

Example (3).

LB	=	3				
B	=	1.56250	−1.76770	1.00000	(minimum-delay)	
LD	=	5				
D	=	2.00	1.00	0.00	1.00	2.00
LA	=	2				
A	=	0.90811	1.01672			
LC	=	4				
C	=	1.41891	−0.01663	−0.88915	1.01672	
ASE	=	0.61621				
SPACE	=	6.56617	−4.52973	−1.35730	2.56250	

Example (4).

LB	=	3				
B	=	1.00000	−1.76770	1.56250	(maximum-delay)	
LD	=	5				
D	=	2.00	1.00	0.00	1.00	2.00
LA	=	2				
A	=	0.58119	0.79120			
C	=	0.58119	−0.23617	−0.49049	1.23625	
ASE	=	0.78375				
SPACE	=	6.56617	−4.52973	0.23230	2.56250	

The above two examples show again that the minimum-delay input does a better job of shaping this particular desired output.

Example (5).

LB	=	5					
B	=	2.00	1.00	0.00	−1.00	−2.00	
LD	=	5					
D	=	2.00	1.00	0.00	1.00	2.00	
LA	=	2					
A	=	0.	0.				
LC	=	6					
C	=	0.	0.	0.	0.	0.	0.
ASE	=	1.00000					
SPACE	=	10.00000	4.00000	0.	0.		

Here it is not possible to shape the input into the desired output; the average-squared error is 100%.

Subroutine SPIKE, which computes spiking filters, also makes use of subroutine EUREKA through subroutine SHAPE. The calling statement is:

CALL SPIKE (LB, B, LA, A, INDEX, ASE, S)

where the subroutine inputs are:

LB = length of B
B = wavelet that is input to spiking filter
LA = length of A

and the subroutine outputs are:

A = coefficients of spiking filter
INDEX = optimum spike position
ASE = average-squared error
S = working space

The program is:

```
      SUBROUTINE SPIKE(LB,B,LA,A,INDEX,ASE,S)
C     SPIKE IS SPIKING FILTER
C     WORKING SPACE IS S(I) , I=1,2*(LA+LB-1)+3*LA
      DIMENSION B(2),A(2),S(14)
      LD=LA+LB-1
      LT1=LD+1
      LT2=2*LD+1
      DO 10 I=1,LD
      CALL IMPULS(LD,S(LT1),I)
      CALL SHAPE(LB,B,LD,S(LT1),LA,A,LD,S(LT1),S(I),S(LT2))
   10 CONTINUE
      CALL MINSN(LD,S,ASE,INDEX)
      CALL IMPULS(LD,S(LT1),INDEX)
      CALL SHAPE(LB,B,LD,S(LT1),LA,A,LD,S(LT1),ASE,S(LT2))
      RETURN
      END
```

The values of ASE for each spike position are found in the first LD = LA + LB − 1 cells of S.

Two numerical examples computed by subroutine SPIKE are:

Example (1).

LB	=	5				
B	=	2.00000	1.00000	0.	1.00000	2.00000
LA	=	8				
A	=	0.10995	−0.03986	0.09315	−0.10514	−0.15317
		0.07600	−0.15306	0.35695		
INDEX	=	12				
ASE	=	0.28610				
S	=	0.28610	0.35022	0.35150	0.35908	0.30458
		0.34852	0.34852	0.30458	0.35908	0.35150
		0.35022	0.28610			

Example (2).

LB	=	5				
B	=	2.00000	1.00000	0.	−1.00000	−2.00000
LA	=	8				
A	=	−0.02714	−0.11459	0.11508	−0.11487	−0.09633
		−0.07348	0.16259	−0.35686		
INDEX	=	12				
ASE	=	0.28627				
S	=	0.28628	0.28899	0.38070	0.38788	0.32764
		0.32851	0.32851	0.32764	0.38788	0.38070
		0.28899	0.28627			

A computationally more efficient spiking-filter subroutine is SPIKER; it makes use of the Simpson sideways recursion (Wiggins and Robinson, 1965), as performed by subroutine SIDE. Subroutine SIDE is used in conjunction with EUREKA to recur sideways. The program is:

```
      SUBROUTINE SIDE(H,LF,F,A,R)
      DIMENSION F(LF),A(LF),R(LF)
      V=R(1)
      S=0.0
      T=0.0
      IF(LF.EQ.1)GO TO 2
      DO 1 I=2,LF
      J=LF+2-I
      S=S+F(I-1)*R(I)
      T=T+A(J)*R(I)
    1 V=V+A(I)*R(I)
    2 FLF=F(LF)
      W=(H-S+FLF*T)/V
      IF(LF.EQ.1)GO TO 4
      DO 3 I=2,LF
      J=LF-I+2
    3 F(J)=F(J-1)+W*A(J)-FLF*A(I)
    4 F(1)=W
      RETURN
      END
```

The program for subroutine SPIKER, with the explanation given by comment cards, is:

```
      SUBROUTINE SPIKER(LB,B,LA,A,LC,C,INDEX,ERRORS,SPACE)
C     INPUTS ARE
C         LB = LENGTH OF INPUT WAVELET
C          B = INPUT WAVELET
C         LA = LENGTH OF FILTER
C     OUTPUTS ARE
C          A = FILTER FOR OPTIMUM SPIKE POSITION
C         LC = LA+LB-1 = LENGTH OF ACTUAL OUTPUT
C          C = ACTUAL OUTPUT FOR OPTIMUM SPIKE POSITION
C      INDEX = OPTIMUM SPIKE POSITION AS FORTRAN SUBSCRIPT
C     ERRORS = ACTUAL SUMS OF SQUARED ERRORS,ONE
C         SUM FOR EACH SPIKE POSITION FROM 1 TO LC.
C     SPACE TAKES 3*LA CELLS
      DIMENSION B(1),A(1),C(1),ERRORS(1),SPACE(1)
      LC=LA+LB-1
      LT1=LA+1
      LT2=2*LA+1
      CALL CROSS(LB,B,LB,B,LA,SPACE)
      DO 4 I=1,LC
      CALL IMPULS(LC,C,I)
      CALL CROSS(LC,C,LB,B,LA,SPACE(LT1))
      IF(I-1)1,1,2
    1 CALL EUREKA(LA,SPACE,SPACE(LT1),A,SPACE(LT2))
      GO TO 3
    2 CALL SIDE(SPACE(LT1),LA,A,SPACE(LT2),SPACE)
    3 CALL DOT(LA,A,SPACE(LT1),Q)
      CALL FOLD(LA,A,LB,B,LC,C)
    4 ERRORS(I)=1.0-Q
      CALL MINSN(LC,ERRORS,EMIN,INDEX)
      CALL IMPULS(LC,C,INDEX)
      CALL SHAPE(LB,B,LC,C,LA,A,LC,C,EMIN,SPACE)
      RETURN
      END
```

Tables 5-1 and 5-2 give a complete set of spike filters and their actual outputs.

The effectiveness of a waveshaping filter depends upon the relative position in time between the input and desired output. We fix our time origin $k = 0$ at the initial value of the finite-length input sequence $(b_0, b_1, \ldots b_n)$. We restrict ourselves to a finite-length causal filter $(a_0, a_1, \ldots a_m)$ so the actual output can only have non-zero values for the points in the sequence $(c_0, c_1, \ldots c_{m+n})$. If all the points of the desired output lie outside of the interval $0 \leqslant k \leqslant m + n$, then the actual output cannot reach the desired output, so the entire desired output represents irreducible error. Thus, we consider the cases where there is an overlap between actual and desired outputs.

Suppose that the desired output waveform is $(d_0, d_1, \ldots d_p)$. We can position the desired output waveform with respect to the time interval $0 \leqslant k \leqslant m + n$ as follows:

TABLE 5-1

A complete set of digital spike filters. The filters are designed so as to transform the input wavelet $b = (-1.25, 1.00)$ as well as possible into an output spike. Each filter has eight coefficients, i.e. $(a_0, a_1, a_2, \ldots a_7)$, and appears as a row of the table. The first row is the spike filter designed to produce a spike at time $k = 0$; the second row, at time $k = 1$; . . . the last row at time $k = 8$. The actual outputs of the spike filters are shown in Table 5-2

−0.79174	−0.62307	−0.48556	−0.37232	−0.27771	−0.19697	−0.12609	−0.06151
0.01032	−0.77884	−0.60695	−0.46540	−0.34713	−0.24621	−0.15761	−0.07688
0.01290	0.02645	−0.75869	−0.58176	−0.43391	−0.30777	−0.19701	−0.09610
0.01612	0.03306	0.05164	−0.72719	−0.54239	−0.38471	−0.24626	−0.12013
0.02016	0.04132	0.06455	0.09100	−0.67799	−0.48089	−0.30783	−0.15016
0.02519	0.05165	0.08068	0.11375	0.15251	−0.60111	−0.38478	−0.18770
0.03149	0.06456	0.10085	0.14219	0.19063	0.24861	−0.48098	−0.23462
0.03936	0.08070	0.12606	0.17773	0.23829	0.31076	0.39877	0.29328
0.04920	0.10087	0.15758	0.22216	0.29786	0.38845	0.49846	0.63339

TABLE 5-2

The complete set of actual outputs for the digital spike filters given in Table 5-1. Each actual output has nine coefficients, i.e. (c_0, c_1, c_2, ... c_8), and appears as a row in the table. The first row is the actual output of the spike filter designed to produce a spike at time $k = 0$; the second row, at time $k = 1$; ... the last row, at time $k = 8$. The input is always the minimum-delay wavelet $b = (-1.25, 1.00)$; hence, the best job is done by the spike filter for time $k = 0$, as seen by the actual spike 0.98968, and the worst job is done by the spike filter for time $k = 8$, as seen by the actual spike 0.63339. Table 5-3 gives the amplitude and phase spectrum of the input wavelet $b = (-1.25, 1.00)$

0.98968	−0.01290	−0.01612	−0.02015	−0.02519	−0.03149	−0.03936	−0.04920	−0.06151
−0.01290	0.98387	−0.02015	−0.02519	−0.03149	−0.03936	−0.04920	−0.06151	−0.07688
−0.01613	−0.02016	0.97480	−0.03149	−0.03936	−0.04920	−0.06151	−0.07688	−0.09610
−0.02016	−0.02519	−0.03149	0.96063	−0.04920	−0.06151	−0.07688	−0.09610	−0.12013
−0.02519	−0.03149	−0.03937	−0.04921	0.93849	−0.07688	−0.09610	−0.12013	−0.15016
−0.03149	−0.03936	−0.04921	−0.06151	−0.07688	0.90389	−0.12013	−0.15016	−0.18770
−0.03936	−0.04920	−0.06151	−0.07688	−0.09610	−0.12013	0.84983	−0.18770	−0.23462
−0.04920	−0.06151	−0.07688	−0.09610	−0.12013	−0.15016	−0.18770	0.76537	−0.29328
−0.06151	−0.07688	−0.09610	−0.12013	−0.15016	−0.18770	−0.23463	−0.29328	0.63339

Case	0, 1, 2, . . .				$m + n$
1	$(d_0, d_1 . . . d_p)$				
2	$(d_0, d_1, . . . d_p)$				
3		$(d_0, d_1, . . . d_p)$			
⋮					
$p + 1$			$(d_0, d_1, . . . d_p)$		
$p + 2$				$(d_0, d_1, . . . d_p)$	
⋮					
$m + n + p + 1$					$(d_0, d_1, . . . d_p)$

That is, each case is obtained from the previous one by lagging the desired output waveform one time unit from its time position in the preceding case. The first case is when the last coefficient d_p of the desired output occurs at the same time as the first coefficient c_0 of the actual output. The final case is when the first coefficient d_0 of the desired output occurs at the same time as the last coefficient c_{m+n} of the actual output. In all, there are $m + n + p + 1$ cases; one of these cases represents the optimum positioning of the desired output with respect to input. Subroutine SHAPER computes the waveshaping filter for this optimum position. (Note that subroutine SHAPE computes the least-squares waveshaping filter for the position corresponding to case $p + 1$; subroutine SHAPER computes all the cases and picks out the least-squares shaping filter for the best case.) The calling statement is:

CALL SHAPER (LB, B, LD, D, LA, A, LC, C, INDEX, ERRORS, S)

where the subroutine inputs are:

LB = length of B = $n + 1$
B = input waveform = $(b_0, b_1, . . . b_n)$
LD = length of D = $p + 1$
D = desired output waveform = $(d_0, d_1, . . . d_p)$
LA = length of A = $m + 1$

and the subroutine outputs are:

A = least-squares shaping filter for optimum positioning
= $(a_0, a_1, . . . a_m)$
LC = length of C = $m + n + 1$
C = actual output of above shaping filter A
INDEX = number of the optimum case
ERRORS = average-squared error for each case

The program is:

```
      SUBROUTINE SHAPER(LB,B,LD,D,LA,A,LC,C,INDEX,ERRORS,S)
      DIMENSION B(LB),D(LD),A(LA),C(2),ERRORS(2),S(2)
      DIMENSION C(LCD),ERRORS(LCD),S(3*LA)
      LC=LA+LB-1
      LCD=LC+LD-1
      LT1=LA+1
      LT2=2*LA+2
      CALL DOT(LD,D,D,DD)
      CALL CROSS(LB,B,LB,B,LA,S)
      DO 4 I=1,LCD
      CALL ZERO(LCD,C)
      LDI=LD-I+1
      IF(I.LE.LD)CALL MOVE(I,D(LDI),C)
      ILD=I-LD+1
      IF(I.GT.LD)CALL MOVE(LD,D,C(ILD))
      CALL CROSS(LC,C,LB,B,LA,S(LT1))
      IF(I.GE.2)GO TO 2
      CALL EUREKA(LA,S,S(LT1),A,S(LT2))
      GO TO 3
    2 CALL SIDE(S(LT1),LA,A,S(LT2),S)
    3 CALL DOT(LA,A,S(LT1),AG)
    4 ERRORS(I)=(DD-AG)/DD
      CALL MINSN(LCD,ERRORS,EMIN,INDEX)
      CALL ZERO(LCD,C)
      LDIND=LD-INDEX+1
      IF(INDEX.LE.LD)CALL MOVE(INDEX,D(LDIND),C)
      INDLD=INDEX-LD+1
      IF(INDEX.GT.LD)CALL MOVE(LD,D,C(INDLD))
      CALL SHAPE(LB,B,LC,C,LA,A,LC,C,EMIN,S)
      RETURN
      END
```

Subroutine WVPRED performs wavelet prediction. The calling statement is:

CALL WVPRED (LB, B, LALPHA, LA, A, LC, C, S)

where subroutine inputs are:

LB = length of B
B = wavelet to be predicted (input to prediction filter)
LALPHA = prediction distance
LA = length of A

and where the subroutine outputs are:

A = coefficients of prediction filter
LC = length of C
C = actual output of prediction filter
S = working space, with
S(1) = autocorrelation at lag zero
S(2) = ASE = average-squared error
S(3) = ASFE = average-squared future error
S(4) = ASPE = average-squared past error

The program is:

```
      SUBROUTINE WVPRED(LB,B,LALPHA,LA,A,LC,C,S)
C     WORKING SPACE S(I) , I=1,LA+LALPHA+2*LA
C     LALPHA MUST BE GREATER THAN ZERO.
C     C(J) IS THE PREDICTED VALUE OF B(J+LALPHA).
      DIMENSION B(2),A(2),C(2),S(2)
      J=LA+LALPHA
      CALL CROSS(LB,B,LB,B,J,S)
      CALL EUREKA(LA,S,S(LALPHA+1),A,S(J+1))
      CALL FOLD(LB,B,LA,A,LC,C)
      CALL DOT(LA,A,S(LALPHA+1),AG)
      S(2)=1.0-AG/S(1)
      CALL DOT(LALPHA,B,B,S(3))
      S(3)=S(3)/S(1)
      S(4)=S(2)-S(3)
      RETURN
      END
```

Subroutine BNDPAS computes a band-pass filter. The program with its explanation is:

```
      SUBROUTINE BNDPAS(N,DT,FL,FH,FILT)
C     BNDPAS IS BANDPASS FILTER
C     PRODUCES ONE LOBE OF A SYMMETRICAL NONCAUSAL FILTER WHICH
C       PASSES ONLY FREQUENCIES BETWEEN FL AND FH, AND PASSES
C       THEM WITH ZERO PHASE SHIFT
C     N = LENGTH OF ONE LOBE PLUS ONE CENTER POINT
C     FILT = THE OUTPUT FILTER OF LENGTH N (CAUSAL PART)
C     FL = THE LOWER FREQUENCY PASSED - IN HERTZ
C     FH = THE HIGHER FREQUENCY PASSED - IN HERTZ
C     DT = THE TIME SPACING BETWEEN DATA POINTS - IN SECONDS
C     IF THE USER SPECIFIES FL=FH OR (FH-FL) LESS THAN 1.0/(DT*N),
C       HE WILL BE RETURNED THE NARROWEST FILTER CONSISTENT WITH
C       THE UNCERTAINTY PRINCIPLE,CENTERED AT (FH+FL)/2.0.
C     IF FH IS HIGHER THAN THE FOLDING FREQUENCY (FH*DT
C       GREATER THAN 0.5),THE SUBROUTINE RETURNS
      DIMENSION FILT(10)
      IF(FH*DT-0.5)8,8,70
    8 FN=N
      IF((FH-FL)-1.0/(DT*FN))10,30,30
   10 FC=(FH+FL)/2.0
      WL=6.2831853*(FC*DT-0.5/FN)
      WH=6.2831853*(FC*DT+0.5/FN)
      GO TO 40
   30 WL=FL*DT*6.283185
      WH=FH*DT*6.2831853
   40 FILT(1)=WH-WL
      DO 50 I=2,N
      FI=I-1
      FILT(I)=(SIN(WH*FI)-SIN(WL*FI))/FI
   50 CONTINUE
      DO 60 I=1,N
      FILT(I)=FILT(I)/3.14159265
   60 CONTINUE
   70 RETURN
      END
```

A related subroutine is BNDPS2, which is:

```
      SUBROUTINE BNDPS2(N,A,B,C,D)
C     SAME AS BNDPAS EXCEPT GETS TWO-SIDED FILTER CENTERED AT (N+1)/2
      DIMENSION D(100)
      D(N)=0.0
      M=(N+1)/2
      CALL BNDPAS(M,A,B,C,D(M))
      MM=M+M
```

```
      DO 10 I=1,M
      J=MM-I
   10 D(I)=D(J)
      RETURN
      END
```

Two similar subroutines are TREN (traveling energy) and TRAP (traveling power). The programs with their explanations are:

```
      SUBROUTINE TREN(LY,Y,L,LE,E)
C     TREN IS TRAVELING ENERGY
C     INPUTS ARE LY,Y,AND L WHICH IS SPECIFIED LENGTH OF ENERGY OPERATOR.
C     OUTPUTS ARE LE,E - THE TRAVELING SUM OF SQUARES OF L TERMS.
C     BECAUSE OF END EFFECTS,L MUST NOT EXCEED LY.
      DIMENSION Y(2),E(2)
      E(1)=0.0
      DO 10 I=1,L
   10 E(1)=E(1)+Y(I)*Y(I)
      IF(LY-L)25,25,15
   15 LE=LY-L+1
      DO 20 I=2,LE
      J=I+L-1
   20 E(I)=E(I-1)-Y(I-1)*Y(I-1)+Y(J)*Y(J)
   25 RETURN
      END
```

```
      SUBROUTINE TRAP(LY,Y,L,LP,P)
C     TRAP IS TRAVELING POWER
C     INPUTS ARE LY,Y,AND L WHICH IS SPECIFIED LENGTH OF
C       POWER OPERATOR.
C     P IS OUTPUT - TRAVELING SUM OF SQUARES OF L TERMS DIVIDED BY L.
C     P(I) =(SUM OF SQUARES FROM I TO I+L)/L, L MUST NOT BE
C       GREATER THAN LY.
C     BECAUSE OF END EFFECTS, LY-L+1=LP=LENGTH OF P
C     BOX-CAR SPECTRAL WINDOW FOR DC POWER.
      DIMENSION Y(2),P(2)
      P(1)=0.0
      DO 10 I=1,L
   10 P(1)=P(1)+Y(I)*Y(I)
      IF(LY-L)25,25,15
   15 LP=LY-L+1
      DO 20 I=2,LP
      J=I+L-1
   20 P(I)=P(I-1)-Y(I-1)*Y(I-1)+Y(J)*Y(J)
   25 FL=L
      DO 30 I=1,LP
   30 P(I)=P(I)/FL
      RETURN
      END
```

A very useful subroutine is TRAF, which computes a traveling correlation filter. It is:

```
      SUBROUTINE TRAF(N,DATA,L,FILTER,ISHIFT,ISUM,OUTPUT)
C     TRAF IS TRAVELING CORRELATION FILTER
C     N=DATA LENGTH
C     DATA=SERIES TO BE FILTERED
C     L=LENGTH OF FILTER
C     FILTER=THE CORRELATION FILTER
C     ISHIFT=SHIFTING INCREMENT
```

```
C     ISUM=DENSITY OF SUMMATION , ISUM=1 MEANS EVERY POINT,
C       ISUM=2 MEANS EVERY OTHER POINT,ETC.
C     OUTPUT=CROSS CORRELATION OF FILTER WITH DATA.
C     OUTPUT IS DENSELY PACKED REGARDLESS OF VALUES OF ISHIFT AND ISUM.
C     OUTPUT IS OF LENGTH (N-L+1)/ISHIFT,ROUNDED DOWN.
C     EQUIVALENCE (DATA,OUTPUT) IS ALLOWED
      DIMENSION DATA(2),FILTER(2),OUTPUT(2)
      K=1
      LS=N-L+1
      DO 20 I=1,LS,ISHIFT
      TEMP=0.0
      DO 10 J=1,L,ISUM
      II=I+J
   10 TEMP=TEMP+DATA(II-1)*FILTER(J)
      OUTPUT(K)=TEMP
   20 K=K+1
      RETURN
      END
```

The purpose of subroutine REMAV is to remove the arithmetic average from an array Y of length LY. The average is stored in AVERAG. The program is:

```
      SUBROUTINE REMAV(LY,Y,AVERAG)
C     REMAV REMOVES THE ARITHMETIC AVERAGE
      DIMENSION Y(2)
      S=0.0
      DO 10 I=1,LY
   10 S=S+Y(I)
      AVERAG=S/FLOAT(LY)
      DO 20 I=1,LY
   20 Y(I)=Y(I)-AVERAG
      RETURN
      END
```

The purpose of subroutine RMSDEV is to compute the RMS (root-mean-square) deviation of an array Y of length N. The RMS deviation is stored in cell DEV. The program is:

```
      SUBROUTINE RMSDEV(N,Y,DEV)
C     THIS ROUTINE RETURNS SQRT(SUM(Y-YMEAN)**2)/N)
      DIMENSION Y(2)
      S=0.0
      DO 10 I=1,N
   10 S=S+Y(I)
      YMEAN=S/FLOAT(N)
      SS=0.0
      DO 20 I=1,N
   20 SS=SS+(Y(I)-YMEAN)*(Y(I)-YMEAN)
      DEV=SQRT(SS/FLOAT(N))
      RETURN
      END
```

Subroutine BURG computes the coefficients (a_1, a_2, . . . a_n) of a finite-length causal forward or backward prediction filter. The key feature of this technique is that it does not explicitly depend on the autocorrelation of the available data. The method utilizes both forward and backward predictions in a symmetric manner in the Toeplitz recursion. (Refer to pp. 130-133 for the mathematical details of the Burg method). This subroutine also generates the "maximum entropy" spectrum.

CALL BURG (LX, X, F, B, LA, A, M, S)

where the subroutine inputs are:

LX = length of data record
X = input data record
LA = length of forward or backward prediction filter
M = integer exponent which controls spectral resolution

and the subroutine outputs are:

A = forward or backward prediction filter; − A(2), − A(3), . . . − A(LA + 1)
S = maximum entropy spectrum at equally spaced frequencies.

The program is:

```
      SUBROUTINE BURG(LX,X,F,B,LA,A,M,S)
C     NP=2**M CONTROLS SPECTRAL RESOLUTION.
C     DIMENSION A(2**M),Y(2**M),S(2**(M-1)+1)
      DIMENSION X(LX),F(LX),B(LX),A(64),S(33),Y(64),TABLE(17)
      NP=2**M
      V=0.
      DO 1 I=1,LX
      V=V+X(I)**2
      F(I)=X(I)
    1 B(I)=X(I)
      V=V/FLOAT(LX)
      DO 5 N=1,LA
      SN=0.
      SD=0.
      L=N+1
      DO 2 J=L,LX
      SN=SN+F(J)*B(J-1)
    2 SD=SD+F(J)*F(J)+B(J-1)*B(J-1)
      G=2.*SN/SD
      V=V*(1.-G*G)
      A(N)=G
      IF(N.EQ.1) GO TO 4
      NH=N/2
      DO 3 K=1,NH
      KK=N-K
      AA=A(KK)-G*A(K)
      A(K)=A(K)-G*A(KK)
    3 A(KK)=AA
    4 M1=N+1
      BS=B(N)
      DO 5 K=M1,LX
      FF=F(K)
      BB=BS
      BS=B(K)
      F(K)=FF-G*BB
    5 B(K)=BB-G*FF
      DO 6 I=1,LA
      K=LA-I+1
    6 A(K+1)=-A(K)
      A(1)=1.
      DO 7 I=1,NP
    7 Y(I)=0.
      LP=LA+2
      DO 8 I=LP,NP
```

```
    8 A(I)=0.
      CALL COSQT(M,TABLE)
      XX=1.442695*ALOG(FLOAT(LA+1))
      LL=IFIX(XX)
      IF(XX/FLOAT(LL).GT.1.)LL=LL+1
      CALL FFT(A,Y,TABLE,M,LL,-1)
      NF=NP/2+1
      DO 9 I=1,NF
    9 S(I)=V/(A(I)*A(I)+Y(I)*Y(I))
      RETURN
      END
```

5.4. SPECTRAL PACKAGE

Here we will give some elementary spectral routines. The subroutines in our spectral package are:

COSTAB	(cosine table)
SINTAB	(sine table)
COSP	(cosine or sine spectrum)
SPLIT	(split data into even and odd parts)
FTRAN	(Fourier transform)
DANWT	(Daniell weighting of autocorrelation)
ASPECT	(auto-spectrum)
ARCTAN	(arctangent)
XPHAZ	(cross-phase spectra)
FT	(Fourier transform)
COSTR	(cosine transform)
SMOOTH	(smooth by Tukey-Hamming formula)
DRUM	(make phase curve continuous, as around a drum)
POLAR	(polar coordinates)
CAST	(cosine and sine transform)
FFT	(fast Fourier transform)

Two basic subroutines are COSTAB, which generates a cosine table, and SINTAB, which generates a sine table. The programs are:

```
      SUBROUTINE COSTAB(M,TABLE)
      COSTAB GENERATES FULL WAVELENGTH COSINE TABLE.
      TABLE LENGTH = 2*M-1,AND TABLE(1)=TABLE(2*M-1) = 1.0
      DIMENSION TABLE(2)
      FM=M+M-2
      MM=M+M-1
      DO 10 I=1,MM
   10 TABLE(I)=COS(FLOAT(I-1)*6.2831853/FM)
      RETURN
      END


      SUBROUTINE SINTAB(M,TABLE)
C     SINTAB GENERATES FULL WAVELENGTH SINE TABLE.
C     TABLE LENGTH = 2*M-1,AND TABLE(1) = TABLE(2*M-1) = 0.0
      DIMENSION TABLE(2)
      FM=M+M-2
      MM=M+M-1
```

```
      DO 10 I=1,MM
   10 TABLE(I)=SIN(FLOAT(I-1)*6.2831853/FM)
      RETURN
      END
```

The purpose of subroutine COSP is to compute the kth value of either a cosine transform or a sine transform. The calling statement is:

CALL COSP (N, DATA, TABLE, M, K, C)

The subroutine inputs are:

N	=	$n + 1$ = number of data points
DATA	=	$X_0, X_1, X_2, \ldots X_n$
TABLE	=	either a full wavelength cosine table of length 2*M−1 with TABLE(1) = TABLE(2*M−1)=1.0, or a full wavelength sine table of length 2*M−1 with TABLE(1) = TABLE(2*M−1)=0.0
M	=	$m + 1$ = number of cosine (or sine) table values
K	=	$k + 1$, where $0 \leqslant k \leqslant m$

The subroutine output is either:

$$\mathrm{C} = c_k = \sum_{i=0}^{n} x_i \cos\left(\frac{\pi i k}{m}\right)$$

if TABLE is the cosine table, or:

$$\mathrm{C} = s_k = \sum_{i=0}^{n} x_i \sin\left(\frac{\pi i k}{m}\right)$$

if TABLE is the sine table. TABLE is conveniently generated by subroutine COSTAB in the case of the cosine table, or by subroutine SINTAB in the case of the sine table. The program for subroutine COSP is:

```
      SUBROUTINE COSP(N,DATA,TABLE,M,K,C)
C     COSP COMPUTES (FOR A GIVEN K BETWEEN 1 AND M INCLUSIVE)
C         N
C     C = SUM(COS(PI*(I-1)*(K-1)/(M-1))*DATA(I))
C         I=1
C     INPUT IS A FULL WAVELENGTH COSINE TABLE OF LENGTH 2*M-1, FOR
C       WHICH TABLE(1) = TABLE(2*M-1) = 1.0 , OR ALTERNATIVELY BY
C       THE CORRESPONDING FULL WAVELENGTH SINE TABLE.(THE ABOVE
C       FORMULA THEN IS COMPUTED WITH SIN REPLACING COS.)
      DIMENSION DATA(2),TABLE(2)
      J=1
      C=0.0
      KK=K-1
      MM=M+M-1
      MMM=MM-1
      DO 20 I=1,N
      C=C+DATA(I)*TABLE(J)
      J=J+KK
      IF(J-MM)20,20,10
```

```
10 J=J-MMA
20 CONTINUE
   RETURN
   END
```

The basic loop requires 1 floating multiplication, 1 floating addition, 1 fixed addition, and 1 fixed subtraction.

The purpose of subroutine SPLIT is to split data into its even and odd parts. The program is:

```
      SUBROUTINE SPLIT(N,DATA,EVEN,ODD)
C     GIVEN DATA(I) FOR I=1 TO N,SPLIT FINDS FOR I=1,N
C       EVEN(I) = 0.5*(DATA(N-1+I)+DATA(N+1-I))
C       ODD(I) = 0.5*(DATA(N-1+I)-DATA(N+1-I))
      DIMENSION DATA(2),EVEN(2),ODD(2)
      DO 10 I=1,N
      K=N-1+I
      L=N+1-I
      EVEN(I)=0.5*(DATA(K)+DATA(L))
      ODD(I)=0.5*(DATA(K)-DATA(L))
   10 CONTINUE
      RETURN
      END
```

Subroutine FTRAN performs a Fourier transform. The program with its explanation is:

```
      SUBROUTINE FTRAN(N,DATA,C,S,SP)
C     FTRAN COMPUTES THE FOURIER TRANSFORM OF AN ODD NUMBER OF POINTS
C       WHERE THE TIME ORIGIN IS AT THE MIDDLE POINT.
C     THE SUBROUTINE INPUTS ARE N AND DATA(I) I=1,2*N-1 (THE INPUT DATA).
C     THE SUBROUTINE OUTPUTS ARE C(I) ,
C       I=1,N (THE COSINE TRANSFORM ABOUT THE MIDDLE)
C       S(I) , I=1,N (THE SINE TRANSFORM ABOUT THE MIDDLE,
C       WHERE S(1) = S(N) = 0.0)
C       SP(I) , I=1,4*N-1 (WORKING SPACE)
      DIMENSION DATA(2),C(2),S(2),SP(2)
      CALL SPLIT(N,DATA,SP(2*N),SP(3*N))
      CALL COSTAB(N,SP)
      DO 10 I=1,N
   10 CALL COSP(N,SP(2*N),SP,N,I,C(I))
      C(I)=C(I)/2.0
      C(N)=C(N)/2.0
      CALL SINTAB(N,SP)
      DO 20 I=1,N
   20 CALL COSP(N,SP(3*N),SP,N,I,S(I))
      RETURN
      END
```

The Daniell weighting function is:

$$w_k = \frac{\sin(\pi k/m)}{(\pi k/m)} \quad \text{for } |k| = 0, 1, 2, \ldots$$

which has no truncation point, so all the empirical autocorrelation coefficients R_k for $k = 0, 1, 2, \ldots n-1$ need to be computed for a time series

$x_1, x_2, \ldots x_n$ of n observations. The parameter m controls the resolution of the spectral estimate; the greater the value of m, the greater will be the resolution. However, the greater the value of m, the greater will be the variance of the spectral estimate, for in the case of a Gaussian process as $n \to \infty$ and $(m/n) \to 0$, the variance of the Daniell estimate is asymptotically equal to $(m/n)[\Phi(f)]^2$. Usually m will be chosen to be a fraction, say 10 or 20%, of the number n of observations.

The purpose of subroutine DANWT is to prepare an autocorrelation function for subroutine COSP to compute a Daniell spectral estimate. The program with its explanation is:

```
      SUBROUTINE DANWT(N,AUTO,M)
C     DANWT TAKES AUTOCORRELATION OF HALF LENGTH N,DIVIDES
C       AUTO(1) BY 2.0, THEN WEIGHTS IT BY SIN(X)/X WHERE
C       X = 0.5*(I-1)*PI/(M-1),AND THEN COLLAPSES IT LIKE AN
C       ACCORDIAN TO LENGTH M.
C     AUTO IS THUS PREPARED FOR SUROUTINE COSP TO COMPUTE DANIELL
C       ESTIMATE.
      DIMENSION AUTO(2)
      AUTO(1)=AUTO(1)/2.0
      FM=M-1
      SC=1.57079632/FM
      DO 10 I=2,N
      FI=I-1
   10 AUTO(I)=AUTO(I)*SIN(SC*FI)/(SC*FI)
      IF(M-N)20,60,60
   20 L=-1
      K=M
      MP=M+1
      DO 50 I=MP,N
      K=K+L
      AUTO(K)=AUTO(K)+AUTO(I)
      IF(K-M)30,40,40
   30 IF(K-I)40,40,50
   40 L=-L
   50 CONTINUE
   60 RETURN
      END
```

The purpose of subroutine ASPECT is to compute the Daniell spectral estimate at M equi-spaced frequencies between zero and the folding (i.e. Nyquist) frequency inclusive. The program with its explanation is:

```
      SUBROUTINE ASPECT(N,DATA,M,SPECT,SN,SM)
C     ASPECT COMPUTES THE DANIELL ESTIMATE AT M EQUISPACED
C       FREQUENCIES BETWEEN ZERO AND FOLDING FREQUENCY INCLUSIVE.
C     THE WINDOWS ARE SIDE BY SIDE (NO 50 PERCENT OVERLAP).
C     INPUTS ARE N,M,DATA(I) , I=1,N(SAMPLE FUNCTION).
C     OUTPUT IS SPECT(I) , I=1,M (DANIELL ESTIMATE)
C     WORKING SPACE IS SN(I) , I=1,N AND SM(I) , I=1,2*M
      DIMENSION DATA(2),SPECT(2),SN(2),SM(2)
      CALL CROSS(N,DATA,N,DATA,N,SN)
      CALL DANWT(N,SN,M)
      CALL COSTAB(M,SM)
      K=MINO(M,N)
      DO 10 I=1,M
   10 CALL COSP(K,SN,SM,M,I,SPECT(I))
      RETURN
      END
```

Subroutine ARCTAN computes the arctangent THETA of the ratio Y/X; the result lies in the correct quadrant between $-\pi$ and π. The program is:

```
      SUBROUTINE ARCTAN(X,Y,THETA)
      IF(X)5,1,5
    1 IF(Y)4,3,2
    2 THETA=1.570796327
      RETURN
    3 THETA=0.0
      RETURN
    4 THETA=-1.570796327
      RETURN
    5 THETA=ATAN(ABS(Y/X))
      IF(X)9,6,6
    6 IF(Y)7,8,8
    7 THETA=-THETA
    8 RETURN
    9 IF(Y)11,10,10
   10 THETA=3.141592654-THETA
      RETURN
   11 THETA=-3.141592654+THETA
      RETURN
      END
```

Subroutine XPHAZ computes the cross-phase spectra of two data series. The program with explanation is:

```
      SUBROUTINE XPHAZ(N,X,Y,PHASE,SPACE)
C     XPHAZ COMPUTES CROSS PHASE SPECTRA OF TWO SERIES.
C     INPUTS ARE X(I),Y(I) , I=1,N
C     OUTPUT IS PHASE(I) , I=1,N (PHASE SPECTRA OF
C       CROSS CORRELATION OF X AND Y).
C     SPACE(I) , I=1,8*N IS WORKING SPACE.
      DIMENSION X(2),Y(2),PHASE(2),SPACE(2)
      CALL REVERS(N,X)
      CALL FOLD(N,X,N,Y,LC,SPACE)
      CALL FTRAN(N,SPACE,SPACE(2*N),SPACE(3*N),SPACE(4*N))
      ICOS=2*N-1
      ISIN=3*N-1
      DO 10 I=1,N
      J=ICOS+I
      K=ISIN+I
   10 CALL ARCTAN(SPACE(J),SPACE(K),PHASE(I))
      CALL REVERS(N,X)
      RETURN
      END
```

Subroutine FT computes one value of the Fourier transform by the sum of angles formulas for sine and cosine. The subroutine inputs are:

N = length of DATA
DATA = series whose Fourier transform is desired
W = angular frequency at which the Fourier transform is evaluated

and the subroutine outputs are:

S = sine transform at angular frequency W
C = cosine transform at angular frequency W

The program is:

```
      SUBROUTINE FT(N,DATA,W,S,C)
C     FT COMPUTES ONE VALUE OF FOURIER TRANSFORM OF DATA AT
C       ANGULAR FREQUENCY W BY MEANS OF SUM OF ANGLES FORMULAS
C       FOR SINE AND COSINE.
C     OUTPUTS ARE
C         N-1                                 N-1
C     S = SUM(DATA(K+1)*SIN(W*K))  AND C = SUM(DATA(K+1)*COS(W*K))
C         K=0                                 K=0
      DIMENSION DATA(2)
      COSNW=1.0
      SINNW=0.0
      SINW=SIN(W)
      COSW=COS(W)
      S=0.0
      C=0.0
      DO 10 I=1,N
      C=C+COSNW*DATA(I)
      S=S+SINNW*DATA(I)
      T=COSW*COSNW-SINW*SINNW
      SINNW=COSW*SINNW+SINW*COSNW
      COSNW=T
   10 CONTINUE
      RETURN
      END
```

Subroutine COSTR computes the cosine transform of the autocorrelation; it is also based on the sum of angles formula. The program with its explanation is:

```
      SUBROUTINE COSTR(N,R,W,S)
C     COSTR IS COSINE TRANSFORM OF AUTOCORRELATION
C     COSTR COMPUTES
C                           N-1
C        S = R(1)+2.0*SUM(R(K+1)*COS(K*W))
C                           K=1
      DIMENSION R(500)
      COSNW=1.0
      SINNW=0.0
      COSW=COS(W)
      SINW=SIN(W)
      S=R(1)
      DO 10 I=2,N
      T=COSW*COSNW-SINW*SINNW
      SINNW=COSW*SINNW+SINW*COSNW
      COSNW=T
   10 S=S+2.0*R(I)*COSNW
      RETURN
      END
```

Subroutine SMOOTH computes the smoothed spectrum from the cosine transform by the Tukey-Hamming formula. The program is:

```
      SUBROUTINE SMOOTH(SPECT,LS)
C     SMOOTH COMPUTES SMOOTHED SPECTRUM FROM THE
C       COSINE TRANSFORM USING TUKEY-HAMMING FORMULA.
C     THE SUBROUTINE INPUTS ARE
C       SPECT = UNSMOOTHED SPECTRUM
C       LS = ITS LENGTH
C     THE SUBROUTINE OUTPUT IS
```

```
C          SPECT = SMOOTHED SPECTRUM
        DIMENSION SPECT(2)
        MM=LS-1
        A=0.54*SPECT(1)+0.46*SPECT(2)
        B=0.54*SPECT(LS)+0.46*SPECT(MM)
        SJ=SPECT(1)
        SK=SPECT(2)
        DO 10 J=2,MM
        SI=SJ
        SJ=SK
        SK=SPECT(J+1)
     10 SPECT(J)=0.54*SJ+0.23*(SI+SK)
        SPECT(1)=A
        SPECT(LS)=B
        RETURN
        END
```

Subroutine DRUM makes a phase curve continuous; it is used by subroutine CAST. The program for DRUM is:

```
        SUBROUTINE DRUM(LPHZ,PHZ)
C       DRUM MAKES PHASE CURVE CONTINUOUS.
        DIMENSION PHZ(2)
        PJ=0.0
        DO 40 I=2,LPHZ
        IF(ABS(PHZ(I)+PJ-PHZ(I-1))-3.14159265)40,40,10
     10 IF(PHZ(I)+PJ-PHZ(I-1))20,40,30
     20 PJ=PH+6.283185307
        GO TO 40
     30 PH=PJ-6.283185307
     40 PHZ(I)=PHZ(I)+PJ
        RETURN
        END
```

Subroutine POLAR computes polar coordinates; it is also used by subroutine CAST. The program is:

```
        SUBROUTINE POLAR(L,RE,XIM,AMP,PHZ)
C       POLAR COMPUTES POLAR COORDINATES
        DIMENSION RE(2),XIM(2),AMP(2),PHZ(2)
        PI=3.14159265
        DO 110 I=1,L
        AMP(I)=SQRT(RE(I)**2+XIM(I)**2)
        IF(XIM(I))10,20,30
     10 IF(RE(I))40,50,60
     20 IF(RE(I))70,80,60
     30 IF(RE(I))90,100,60
     40 PHZ(I)=ATAN(XIM(I)/RE(I))-PI
        GO TO 110
     50 PHZ(I)=-0.5*PI
        GO TO 110
     60 PHZ(I)=ATAN(XIM(I)/RE(I))
        GO TO 110
     70 PHZ(I)=-PI
        GO TO 110
     80 PHZ(I)=0.0
        GO TO 110
     90 PHZ(I)=ATAN(XIM(I)/RE(I))+PI
        GO TO 110
    100 PHZ(I)=0.5*PI
    110 CONTINUE
        RETURN
        END
```

Subroutine CAST computes the cosine and sine transform, and also the amplitude and phase spectra. The subroutine inputs are:

LW = length of W
W = data whose transform is desired
LT = length of TCOS, TSIN, AMP, or PHZ

and the subroutine outputs are:

TCOS = cosine transform at equally spaced frequencies from zero to Nyquist
TSIN = sine transform at equally spaced frequencies from zero to Nyquist
AMP = amplitude spectrum at equally spaced frequencies from zero to Nyquist
PHZ = phase spectrum at equally spaced frequencies from zero to Nyquist

The program is:

```
      SUBROUTINE CAST(LW,W,LT,TCOS,TSIN,AMP,PHZ)
C     CAST IS COSINE AND SINE TRANSFORM
      DIMENSION W(2),TCOS(3),TSIN(3),AMP(3),PHZ(3)
      COMPLEX X,A
      DANG=3.141592651/FLOAT(LT-1)
      ANG=0.0
      DO 10 I=1,LT
      X=CMPLX(COS(ANG),SIN(ANG))
      CALL POLYEV(LW,W,X,A)
      TCOS(I)=REAL(A)
      TSIN(I)=AIMAG(A)
   10 ANG=ANG+DANG
      CALL POLAR(LT,TCOS,TSIN,AMP,PHZ)
      CALL DRUM(LT,PHZ)
      RETURN
      END
```

The discrete Fourier transform (DFT) of the signal x_k is defined by:

$$X_n = \sum_{k=0}^{N-1} x_k \, e^{-i2\pi kn/N} \quad \text{for } n = 0, 1, 2, \ldots N-1$$

where both x_k and X_n may be complex quantities.

The inverse discrete Fourier transform (IDFT) of X_n is x_k where:

$$x_k = \frac{1}{N} \sum_{n=0}^{N-1} X_n \, e^{i2\pi kn/N} \quad \text{for } k = 0, 1, 2, \ldots N-1$$

We note that the above expressions for the DFT and IDFT differ only in the sign of the exponent of the phasor exp $(i2\pi kn/N)$ and in a scale factor $1/N$. Hence, computational procedures for the DFT can be easily modified to compute the IDFT.

The most straightforward method of calculating an N-point discrete Fourier transform, using the definition directly, requires N^2 multiplications; N multiplications for each of the N frequency points, X_0, X_1, ... X_{N-1}.

However, since the amount of computation (and thus computation time) is proportional to N^2, the number of computations (and computation time) required to compute the DFT by such a direct method becomes very large for large values of N. Subroutine FFT is an algorithm which reduces the number of computations required for the DFT from N^2 to $N \log_2 N$ when N is a power of two. Thus, the letters FFT stand for fast Fourier transform.

The fundamental principle behind subroutine FFT is to decompose the computation of the DFT of a sequence $x_0, x_1, \ldots x_{N-1}$ of length N into successively smaller discrete Fourier transforms. For example, we can compute an N-point DFT by combining two $N/2$-point DFTs. The $N/2$-point DFTs can be evaluated by combining two $N/4$-point DFTs and so on. The FFT is then naturally computed by adding sums pair-wise, then adding the pairs of sums pair-wise, and so on. This would appear to be a very unnatural way of computing the DFT from the defining expression, but it is the most natural way to compute the FFT.

There are many suitable ways of understanding what is actually going on in the FFT computation, i.e. in terms of Fourier series, matrix theory, number theory, multidimensional transforms, flow graphs, and others. A physical description of the FFT is as follows. Consider a line array of equally spaced radar receivers and a plane wave signal is approaching the array at an angle. The phasor measured by each sensor will be out of phase with the phasor at each other sensor, in general, and the sum of all the phasors will be approximately zero (or small) because of the different phases. However, if each sensor output is subjected to an appropriate phase shift, they can all be added together to produce a relatively large output. A radar beam-forming network is a collection of phase shifters and adders which produce outputs which are maximum when the wavefront is impinging on the array from a given direction. This beam-forming operation turns out to be a Fourier transform.

The conventional method of forming beams is the so-called Blass array — a collection of N^2 phase shifters and adders, which is analogous to the direct method of computing the DFT. A phase shifter operating on a sinusoidal waveform is equivalent to multiplying a complex number by a root of unity. J. Butler and, independently J.P. Shelton, had figured out a way of connecting together phase shifters and adders to the sensors so as to require only $N \log_2 N$ elements to generate N beams for N antennas. The device is called a Butler matrix, and is completely analogous to the fast Fourier transform.

In the process of decomposing an N-point DFT into successively smaller-point DFTs, the FFT algorithm "shuffles" the input data $x_0, x_1, \ldots x_{N-1}$. The FFT then combines the data by adding sums pair-wise, then adding pairs of sums pair-wise, and so on. This merging procedure involves the use of the quarter-length cosine table and repeatedly involves the so-called butterfly operation. Because the data have been shuffled, we must reshuffle the data by a procedure usually referred to as a bit reversal operation. An important feature of the FFT algorithm is that the merging and bit reversal can be done "in place", without redefining the whole array during each butterfly operation.

If we wish to calculate the FFT repeatedly, we can compute the quarter-length cosine table once and store these values in a permanent array. Also, it is sometimes desirable to perform interpolation in the frequency or time domain. This is usually done by appending a sequence of zeroes to the actual data sequence. Thus, if 2^L points are to be interpolated in the transform domain (resulting in $N = 2^M$ points, for example, where $M > L$) this can be easily done by applying a 2^M-point FFT to the input data where 2^{M-L} zeroes are appended to the first 2^L points. To avoid the added computation introduced by appending zeroes to our input data record, we can eliminate the operations involving zeroes by the process of pruning. With the particular FFT algorithm developed by Sande (see Gentleman and Sande, 1966), the pruning procedure due to Markel (1971) is straightforward.

The calling statement for subroutine FFT is:

CALL FFT (X, Y, TABLE, M, LL, ISN)

where the subroutine inputs are:

X	=	array of length N used to hold the real part of complex input
Y	=	array of length N used to hold the imaginary part of complex input
TABLE	=	array of length N/4 + 1 used to hold the quarter-length cosine table
M	=	integer power of two which determines the size of the FFT to be performed ($N = 2^M$)
LL	=	number of stages in which no pruning is allowable; the actual number of data points
ISN	=	DFT or IDFT indicator; set
ISN	=	−1 for DFT (forward FFT) and set
ISN	=	+1 for IDFT (inverse FFT)

and the subroutine outputs are:

X = array of length N used to store real part of complex output
Y = array of length N used to store imaginary part of complex output

The program is:

```
      SUBROUTINE FFT(X,Y,TABLE,M,LL,ISN)
C     FFT IS IN PLACE DFT COMPUTATION USING SANDE ALGORITHM
C       AND MARKEL PRUNING MODIFICATION.
C     X = ARRAY OF LENGTH 2**M USED TO HOLD REAL PART OF COMPLEX INPUT.
C     Y = ARRAY OF LENGTH 2**M USED TO HOLD IMAGINARY PART OF COMPLEX INPUT=
C     TABLE = ARRAY OF LENGTH N/4+1 (N=2**M) , CONTAINS
C       QUARTER-LENGTH COSINE TABLE.
C     M = INTEGER POWER OF TWO WHICH DETERMINES SIZE OF
C       FFT TO BE PERFORMED. (BIT REVERSE TABLE IS SET FOR A
C       MAXIMUM OF N=2**12=4096)
```

```
C      LL = INTEGER POWER OF TWO WHICH DETERMINES 2**LL ACTUAL
C        DATA POINTS , NUMBER OF STAGES IN WHICH NO PRUNING IS ALLOWABLE.
C      ISN - SET TO -1 FOR FOWARD FFT (DFT)
C      ISN - SET TO +1 FOR INVERSE FFT (IDFT)
C      X = REAL PART OF COMPLEX OUTPUT
C      Y = IMAGINARY PART OF COMPLEX OUTPUT
       DIMENSION X(1096),Y(1096),TABLE(1025),L(12)
       EQUIVALENCE (L12,L(1)),(L11,L(2)),(L10,L(3)),(L9,L(4)),(L8,L(5)),
      1(L7,L(6)),(L6,L(7)),(L5,L(8)),(L4,L(9)),(L3,L(10)),(L2,L(11)),
      2(L1,L(12))
       N=2**M
       ND4=N/4
       ND4P1=ND4+1
       ND4P2=ND4P1+1
       ND2P2=ND4+ND4P2
       LLL=2**LL
       DO 8 LO=1,M
       LMX=2**(M-LO)
       LMM=LMX
       LIX=2*LMX
       ISCL=N/LIX
C      TEST FOR PRUNING
       IF(LO-M+LL) 1,2,2
     1 LMM=LLL
     2 DO 8 LM=1,LMM
       IARG=(LM-1)*ISCL+1
       IF(IARG.LE.ND4P1) GO TO 4
       K1=ND2P2-IARG
       C=-TABLE(K1)
       K3=IARG-ND4
       S=ISN*TABLE(K3)
       GO TO 6
     4 C=TABLE(IARG)
       K2=ND4P2-IARG
       S=ISN*TABLE(K2)
     6 CONTINUE
       DO 8 LI=LIX,N,LIX
       J1=LI-LIX+LM
       J2=J1+LMX
       T1=X(J1)-X(J2)
       T2=Y(J1)-Y(J2)
       X(J1)=X(J1)+X(J2)
       Y(J1)=Y(J1)+Y(J2)
       X(J2)=C*T1-S*T2
       Y(J2)=C*T2+S*T1
     8 CONTINUE
C
C      PERFORM BIT REVERSAL
C
       DO 40 J=1,12
       L(J)=1
       IF(J-M) 31,31,40
    31 L(J)=2**(M+1-J)
    40 CONTINUE
       JN=1
       DO 60 J1=1,L1
       DO 60 J2=J1,L2,L1
       DO 60 J3=J2,L3,L2
       DO 60 J4=J3,L4,L3
       DO 60 J5=J4,L5,L4
       DO 60 J6=J5,L6,L5
       DO 60 J7=J6,L7,L6
       DO 60 J8=J7,L8,L7
       DO 60 J9=J8,L9,L8
       DO 60 J10=J9,L10,L9
       DO 60 J11=J10,L11,L10
```

```
      DO 60 JR=J11,L12,L11
      IF(JN-JR) 51,51,54
   51 R=X(JN)
      X(JN)=X(JR)
      X(JR)=R
      FI=Y(JN)
      Y(JN)=Y(JR)
      Y(JR)=FI
   54 IF(ISN)53,53,52
   52 X(JR)=X(JR)/FLOAT(N)
      Y(JR)=Y(JR)/FLOAT(N)
   53 JN=JN+1
   60 CONTINUE
      RETURN
      END
```

The quarter-length cosine table used in subroutine FFT may be generated by subroutine COSQT. The program is:

```
      SUBROUTINE COSQT(M,TABLE)
C     COSQT GENERATES QUARTER-LENGTH COSINE TABLE
      DIMENSION TABLE(2)
      N=2**M
      ND4P1=N/4+1
      SCL=6.283185307/FLOAT(N)
      DO 10 I=1,ND4P1
      ARG=FLOAT(I-1)*SCL
   10 TABLE(I)=COS(ARG)
      RETURN
      END
```

where the subroutine input is:

M = integer power of 2 which determines the size of the FFT to be performed (N = 2**M)

and the subroutine output is:

TABLE = quarter-length cosine table of length N/4 + 1.

On the following pages we present Tables 5-3 through 5-10, which (except for Table 5-8) show some computed spectra. Table 5-8 gives spike filters whose spectra are given in Tables 5-9 and 5-10.

5.5. CONCLUDING REMARKS

In the preceding chapters, we have devoted a great deal of effort to the mathematical details and models associated with the deconvolution of reflection seismograms. Although deconvolution is an important part of the exploration process, it is only one stage of an extensive engineering effort. The geophysical time series necessary for deconvolution analysis are the result of land and marine seismic prospecting conducted under extremely varied climatic conditions. The geophysical prospecting crews are generally led by engineers and are composed of a varied number of technicians (between one and thirty) and some hundred local workers. Such field

TABLE 5-3

The amplitude and phase spectrum of the minimum-delay wavelet $b = (-1.25, 1.00)$. A minimum-delay wavelet has minimum phase-lag characteristic; the total phase-lag displacement for this wavelet from frequency 0 to 180 degrees to $180 - 180 = 0$ degrees. The amplitude and phase spectrum of the reverse of this wavelet, namely $(1.00, -1.25)$ is given in Table 5-4

Frequency (degrees)	Amplitude spectrum	Phase-lag (degrees)
0	0.25000	180
5	0.26835	161
10	0.31699	147
15	0.38430	138
20	0.46181	132
25	0.54473	129
30	0.63042	128
35	0.71737	127
40	0.80460	127
45	0.89148	128
50	0.97751	128
55	1.06234	130
60	1.14564	131
65	1.22717	132
70	1.30669	134
75	1.38400	136
80	1.45889	138
85	1.53121	139
90	1.60078	141
95	1.66745	143
100	1.73107	145
105	1.79152	147
110	1.84866	149
115	1.90237	152
120	1.95256	154
125	1.99910	156
130	2.04192	158
135	2.08092	160
140	2.11603	162
145	2.14717	165
150	2.17429	167
155	2.19733	169
160	2.21624	171
165	2.23098	173
170	2.24154	176
175	2.24788	178
180 (Nyquist frequency)	2.25000	180

TABLE 5-4

The amplitude and phase spectrum of the maximum-delay wavelet $b = (1.00, -1.25)$. A maximum-delay wavelet has maximum phase-lag characteristic; the total phase-lag displacement for this wavelet from 0 to 180 degrees is $0 - (-180) = 180$ degrees. Note that the amplitude spectrum is the same as the amplitude spectrum of the wavelet $(-1.25, 1.00)$

Frequency (degrees)	Amplitude spectrum	Phase-lag (degrees)
0	0.25000	−180
5	0.26835	−156
10	0.31699	−137
15	0.38430	−123
20	0.46181	−112
25	0.54473	−104
30	0.63042	− 98
35	0.71737	− 92
40	0.80460	− 87
45	0.89148	− 83
50	0.97751	− 78
55	1.06234	− 75
60	1.14564	− 71
65	1.22717	− 67
70	1.30669	− 64
75	1.38400	− 61
80	1.45890	− 58
85	1.53121	− 54
90	1.60078	− 51
95	1.66745	− 48
100	1.73107	− 45
105	1.79152	− 42
110	1.84866	− 39
115	1.90237	− 37
120	1.95256	− 34
125	1.99910	− 31
130	2.04192	− 28
135	2.08092	− 25
140	2.11603	− 22
145	2.14717	− 20
150	2.17429	− 17
155	2.19733	− 14
160	2.21624	− 11
165	2.23098	− 8
170	2.24154	− 6
175	2.24788	− 3
180 (Nyquist frequency)	2.2500	− 0

TABLE 5-5

Amplitude and phase spectrum of minimum-delay wavelet (2, 1)

Frequency	Amplitude spectrum	Phase-lag
0.000	3.00	0.00
0.025	2.99	0.05
0.050	2.96	0.10
0.075	2.92	0.15
0.100	2.86	0.20
0.125	2.79	0.25
0.150	2.71	0.30
0.175	2.61	0.34
0.200	2.49	0.39
0.225	2.37	0.42
0.250	2.23	0.46
0.275	2.09	0.49
0.300	1.94	0.51
0.325	1.78	0.52
0.350	1.62	0.52
0.375	1.47	0.50
0.400	1.32	0.45
0.425	1.19	0.38
0.450	1.09	0.28
0.475	1.02	0.15
0.500 (Nyquist)	1.00	0.00

personnel are responsible for the positioning of seismic arrays, hole boring, seismic shooting, digital recording of data and many engineering functions required to properly conduct a seismic survey. In addition to such field procedures, the recorded data are preprocessed and prepared for deconvolution and seismic interpretation. The overall effort is quite an engineering accomplishment and deserves some elaboration.

Seismic crews work in the deserts of Africa, the jungles of South America, mountains and arctic regions of North America as well as in a wide variety of situations in the United States ranging from plains to forests to swamps. These crews must be prepared to operate in jungles where every foot of access must be hacked out by machete, in marshes where special flotation vehicles are required, and on snow, ice and tundra where insulated tracked vehicles are essential. Sometimes the terrain problems require the construction of a road for trucks to follow, or to transport the entire seismic operation by off-road vehicles, aircraft, or manpower as necessary.

For land surveys, crews use both the traditional explosive charges (dynamite) in shot holes and the newer surface energy sources such as Vibroseis and Dinoseis. Unlike dynamite, which imparts an impulsive seismic signal into the earth, the wave put into the earth by a Vibroseis source is oscillatory

TABLE 5-6

The first three columns give the amplitude and phase spectrum of the maximum-delay wavelet (1, 2). The amplitude spectrum is the same as that of the minimum-delay wavelet (2, 1) of Table 5-5. The fourth column is the sum of the phase-lags (at a given frequency) of the minimum-delay and maximum-delay wavelets. The fifth column gives the first differences of the fourth column. It is seen that the sum of phase-lags is equal to $2\pi f$

Frequency F	Amplitude spectrum	Phase-lag	Sum of phase-lag	First differences
0.000	3.00	0.00	0.00	0.15
0.025	2.99	0.10	0.15	0.15
0.050	2.96	0.20	0.30	0.16
0.075	2.92	0.31	0.46	0.16
0.100	2.86	0.42	0.62	0.15
0.125	2.79	0.52	0.77	0.16
0.150	2.71	0.63	0.93	0.16
0.175	2.61	0.75	1.09	0.16
0.200	2.49	0.86	1.25	0.15
0.225	2.37	0.98	1.40	0.16
0.250	2.23	1.10	1.56	0.16
0.275	2.09	1.23	1.72	0.16
0.300	1.94	1.37	1.88	0.15
0.325	1.78	1.51	2.03	0.16
0.350	1.62	1.67	2.19	0.16
0.375	1.47	1.85	2.35	0.15
0.400	1.32	2.05	2.50	0.16
0.425	1.19	2.28	2.66	0.16
0.450	1.09	2.54	2.82	0.16
0.475	1.02	2.83	2.98	0.16
0.500 (Nyquist)	1.00	3.14	3.14	

and persists for many seconds, the frequency changing slowly over the duration of the signal. The source system designated as Dinoseis utilizes a gas (propane and oxygen) explosion, detonated inside a closed chamber in contact with the ground, to generate an impulsive signal. Vibroseis and Dinoseis equipment are usually mounted on conventional trucks and on specialized off-road vehicles for use in rough or forested terrain. A full range of capabilities is usually utilized with these surface sources in order to maximize the seismic record quality in any area with cultural noise (Vibroseis) or with prohibitively expensive shot-hole drilling costs (either Vibroseis or Dinoseis). For specialized source requirements in marshy areas, the use of high-pressure air guns as in-hole seismic sources is employed.

The seismic receiver is usually an array (linear tapered or other areal patterns optimally designed for noise discrimination) of digital-grade geophones (standard; 10 Hz, individually damped). Nearly all geophones used for seismic recording on land are of the electromagnetic type. This kind of detector (geophone) produces a voltage which is proportional to the particle velocity of wave motion and is about the size of a golf ball.

TABLE 5-7

The second column gives the common amplitude spectrum of the wavelets (1.00, 0.00, 0.25) and (0.25, 0.00, 1.00). The third column gives the phase-lag of the minimum-delay wavelet (1.00, 0.00, 0.25) and the fourth column gives the phase-lag of the maximum-delay wavelet (0.25, 0.00, 1.00). The fifth column gives the sum of the phase-lags, which is equal to $4\pi f$

Frequency F	Amplitude spectrum	Phase-lag for minimum-delay wavelet	Phase-lag for maximum-delay wavelet	Sum of phase-lags	First difference
0.000	1.25	0.00	0.00	0.00	0.31
0.025	1.24	0.06	0.25	0.31	0.31
0.050	1.21	0.12	0.50	0.62	0.31
0.075	1.16	0.17	0.76	0.93	0.31
0.100	1.10	0.21	1.03	1.24	0.32
0.125	1.03	0.24	1.32	1.56	0.32
0.150	0.95	0.25	1.63	1.88	0.31
0.175	0.87	0.23	1.96	2.19	0.32
0.200	0.81	0.18	2.33	2.51	0.31
0.225	0.76	0.10	2.72	2.82	0.32
0.250	0.75	0.00	3.14	3.14	0.31
0.275	0.76	−0.10	3.55	3.45	0.32
0.300	0.81	−0.18	3.95	3.77	0.31
0.325	0.87	−0.23	4.31	4.08	0.32
0.350	0.95	−0.25	4.65	4.40	0.31
0.375	1.03	−0.24	4.95	4.71	0.32
0.400	1.10	−0.21	5.24	5.03	0.31
0.425	1.16	−0.17	5.51	5.34	0.31
0.450	1.21	−0.12	5.77	5.65	0.32
0.475	1.24	−0.06	6.03	5.97	0.31
0.500 (Nyquist)	1.25	0.00	6.28	6.28	0.31

An array may have up to 100 geophones with the output of each geophone combined to produce the array signal or trace, which is recorded on a single channel. Today's seismic crews use the latest binary gain digital equipment with computer-compatible magnetic tape recording. Depending upon the nature of the exploration problem, 24, 48, and up to 96 array signals are multiplexed and recorded digitally on magnetic tape. This procedure generates a *single* seismic record.

Digital seismic recording systems have a typical dynamic range of 80 db and record with a 16-bit capability. Reflected signals contain frequencies from a few hertz to several hundred hertz. Pertinent seismic formation is usually contained in the 10—250-Hz frequency band and signals are sampled at 1 ms, 2 ms, or 4 ms rates. For land surveys, a 2-ms sampling rate is commonly used. This entire operation is conducted at the site of the seismic survey with the data recorded and transmitted to a nearby instrument truck or other facility.

TABLE 5-8

The complete set of spike filters and their actual outputs for the minimum-delay input wavelet $(b_0, b_1) = (-1.25, 1.00)$

	Spike at			
	$k = 0$	$k = 1$	$k = 2$	$k = 3$
Spike filter coefficients				
a_0	−0.70	0.11	0.14	0.17
a_1	−0.45	−0.56	0.29	0.36
a_2	−0.22	−0.27	−0.34	0.56
Actual output				
c_0	0.88	−0.14	−0.17	−0.22
c_1	−0.14	0.82	−0.22	−0.27
c_2	−0.17	−0.22	0.72	−0.34
c_3	−0.22	−0.27	−0.34	0.56
Desired output				
d_0	1.00	0.00	0.00	0.00
d_1	0.00	1.00	0.00	0.00
d_2	0.00	0.00	1.00	0.00
d_3	0.00	0.00	0.00	1.00

TABLE 5-9

The phase-lag spectra for the spike filters of Table 5-8. Note how the phase-lag increases monotonically as the spike position moves from $k = 0$ to $k = 3$. The spike filter for $k = 0$ is minimum-delay; for $k = 1$ and $k = 2$, mixed-delay; and for $k = 3$, maximum-delay

Frequency F	Phase-lag for spike filters for spike at			
	$k = 0$	$k = 1$	$k = 2$	$k = 3$
0.000	−3.14	−3.14	0.00	0.00
0.025	−3.04	−2.90	−0.55	0.21
0.050	−2.93	−2.66	−0.68	0.42
0.075	−2.84	−2.42	−0.64	0.64
0.100	−2.74	−2.19	−0.53	0.86
0.125	−2.65	−1.97	−0.38	1.08
0.150	−2.57	−1.75	−0.22	1.31
0.175	−2.50	−1.54	−0.04	1.55
0.200	−2.44	−1.34	0.14	1.81
0.225	−2.40	−1.15	0.33	2.09
0.250	−2.39	−0.96	0.53	2.39
0.275	−2.41	−0.79	0.74	2.72
0.300	−2.48	−0.63	0.96	3.11
0.325	−2.59	−0.48	1.19	3.54
0.350	−2.74	−0.35	1.43	4.00
0.375	−2.89	−0.23	1.68	4.46
0.400	−3.00	−0.14	1.95	4.89
0.425	−3.07	−0.07	2.23	5.27
0.450	−3.11	−0.02	2.52	5.63
0.475	−3.13	−0.00	2.83	5.96
0.500 (Nyquist)	−3.14	−0.00	3.14	6.28

TABLE 5-10

The amplitude spectra for the spike filters of Tables 5-8 and 5-9

Frequency F	Amplitude spectra for spike filters for spike at			
	$k = 0$	$k = 1$	$k = 2$	$k = 3$
0.000	1.38	0.73	0.08	1.10
0.025	1.37	0.73	0.11	1.10
0.050	1.34	0.73	0.17	1.07
0.075	1.30	0.73	0.24	1.04
0.100	1.24	0.73	0.31	0.99
0.125	1.16	0.73	0.37	0.93
0.150	1.07	0.73	0.42	0.86
0.175	0.97	0.72	0.47	0.78
0.200	0.87	0.72	0.51	0.69
0.225	0.76	0.70	0.54	0.61
0.250	0.66	0.68	0.56	0.53
0.275	0.57	0.66	0.57	0.45
0.300	0.49	0.63	0.58	0.39
0.325	0.43	0.60	0.57	0.34
0.350	0.40	0.56	0.56	0.32
0.375	0.40	0.52	0.55	0.32
0.400	0.41	0.49	0.53	0.33
0.425	0.43	0.45	0.52	0.34
0.450	0.45	0.42	0.50	0.36
0.475	0.47	0.41	0.49	0.37
0.500 (Nyquist)	0.47	0.40	0.49	0.38

To appreciate the capacity of this entire operation, let us consider the following figures. Suppose that a seismic trace is obtained from an array consisting of 100 geophones. If the output of each geophone is 6 seconds in duration and sampled at a 2-ms rate, then each geophone contributes 3000 samples of data or 48,000 bits of information. Further, the seismic trace involves 4.8 million bits of information. Now if a single record contains 48 traces, then a single 6-second record of seismic data involves about 250 million bits of data. If a land crew is able to generate 10 records per day, one is faced with the enormous task of processing approximately 2.5 billion (2.5×10^9) bits of information for a single day's work!

The first marine seismic survey ever made was in 1934 and it resulted in the discovery of a Gulf of Mexico offshore oil field. Since that time, marine surveys have recorded hundreds of thousands of miles of data in many of the world's oceans, Atlantic, Pacific, Indian and Arctic; into the Mediterranean Sea, North Sea, the Great Lakes, above the Arctic circle in Canadian lakes and into the Straits of Magellan, the Caribbean and the Gulf of Mexico.

A marine seismic crew for oil prospecting would make use of a boat 50—70 m long, complete with up-to-date navigational and geophysical equipment. Some marine recording makes use of cables 3200 m long, which contain 48 recording sections, each feeding a separate channel. Each section may contain 20—32 hydrophones. The seismic receivers or hydrophones in marine work use piezo-electric crystals or comparable ceramic elements as pressure sensors. The hydrophones are usually the acceleration cancelling type with high pressure sensitivity.

Almost every marine reflection survey today is carried out as a single-ship operation. The same ship tows both the recording cable and energy source, commonly an air gun or tuned air gun array. Some marine surveys use a total of 16 air guns of various sizes which are suspended from two frames on respectively the starboard and port sides of the ship. Air gun arrays of up to six guns and as much as 31,000 cm^3 capacity have been fired at repetitive rates as frequently as one shot every 15 seconds. It is extremely important that the marine survey ship accurately know its position, course, and speed and the position of the cable as it is towed through the water. Hence, the latest navigational equipment, including satellite navigation systems, are used in conducting the marine survey.

The marine vessel has its recording instruments custom-installed in shock-mounted cabinets in the instrument room along with other peripheral equipment. Most recording systems have sampling rates of 1 ms, 2 ms or 4 ms, with the latter sampling rate most commonly used in marine work. Marine recording systems have the capability of 24, 48 and up to 96 recording channels and a time-coordinator to start the recording system, mark the ship's position and fathometer, and so on. Unlike the case of land surveys, a marine crew can acquire much more data for a single day's work. For example, under good weather conditions, a marine survey could collect over 200 seismic records per day with each record involving about 40 million bits of information.

Both land and marine surveys utilize common depth point recording, in which each geophone (or hydrophone) receives signals from different shot points, the individual recordings being subsequently composited for reduction of noise. In either case, seismic reflection surveys involve a great deal of engineering effort. The operation and maintenance of Vibroseis and air gun sources, the computer specialists and technicians who operate and repair the recording apparatus, the ship's navigators in marine work, and so on, are all part of a complex operation. However, all this work is necessary before any software or deconvolution analysis can take place. Let us now consider the digital processing of these reflection data.

Before the application of any deconvolution techniques, the field data acquired from land and marine surveys are subjected to preprocessing operations. The preprocessing has the task of inputting all usual field tape formats, of demultiplexing, and of sorting the seismic traces according to subsurface points. After demultiplexing and gain recovery, every seismic record is usually displayed and inspected for noisy traces, reversed traces, spikes, or other anomalous conditions that require editing, a time-consuming but essential step to producing the best possible final product. In addition to the editing of actual seismic traces, all field data can be read in as input, sorted and written out as output onto tape as basic data for subsequent standard processing, e.g. basic static and dynamic corrections, elevation data, shot and geophone positions, normal movement corrections, stacking, velocity analyses, migration analyses, etc. Some processing schemes deconvolve every seismic trace after the editing phase, whereas other methods of analysis perform deconvolution after the traces have been gathered into common depth point sets. In any event, the deconvolution of seismic data is performed after editing and subjected to seismic interpretation.

The above discussion is an overview of the basic signal processing operations performed on reflection seismic data. Each individual operation is a study within itself and detailed explanations of such procedures as common depth point (CDP) gathering, normal moveout (NMO) corrections, velocity analyses, and so on, can be found in Dobrin (1976).

To give an idea of the computer usage of a geophysical analysis activity, let us consider a typical case. A certain activity might be able to process easily more than 1,000,000 seismic recordings yearly, each of which contains 50,000 samples with a precision of 15 binary digits. Further, the computer (a CDC 6600, IBM 360/44, or PDP 11/15) can carry out in the same time about 2000 billion (2×10^{12}) multiplications, work which would entail more than one million man-years of work if done manually! Although our package of computer programs and deconvolution procedure are essential parts of the exploration process, we add that the overall engineering effort is a credit to existing technology.

Appendix

THE LAPLACE Z-TRANSFORM

In the bilateral Laplace transform:

$$X(s) = \int_{-\infty}^{\infty} x(t)\, e^{-st}\, dt \qquad [A\text{-}1]$$

e^{-st} represents the kernel of the Laplace transformation and $s = \sigma + i\omega$ the complex frequency variable associated with continuous time t. If $x(t)$ is passed through an ideal sampler, the sampler output is given by:

$$y(t) = \sum_{n=-\infty}^{\infty} x(n\Delta t)\delta(t - n\Delta t) \qquad [A\text{-}2]$$

where Δt is the sampling increment chosen in accordance with the sampling theorem and $\delta(t)$ the unit impulse function defined by:

$$\int_{-\infty}^{t_0} \delta(t - \tau)\, dt = \begin{cases} 1, & t_0 \geqslant \tau \\ 0, & t_0 < \tau \end{cases} \qquad [A\text{-}3]$$

The bilateral Laplace transform $Y(s)$ of the sampler output $y(t)$ then becomes:

$$\begin{aligned} Y(s) &= \int_{-\infty}^{\infty} y(t)\, e^{-st}\, dt \\ &= \int_{-\infty}^{\infty} \left[\sum_{k=-\infty}^{\infty} x(k\Delta t)\delta(t - k\Delta t) \right] e^{-st}\, dt \end{aligned} \qquad [A\text{-}4]$$

Term by term integration of [A-4] with the aid of [A-3] yields:

$$\begin{aligned} Y(s) &= \sum_{k=-\infty}^{\infty} x(k\Delta t)\, e^{-sk\Delta t} \\ &= \sum_{k=-\infty}^{\infty} x(k\Delta t)[e^{-s\Delta t}]^k \end{aligned} \qquad [A\text{-}5]$$

Since e^{-st} represents the kernel of the continuous-time bilateral Laplace transform, we shall let the complex variable $z = e^{-s\Delta t}$ denote an analogous incremental kernel. Thus, we shall define the two-sided *Laplace z-transform* $X(z)$ by:

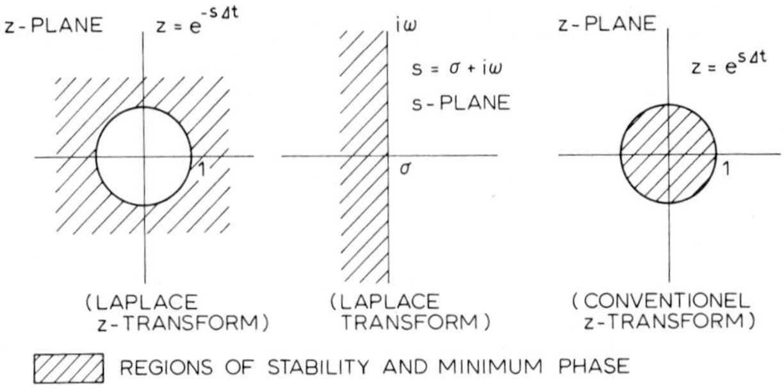

Fig. A-1. The relationship between the Laplace transform, Laplace z-transform, and the conventional z-transform.

$$X(z) = Y(z) = \sum_{k=-\infty}^{\infty} x_k z^k = \text{Laplace } z\text{-transform} \qquad \text{[A-6]}$$

where $x_k = x(k\Delta t)$ and $z = e^{-s\Delta t}$.

In the conventional two-sided z-transform, the complex variable z is defined by $z = e^{s\Delta t}$ where:

$$X(z) = \sum_{k=-\infty}^{\infty} x_k z^{-k} = \text{conventional } z\text{-transform}$$

Although we have approached the Laplace z-transform and the conventional z-transform from a sampling viewpoint, we note that the sequence $\{x_k\}$ need not be the sampled values of a continuous signal and can represent any sequence of numbers. Nevertheless, we include this approach to relate the continuous-time Laplace transform to the discrete-time Laplace z-transform.

The complex variable $z = e^{-s\Delta t} = e^{-\sigma\Delta t}e^{-i\omega\Delta t}$ in the Laplace z-transform maps the right-half s-plane *inside* the unit circle and the left-half s-plane *outside* the unit circle. Similarly, the complex variable $z = e^{s\Delta t} = e^{\sigma\Delta t}e^{i\omega\Delta t}$ in the conventional z-transform maps the right-half s-plane outside the unit circle and the left-half s-plane inside the unit circle. The regions of minimum-phase and stability for the Laplace transform, Laplace z-transform, and conventional z-transform are shown in Fig. A-1.

For causal sequences, i.e. $x_k = 0$ for $k < 0$, we add that the Laplace z-transform is the familiar Maclaurin series in the theory of complex variables. This series contains a polynomial with positive powers of z and conveniently characterizes causal systems. The inverse Laplace z-transform (or inversion integral) can be derived by application of Cauchy's theorem or by a coefficient matching procedure.

REFERENCES

Akaike, H., 1968. On the use of a linear model for the identification of feedback systems. *Ann. Inst. Stat. Math.*, 20: 425—439.

Akaike, H., 1969. Fitting autoregressive models for prediction. *Ann. Inst. Stat. Math.*, 21: 243—247.

Akaike, H., 1971. Autoregressive model fitting for control. *Ann. Inst. Stat. Math.*, 23: 163—180.

Andersen, N., 1974. On the calculation of filter coefficients for maximum entropy spectral analysis. *Geophysics*, 39: 69—72.

Anstey, N.A., 1966. The sectional autocorrelogram and the sectional rectrocorrelogram. *Geophys. Prospect.*, 14: 389—426.

Backus, M.M., 1959. Water reverberations — their nature and elimination. *Geophysics*, 24: 233—261.

Båth, M., 1968. *Mathematical Aspects of Seismology*. Elsevier, Amsterdam.

Båth, M., 1973. *Introduction to Seismology*. Birkhäuser, Basel and Stuttgart.

Båth, M., 1974. *Spectral Analysis in Geophysics*. Elsevier, Amsterdam.

Bateman, H., 1944. *Partial Differential Equations of Mathematical Physics*. Dover, New York, N.Y.

Bayless, J.W. and Brigham, E.O., 1970. Application of the Kalman filter to continuous signal restoration. *Geophysics*, 35: 2—23.

Berkhout, A.J. and Zaanen, P.R., 1976. A comparison between Wiener filtering, Kalman filtering, and deterministic least squares estimation. *Geophys. Prospect.*, 24: 141—197.

Berryman, L.H., Goupillaud, P.L., and Waters, K.H., 1958. Reflections from multiple transition layers. *Geophysics*, 23: 223—243.

Blackman, R.B. and Tukey, J.W., 1958. *The Measurement of Power Spectra*. Dover, New York, N.Y.

Bode, H.W., 1945. *Network Analysis and Feedback Amplifier Design*. D. Van Nostrand, Princeton, N.J.

Bogert, B.P. and Ossanna, J.F. 1966. The heuristics of cepstrum analysis of a stationary complex echoed Gaussian signal in stationary Gaussian noise. *IEEE Trans. Inf. Theory*, 12: 373—380.

Bogert, B.P., Healy, M.J.R. and Tukey, J.W., 1963. The quefrency analysis of time series for echoes. In: M. Rosenblatt (Editor), *Proceedings of the Symposium on Time Series Analysis*. Wiley, New York, N.Y., 209—243.

Box, G.E.P. and Jenkins, G.M., 1970. *Time Series Analysis Forecasting and Control*. Holden-Day, San Francisco, Calif.

Buhl, P., Stoffa, P.L. and Bryan, G.M., 1974. The application of homomorphic deconvolution to shallow-water marine seismology, 2. Real data. *Geophysics*, 39: 417—426.

Burg, J.P., 1968. *A New Analysis Technique for Time Series Data*. Paper presented at NATO Advanced Study Institute on Signal Processing, Enschede.

Burg, K.E., Ewing, M., Press, F. and Stulken, E.J., 1951. A seismic wave guide phenomenon. *Geophysics*, 16: 594—612.

Claerbout, J., 1976. *Fundamentals of Geophysical Data Processing*. McGraw-Hill, New York, N.Y.

Cole, R.H., 1948. *Underwater Explosions.* Princeton University Press, Princeton, N.J.

Dahlman, O. and Israelson, H., 1977. *Monitoring Underground Nuclear Explosions.* Elsevier, Amsterdam.

Davenport, W.B., Jr. and Root, W.L., 1958. *An Introduction to the Theory of Random Signals and Noise.* McGraw-Hill, New York, N.Y.

Deutsch, R., 1965. *Estimation Theory.* Prentice-Hall, Englewood Cliffs, N.J.

Dobrin, M.B., 1976. *Introduction to Geophysical Prospecting.* McGraw-Hill, New York, N.Y., 3rd ed.

Doob, J.L., 1953. *Stochastic Processes.* Wiley, New York, N.Y.

Durbin, J., 1960. The fitting of time series models. *Rev. Inst. Int. Stat.*, 28: 233—243.

Fatou, P., 1906. Séries trigonométriques et séries de Taylor. *Acta Math.*, 30: 335—345.

Frost, O.L., 1976. *Power Spectrum Estimation.* Paper presented at NATO Advanced Study Institute on Signal Processing, Portovenere.

Fryer, G.J., Odegard, M.E. and Sutton, G.H., 1975. Deconvolution and spectral estimation using final prediction error. *Geophysics*, 40: 411—425.

Galbraith, J.N., 1971. Prediction error as a criterion for operator length. *Geophysics*, 35: 251—265.

Gauss, C.F., 1809. *Theoria Motus Corporum Coelestium in Sectionibus Conicis Solem Ambientum.* Perthes and Besser, Hamburg (translation published by Dover, New York, N.Y., 1963).

Gentleman, W.M. and Sande, G., 1966. Fast Fourier transforms for fun and profit. In: 1966 Fall Joint Computer Conference. *AFIPS Conf. Proc.*, 29: 563—578.

Goupillaud, P.L., 1961. An approach to inverse filtering of near-surface layer effects from seismic records. *Geophysics*, 26: 754—760.

Grenander, U. and Szegö, G., 1958. *Toeplitz Forms and Their Applications.* University of California Press, Los Angeles, Calif.

Hagedoorn, B.B., 1954. A process of seismic reflection interpretation. *Geophys. Prospect.*, 2: 85—127.

Hille, E. and Phillips, R.S., 1948. *Functional Analysis and Semigroups.* American Mathematical Society, Providence, R.I.

Hobson, E.W., 1950. *The Theory of Functions of a Real Variable and the Theory of Fourier's Series.* Harren Press, Washington, D.C.

Hofstetter, E.M., 1974. An introduction to the mathematics of linear predictive filtering as applied to speech analysis and synthesis. *Lincoln Lab., MITT, Tech. Note*, 1973-36, Rev. 1.

Jenkins, G.M., and Watts, D.G., 1968. *Spectral Analysis and Its Applications.* Holden-Day, San Francisco, Calif.

Jones, R.H., 1973. Autoregressive spectrum estimation. *Proc. Am. Meteorol. Soc. 3rd Conf. on Probability and Statistics in Atmospheric Science.*

Kailath, T., 1968. An innovations approach to least squares estimation, 1. Linear filtering in additive white noise. *IEEE Trans. Autom. Control*, 13: 646—655.

Kailath, T., 1974. A view of three decades of linear filtering theory. *IEEE Trans. Inf. Theory*, 20: 146—181.

Kalman, R.E., 1960. New methods and results in linear prediction and filtering theory. *Trans. ASME*, 82D: 33—45.

Kalman, R.E. and Bucy, R.S., 1961. New results in linear filtering and prediction theory. *Trans. ASME*, 83D: 35—108.

Kemerait, R.C. and Childers, D.C., 1972. Signal detection and extraction by cepstrum techniques. *IEEE Trans. Inf. Theory*, 18: 745—759.

Khintchine, A., 1934. Korrelationstheorie der stationären stochastischen Prozesse. *Math. Ann.*, 109: 604—615.

Kolmogorov, A.N., 1939. Sur l'interpolation et extrapolation des suites stationnaires. *C.R. Acad. Sci. Paris*, 208: 2043—2045.

Kolmogorov, A.N., 1941a. Interpolation and extrapolation of stationary random sequences. *Izv. Acad. Sci. USSR, Math. Ser.*, 5: 3—14 (in Russian). (Translation by W. Doyle and J. Selin, RM-3090-PR, Rand Corp., Santa Monica, Calif., 1962.)

Kolmogorov, A.N., 1941b. Stationary sequences in Hilbert space. *Bull. State Univ. Moscow, Math.*, 2 (in Russian). (A translation by N. Artin is available in many libraries.)

Kulhánek, O., 1975. *Introduction to Digital Filtering in Geophysics.* Elsevier, Amsterdam.

Kunetz, G., 1964. Génerlization des opérateurs d'antirésonance à un nombre quelconque de réflecteurs. *Geophys. Prospect.*, 12: 283—289.

Kunetz, G. and Fourmann, J.M., 1968. Efficient deconvolution of marine seismic records. *Geophysics*, 33: 412—423.

Lacoss, R.T., 1971. Data adaptive spectral analysis methods. *Geophysics*, 36: 661—675.

Lamb, H., 1904. Propagation of tremors over the surface of an elastic solid. *Philos. Trans., R. Soc. Lond., Ser. A*, 203: 1—42.

Lee, Y.W., 1960. *Statistical Theory of Communication.* Wiley, New York, N.Y.

Levinson, N., 1947. A heuristic exposition of Wiener's mathematical theory of prediction and filtering. *J. Math. Phys.*, 25: 110—119.

Makhoul, J., 1975. Linear prediction: a tutorial review. *Proc. IEEE*, 63: 561—580.

Marden, M., 1966. *Geometry of Polynomials.* American Mathematical Society, Providence, R.I.

Markel, J.D., 1971. FFT pruning. *IEEE Trans. Audio Electroacoust.*, 19: 305—311.

Markel, J.D., 1972. Digital inverse filtering — a new tool for formant trajectory estimation. *IEEE Trans. Audio Electroacoust.*, 20: 129—137.

Markel, J.D. and Gray, A.H., 1976. Linear Prediction of Speech. Springer-Verlag, New York, N.Y.

Mendel, J.M., 1973. *Discrete Techniques of Parameter Estimation.* Marcel Dekker, New York, N.Y.

Mendel, J.M., 1976. Single-channel white noise estimators for predictive deconvolution. Paper presented at the 46th SEG Meeting, Houston, Texas.

Nettleton, L.L., 1940. *Geophysical Prospecting for Oil.* McGraw-Hill, New York, N.Y.

Nuttall, A.H., 1976. Spectral analysis of a univariate process with bad data points, via maximum entropy and linear predictive techniques. *NUSC TR*, No. 5303, Naval Underwater Systems Center, Conn.

Oppenheim, A.V. and Schafer, R.W., 1975. *Digital Signal Processing.* Prentice-Hall, Englewood Cliffs, N.J.

Oppenheim, A.V., Kopec, G.E. and Tribolet, J.M., 1976. Signal analysis by homomorphic prediction. *IEEE Trans. Acoust., Speech, Signal Process.*, 24: 327—332.

Ott, N. and Meder, H.G., 1972. The Kalman filter as a prediction-error filter. *Geophys. Prospect.*, 20: 549—560.

Papoulis, A., 1965. *Probability, Random Variables and Stochastic Processes.* McGraw-Hill, New York, N.Y.

Parzen, E., 1971. Some recent advances in time series analysis. In: M. Rosenblatt (Editor), *Statistical Models and Turbulence.* Springer-Verlag, New York, N.Y.

Peacock, K.L. and Treitel, S., 1969. Predictive deconvolution: theory and practice. *Geophysics*, 34: 155—169.

Peterson, R.A. and Walter, W.C., 1974. *Through the Kaleidoscope.* Lecture notes prepared by United Geophysical Corporation.

Pflueger, J., 1972. Spectra of water reverberations for primary and multiple reflections. *Geophysics*, 37: 788—796.

Poisson, S.D., 1823. Sur la distribution de la chaleur dans les corps solides. *J. Ec. R. Polytech., Ser. I*, 19: 1—162.

Ricker, N., 1953. The form and laws of propagation of seismic wavelets. *Geophysics*, 18: 10—40.

Ricker, N., 1977. *Transient Waves in Visco-Elastic Media.* Elsevier, Amsterdam.

Robinson, E.A., 1954. *Predictive Decomposition of Time Series with Applications to Seismic Exploration.* Ph.D. Thesis, MIT, Cambridge, Mass. Also in *Geophysics*, 1967, 32: 418—484.

Robinson, E.A., 1957. Predictive decomposition of seismic traces. *Geophysics*, 22: 767—778.

Robinson, E.A., 1962. *Random Wavelets and Cybernetic Systems.* Charles Griffin and Co., London.

Robinson, E.A., 1967a. *Multichannel Time Series Analysis with Digital Computer Programs.* Holden-Day, San Francisco, Calif.

Robinson, E.A., 1967b. *Statistical Communication and Detection with Special Reference to Digital Data Processing of Radar and Seismic Signals.* Charles Griffin and Co., London.

Robinson, E.A., 1975. Dynamic predictive deconvolution. *Geophys. Prospect.*, 23: 780—798.

Robinson, E.A. and Treitel, S., 1969. *The Robinson-Treitel Reader.* Compiled by Seismograph Service Corporation, Tulsa, Okla.

Robinson, E.A., and Wold, H., 1963. Minimum-delay structure of least-squares and lo-ipso predicting systems for stationary stochastic processes. In: M. Rosenblatt (Editor), *Proceedings of the Symposium on Time Series Analysis.* Wiley, New York, N.Y., pp. 192—196.

Schafer, R.W., 1969. Echo removal by discrete generalized linear filtering. *Res. Lab. Electron., MIT, Tech. Rep.*, No. 466.

Schwarz, H.A., 1872. Zur Integration der partiellen Differentialgleichung. *Z. Reine Angewandte Math.*, pp. 218—254.

Senmoto, S. and Childers, D.G. 1972. Signal resolution via digital inverse filtering. *IEEE Trans. Aerosp. Electron. Syst.*, 8: 633—640.

Sherwood, J.W.C. and Trorey, A.W., 1965. Minimum-phase and related properties of the response of a horizontally stratified absorptive earth to plane acoustic wave. *Geophysics*, 30: 191—197.

Srinath, M.D. and Viswanathan, M.M., 1975. Sequential algorithm for identification of an autoregressive process. *IEEE Trans. Autom. Control*, 20: 542—546.

Stoffa, P., Buhl, P. and Bryan, G., 1974. The application of homomorphic deconvolution to shallow-water marine seismology, 1. Theory. *Geophysics*, 39: 417—436.

Stolt, R.H., 1978. Migration by Fourier transform. *Geophysics*, 43: 23—48.

Szegö, G., 1915. Ein Grenzwertsatz über die Toeplitzschen Determinanten einer reelen positiven Funktion. *Math. Ann*, 76: 490—503.

Szegö, G., 1939. Orthogonal polynomials. *Am. Math. Soc. Colloq. Publ.*, 23.

Treitel, S., 1966. Seismic wave propagation in layered media in terms of communication theory. *Geophysics*, 25: 106—129.

Treitel, S. and Robinson, E.A., 1966. The design of high resolution digital filters. *IEEE Trans. Geosci. Electron.*, 4: 25—38.

Ulrych, T.J., 1971. Application of homomorphic deconvolution to seismology. *Geophysics*, 36: 650—660.

Ulrych, T.J. and Clayton, R.W., 1976. Time series modelling and maximum entropy. *Phys. Earth Planet. Inter.*, 12: 188—200.

Wadsworth, G.P., Robinson, E.A., Bryan, J.G. and Hurley, P.M., 1953. Detection of reflection on seismic records by linear operators. *Geophysics*, 18: 539—586.

White, R.E. and O'Brien, P.N.S., 1974. Estimation of the primary seismic pulse. *Geophys. Prospect.*, 22: 627—651.

Whittle, P., 1969. A view of stochastic control theory. *J. R. Stat. Soc., Ser. A*, 132: 320—334.

Wiener, N., 1930. Generalized harmonic analysis. *Acta Math.*, 55: 117—258.

Wiener, N., 1942. The extrapolation, interpolation, and smoothing of stationary time series with engineering applications. *MIT DIC Contract*, No. 6037, Cambridge, Mass., National Defense Research Council, Section D2. (Reprinted 1949, Wiley, New York, N.Y.)

Wiggins, R.A. and Robinson, E.A., 1965. Recursive solution to the multichannel filtering problem. *J. Geophys. Res.*, 70: 1885—1891.

Wold, H., 1954. *A Study in the Analysis of Stationary Time Series.* Almqvist and Wiksells, Uppsala, 2nd ed.

Wood, L.C. and Treitel, S., 1975. Seismic signal processing. *Proc. IEEE*, 63: 649—661.

Wuenschel, P.C., 1960. Seismogram synthesis including multiples and transmission coefficients. *Geophysics*, 25: 106—129.

Yule, G.U., 1907. On the theory of correlation for any number of variables treated by a new system of notation. *Proc. R. Stat. Soc.*, 79: 182—193.

Yule, G.U., 1927. On a method of investigating periodicities in disturbed series, with special reference to Wolfer's sunspot numbers. *Phils. Trans., Ser. A*, 226: 267—298.

Zadeh, L.A. and Ragazzini, J.R., 1950. An extension of Wiener's theory of prediction. *J. Appl. Phys.*, 21: 645—655.

SUBJECT INDEX